INTRODUCTION TO
BRYOLOGY

W. B. Schofield
Department of Botany
University of British Columbia

Reprint of First Edition

Introduction to Bryology

ISBN-10: 1-930665-26-1
ISBN-13: 978-1-930665-26-2

Library of Congress Control Number: 2001088824

THE BLACKBURN PRESS
P. O. Box 287
Caldwell, New Jersey 07006 USA
973-228-7077
www.BlackburnPress.com

To
M. I. B.

From these I weave my tapestry:
A dew-jewelled bryophyte fabric
That yields infinite delight.

Contents

Preface

This book is presented as an introduction to a fundamental understanding of the bryophytes. These plants have been seriously neglected in elementary botany textbooks, where treatment is usually restricted to a few representative or conspicuous species which, regrettably, are usually atypical. In biology courses bryophytes are usually ignored entirely.

Concepts and terms are introduced gradually, with the first chapters forming a foundation for those which follow. Extensive illustrations complement the text and show the extraordinary diversity of these plants. A general understanding of the terminology of biology is assumed. A reasonably comprehensive glossary is provided to define all terms that might be unfamiliar to a beginning student.

Two general books concerning bryology have appeared during the past 20 years: E. V. Watson's *The Structure and Life of the Bryophytes* (third edition, 1971) and N. S. Parihar's *Bryophyta* (fifth edition, 1965). The first presupposes a firm understanding in bryology, while the second concerns itself mainly with morphogenetic detail. In consequence, both have limitations as comprehensive introductions to the subject.

The organization of the present book is founded on an establishment of the structure, interrelationships, and possible evolutionary trends among the bryophytes. After this foundation is presented, discussion of the biology of the bryophytes is provided to make the historical development of the science more readily understandable. The appendices give aids for the study of bryophytes.

This book had its genesis as a series of comprehensive notes for student use in a bryology course that the author has taught at the University of British Columbia for more than 20 years. Students have worked with the manuscript for four years, and it has profited from their recommendations.

It would be especially heartening if this book led students into the utilization of bryophytes in diverse areas of biological research, while they become captivated by the fascination of the plants for their own sake. My main ambition has been to understand bryophytes as living plants in their environment. I have been especially intrigued by the diverse ways that bryophytes have evolved physiological, ecological, and morphological features that have enabled them to be such successful plants in seemingly inhospitable environments. I hope that some of my own fascination has been reflected in this book.

ACKNOWLEDGMENTS

An author is never alone when writing a textbook. The ghosts of past researchers hover over him, and their written communications and those of contemporary researchers contain the ideas that form the material from which he must construct his book. The debt I owe is acknowledged, in part, through the citation of published works at the end of each chapter. A number of contemporary researchers have given me ready assistance in a number of chapters in this text. I express my gratitude to them and absolve them from responsibility for any misconceptions that I have inadvertently retained. Those who have read much of the text are John A. Christy, Howard A. Crum, Judith D. Godfrey, Diana G. Horton, Dale M. J. Mueller, Helen P. Ramsay, William C. Steere, Terry T. McIntosh, and Benito C. Tan. Their general and detailed comments have assisted greatly in improvement of the text. Several researchers have commented on particular chapters: Lewis E. Anderson (cytology and genetics), Yoshinori Asakawa (chemistry), Bruce A. Bohm (chemistry), M. Bopp (physiology), Gary E. Bradfield (ecology), Ella O. Campbell (Monocleales, Marchantiales), Anthony D. M. Glass (physiology), Sinske Hattori (Calobryales), Hiroshi Inoue (Calobryales, Marchantiales), Helen P. Ramsay (cytology and genetics), Wolfram Schultze-Motel (Andreaeidae), Nancy Slack (ecology), John Spence (ecology), and G. H. Neil Towers (chemistry).

Several researchers have provided photographs, which greatly enhance the text: Diana G. Horton, Zennoske Iwatsuki, Dale M. J. Mueller, Helen P. Ramsay, and Dale H. Vitt.

The plates so ably prepared by the artists are a valuable asset. My debt to the artists is apparent. Patricia Drukker-Brammall has assisted in all aspects of the illustration of the book, both in the production of original plates and in the designing of many of the others. All artists have profited from her patient and skilled training: Norman Eyolfson, Janet Morgan, Scott Renyard, and Muriel R. Schofield. Two other artists, whose contributions are apparent and whose illustrations we have copied, are Rudolph M. Schuster and P. Janzen. The elegance and accuracy of their drawings and the excellence of the research that accompanied their production cannot be overemphasized.

The figures were drawn by the following artists. Toshio Asakawa: Fig. 22-1; Patricia Drukker-Brammall: Figs. 1-4; 2-2,4; 6-10; 8-1,2,5,15; 9-3,5; 13-1,2; 14-4,5; 8*A,F*; 15-1, 17-1,2,3,15; 19-1. Norman Eyolfson: Figs.

3-1,2,4; 6-1,8*A*–*C*; 7-1. Janet Morgan: Figs. 4-3; 5-1; 8-16; 12-3,5; 15-2,4; 20-1,2. Scott Renyard: Figs. 3-5; 4-7,8; 6-3; 7-3, 12-1,2. Muriel Schofield: Figs. 4-9; 16-1,2. The remaining figures were drawn by the author. Photographs were supplied as follows: T. Hirohama and Z. Iwatsuki: Fig. 8-14. Diana Horton: Figs. 8-3,9*B*. Dale Mueller: Figs. 7-2; 17-10. Helen Ramsay: Fig. 21-1. James Reid: Figs. 24-1,2,3,4. Jane Taylor: Figs. 13-3; 17-4. Dale Vitt: Figs. 4-1; 6-5; 8-8,9*A,C*.

The following publishers requested specific credit lines: Fig. 1-1 from *The Plant Kingdom* by William H. Brown, © Copyright, 1935, by William H. Brown. Used by permission of the publisher, Ginn and Company (Xerox Corporation). Figs. 2-1; 8-13*C,D*; 14-6*C*; 14-8*D-H* from *Nonvascular Plants, An Evolutionary Survey* by R. F. Scagel, R. J. Bandoni, J. R. Maze, G. E. Rouse, W. B. Schofield, and J. R. Stein. © 1982 by Wadsworth, Inc. Reprinted by permission of Wadsworth Publishing Company, Belmont, California 94002. Figs. 2-3*B*; 3-3*E*; 6-7*A*; 8-6; 8-7; 8-10*B*; 8-11*F,H*; 8-13*B*; 9-1*B,C*; 9-2*A* © 1981, Columbia University Press. Reprinted by permission. Fig. 13-3*C,D* © 1980 Columbia University Press. Reprinted by permission. Figs. 12-6; 13-7; 13-8; 13-10; 13-11; 13-12; 13-13*A,B*; 13-14*A,B,F-I*; 13-15; 13-16; 13-17; 13-18; 13-20*A,B*; 14-7*C,E*; 14-9; 17-16 © 1966, Columbia University Press. Reprinted by permission. Figs. 13-13*E-H*; 13-14*C-E* © 1969, Columbia University Press. Reprinted by permission. Fig. 3-3*F,G* reproduced from *Moss Flora of the Maritime Provinces,* by R. R. Ireland (Ottawa, 1982). Courtesy National Museum of Natural Sciences, National Museums of Canada. Fig. 4-4*A,B* from *Illustrated Moss Flora of Fennoscandia* by E. Nyholm, Publishing House of the Swedish Research Councils, Stockholm, Sweden. Figs. 4-7; 13-19 from *Spore Discharge in Land Plants* by C. T. Ingold. © 1939, Oxford University Press. Fig. 6-8*D,E* reprinted by permission of Brigham Young University Press. Fig. 9-1*A* reprinted from *Muelleria* 2: 198 by I. G. Stone, 1973. Fig. 9-2*C-E* reprinted from *Muelleria* 2: 200 by I. G. Stone, 1973. Fig. 9-6 reprinted from *Muelleria* 2: 207 by I. G. Stone, 1973. Figs. 13-21; 19-2*B,C,E,F*; 19-3*C* courtesy of *The Bulletin of the Torrey Botanical Club.* Fig. 14-1*B,C* from *Ann. Missouri Bot. Gard.* 48 (1961). Fig. 15-2*A,B* with permission from *The Botanical Journal of the Linnean Society.* Vol. 51, p. 324. Copyright 1937 by The Linnean Society of London. Fig. 17-11 reproduced by permission from Walker, R. and Pennington, W. "The movements of the air pores of *Preissia quadrata,*" *New Phytologist* 38 (1939). Figs. 17-18, 19-5, 19-6 after *Cryptogamic Botany,* 2 ed, Vol. II by Gilbert M. Smith, © 1955, McGraw-Hill Book Company, reproduced with permission.

I am most grateful to the typists of the Department of Botany at the University of British Columbia, who have borne patiently the handwritten first manuscript and have accurately interpreted numerous revisions: Chris Skelton, Ila Westergard-Thorpe, Peggy Oyama, Susan Berry, and Pat Fornelli.

I am especially indebted to the staff at Macmillan for their careful attention to the editing, typesetting, and design of the book. I express particular thanks to Sarah Greene, Frances Tindall, Milton Horowitz, and Elizabeth Penati. A

Graphic Method Inc., was the typesetter and interior designer. Soho Studio, Inc., designed the cover and the jacket.

Several colleagues have given me encouragement that provided the impetus for me to persist. Robert F. Scagel, Howard A. Crum, Diana G. Horton, Sinske Hattori, Helen P. Ramsay, Nancy G. Slack, William C. Steere, and Dale H. Vitt should be mentioned in particular.

Finally, I acknowledge, with great affection, the support of my wife and family, who have tolerated my agonies over writing and executing many of the illustrations.

W. B. Schofield

1 Introduction

The division Bryophyta includes the mosses and their allies. Bryophytes are generally small but form a striking part of the vegetation in cooler northern and southern latitudes and extremely humid climates of both temperate and tropical regions. Bryophytes sometimes carpet the forest floor with vivid greens and form extensive peatlands with hummocks and depressions of rich greens, browns, and reds. Trees may be sheathed in bryophytes that extend up the trunks and encircle branches. In foggy forests the tree branches are often draped with pendent ropes of bryophytes. These essentially terrestrial plants have achieved the greatest diversity shown by the gametophyte.

The division is generally divided into three classes: Hepaticae (liverworts and scale-mosses), Anthocerotae (hornworts), and Musci (mosses). Although these names are used traditionally, the International Rules of Botanical Nomenclature would recommend Hepaticopsida, Anthocerotopsida, and Bryopsida. Some authors treat each of these classes as a division but, as will be detailed in the following pages, these plants share so many fundamental characteristics that they can be included within a single division.

The bryophytes possess the following features:

1. They are green land plants, possess chlorophyll A and B, starch, cellulose walls, and sometimes possess a cuticle.
2. The sporophyte is short-lived to annual.
3. The sporophyte, although photosynthetic through most of its life span preceding spore dispersal, is always attached to the gametophyte and is at least partially dependent on it. The base of the sporophyte (the foot) penetrates the tissue of the gametophyte.
4. The sporophyte is unbranched and produces a single terminal sporangium; the spore coat is cutinized, and the spores are generally disseminated through the air.
5. The sporophyte and gametophyte possess no lignified tissue.
6. The gametophyte is generally perennial; it often consists of a juvenile,

1

usually filamentous phase (protonema) and a more complex phase (the gametophore) that actually produces the sex organs.

7. The male gametes (sperms) are biflagellate with whiplash flagella and must reach the egg via a water film. They are produced in an antheridium, that is, a stalked sac (Fig. 1-1*B*). This sac is composed of a sterile unistratose (one cell in thickness) jacket enclosing innumerable cells, each of which produces a sperm.

8. The female sex organ (archegonium) is flask-shaped; the neck of the flask is unistratose, while the lower expanded portion (venter) is multistratose; each archegonium encloses a single egg (Fig. 1-1*A*).

9. Growth of the gametophore is by a single apical cell, rather than by meristematic tissue.

FIGURE 1-1
Sex organs of bryophytes. *(A)* Archegonium of a moss (×210). *(B)* Antheridium of a moss (×210). (After Brown, 1935.)

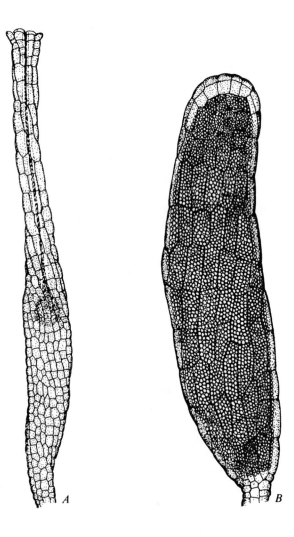

A   *B*

10. Bryophytes are generally small, with the sporophyte usually no more than 3 cm tall and the gametophore usually less than 10 cm tall, although erect forms may exceed 20 cm, and reclining, aquatic, or hanging forms may reach a meter in length.

Bryophytes are generally considered to be primitive plants, and in the early fossil record there is material (*Hepaticites devonicus*), that has been identified as a bryophyte. This is of Devonian age, and consists of small fragments of an apparent hepatic thallus (see p. 181). This material lacks sporophytes and sex organs; thus its interpretation remains tentative, since the gametophytes of some ferns are structurally similar, and it is also possible that this fossil could represent an unlignified portion of a lignified plant.

The ancestry of the bryophytes remains uncertain. Features that bryophytes share with other green plants relate them to the green algae, and the green algae and bryophytes appear to share a common ancestor. The green algae show two evolutionary lines, the Chlorophyceae and Charophyceae. The bryophytes and other archegoniates appear to have been derived from the charophycean line. In common with the green land plants, the charophycean line possesses the same type of cell division with persistent spindles. It may form phragmoplasts, and it produces glycolate oxidase. The chlorophycean line, to which most green algae belong, lacks persistent spindles, forms phycoplasts, and usually possesses glycolate dehydrogenase. Since the ferns and their allies possess gametophytes roughly similar to those of some bryophytes and since the sex organs are also structurally similar, these plants appear to have shared a common origin with the bryophytes, and may have been derived from bryophytic or bryophytelike ancestors.

Numerous theories have been presented to explain the origin of the bryophytes. One suggests that the bryophytes originated from filamentous fresh-water green algae. The fossil record does not support this hypothesis, since material identified as the appropriate algal group appears in the fossil record more than 100 million years after presumptive bryophyte fossils were in existence. This fact alone is insufficient to discard the hypothesis, but the superficial similarity of the filamentous protonema of bryophytes to the algae appears to be a derived feature, in other words, a feature that has originated by reduction from a more elaborate structure. Since protonemata and filamentous algae occupy similar habitats, it is not surprising that they should show a similar general morphology. Furthermore, as detailed earlier, fundamental cytological and biochemical differences separate them from the filamentous green algae.

A second hypothesis suggests that a unicellular green alga may have given rise to a plant that was adapted to a humid soil environment, and this ancestor may have given rise to a terrestrial progenitor of all archegoniate plants as well as the charophyte algae.

A third hypothesis proposes that the bryophytes were derived from primitive vascular plants. This theory places emphasis on the simplicity of the sporophyte that may have originated, through considerable structural reduction, from a more elaborate branched sporophyte that bore many terminal sporangia. Among the archegoniate vascular plants, the extinct class Rhynio-phytina is suggested as a possible ancestral group (Fig. 1-2).

A fourth hypothesis presents a radially symmetrical erect gametophore as the possible ancestral archegoniate from which both leafy and thallose evolutionary lines were derived.

Under any circumstances, the morphological similarities among these archegoniate plants, coupled with detailed features of biochemistry and ultrastructure, point to obvious interrelationships. The origin of the bryophytes is likely to remain mysterious. It is possible that bryophytes are ancestral to the lignified land plants. Certainly the ancestors of all green land plants appear to have been bryophytelike in structure. The bryophytes are now eminently successful plants that can survive under environmental extremes that few other

FIGURE 1-2
Reconstructions of ancient fossil land plants. *(A, B)* Rhyniophytina. *(A) Rhynia major,* showing dichotomizing erect axes with terminal sporangia arising from a creeping axis affixed to the substratum by rhizoids (×1). *(B)* Forked sporangium of *Horneophyton lignieri* showing columella (×5), *(C, D) Sporogonites exuberans.* (*C*) Reconstruction showing many sporophytes arising from a reclining "thallus" (×0.5). *(D)* Sporangium, showing apophysis-like basal portion of sporangium that possessed stomata (×3). *(E) Torticaulis transwalliensis* showing sporangium and twisted axis or seta (×3). (*A,* after Lang, 1937; *B,* after Eggert, 1974; *C,* after Andrews, 1960; *D,* after Lang and Cookson, 1935; *E,* after D. Edwards, 1979.)

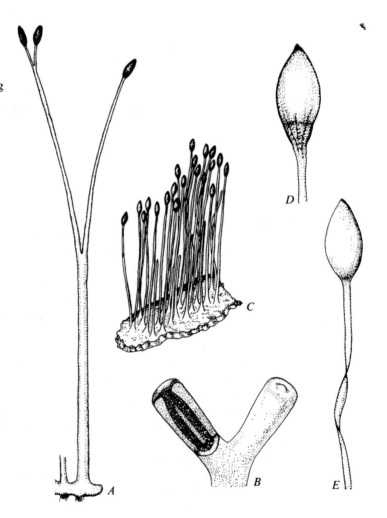

plants can tolerate. This success means that bryophytes are widely distributed in the world, from arctic and alpine environments to the humid tropics, from semi-arid sites to submergence in water.

Bryophytes thrive in humid environments and, like all plants, require water to survive. Some mosses, however, can recover after many years of nearly complete dehydration. Many bryophyte gametophores tolerate periods of freezing, whether under wet or dry conditions. No bryophytes grow in the sea, although a few mosses are confined to seashores. Bryophytes form the dominant vegetation in some parts of the world where mossy peatlands form extensive mats (Fig. 4-1), on which the vascular plants are epiphytic. Bryophytes often form an almost continuous carpet on the floor of humid forests. In arctic, as well as other extreme environments, bryophytes can produce a substratum that initiates conditions favorable for other plants and fungi to colonize and contribute to soil building. In such environments bryophytes are essential in the creation of sites in which most other terrestrial organisms can survive.

Since bryophytes are autotrophic organisms, they must absorb minerals, water, and carbon dioxide. Generally these are absorbed in soluble form directly through the wall of the cell in which photosynthesis occurs, but there is also interchange among cells through cytoplasmic interconnections. Generally the morphology of the gametophore permits external capillary conduction of water up and down the gametophore, and sometimes there is an elaborate internal conducting system (see p. 60).

The life history of bryophytes involves an alternation between sporophytic and gametophytic generations (Fig. 1-3) that are distinctly different in form. The spore is the first cell of the gametophyte generation. This germinates to produce a protonema or sporeling in most cases, and this is attached to the substratum by nonchlorophyllose filamentous branches, termed rhizoids. Ultimately the protonema or sporeling produces a bud that differentiates an apical cell from which the more complex gametophore grows. The gametophore produces the sex organs. Often each gametophore bears only one type of sex organ, and thus there may be extensive colonies with only antheridia or only archegonia. Since this gametophore may be unable to reproduce sexually, it often has various means of asexual reproduction. Often gemmae are produced. These bodies are formed of a cluster of cells that, when they fall from the gametophore, produce a protonema or sporeling that gives rise to a new gametophore. Gemmae are sometimes miniature gametophores that possess an apical cell, and thus when they fall from the parent gametophore, are able to produce a normal gametophore without first producing the protonemal or sporeling stage. The gametophores are sometimes brittle, especially when dry, and isolated fragments can regenerate entire gametophores. For sexual reproduction, water is necessary. Since sperms can swim only short distances, the gametophores that bear the antheridia must be near those bearing archegonia if sexual reproduction is to take place.

When mature, the walls of the neck canal cells of the archegonium disintegrate and release the fluid material of the cell contents, and that fluid exudes

FIGURE 1-3
Life cycle of a bryophyte. All
cell division in the gametophyte is by mitosis; the only
meiotic divisions are within
the sporangium (see text).

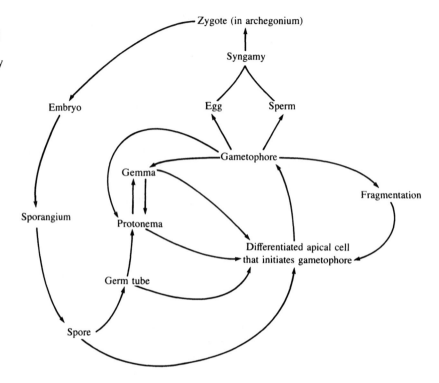

outward through the ruptured apex of the archegonium into water that bathes
it. The mass of sperms is either squeezed out of, or emerges gradually from, the
antheridium. Usually this mass moves upward to the surface film of the water
that bathes the antheridia. Since fatty materials are also released, the mass
often spreads outward rapidly by repulsion of the fatty droplets for each other.
If a sperm encounters the diffused fluid contents from the neck canal cells of an
archegonium, it swims toward the site of the greatest concentration of this fluid,
and thus down the archegonium neck to the egg. The sperm burrows through
the wall of the egg; the sperm nucleus unites with the egg nucleus, and a zygote
results. This is the first cell of the sporophyte.

The zygote, after several mitotic divisions, produces a basal foot that penetrates the gametophore, and the embryonic cells that will undergo further
divisions to produce the remainder of the sporophyte. While the sporophyte
enlarges, the lower cells of the archegonium wall (venter) also undergo cell
divisions and form a protective layer of tissue that protects the developing
embryo; this is the calyptra.

Details of sporophyte development differ in each of the classes of bryophytes, but in all classes a sporangium is finally produced, and spore mother
cells are differentiated within it. The spore mother cells undergo meiosis, and
each produces a tetrad of spores. Rupture of the sporangium jacket also varies

among the different classes of bryophytes; the spores are released and the life cycle is completed.

A number of features that characterize bryophytes are best understood by examination of material or reference to an illustration (Fig. 1-4). These features are mentioned frequently throughout the text.

In the following chapters each of the major and many of the minor evolutionary lines within the bryophytes are discussed. The mosses are treated first. This is done out of convenience, and does not imply that they are the most primitive of the bryophytes. Indeed, within each of the main evolutionary lines (the mosses, hepatics, and hornworts), a number of representatives show equally generalized features and others that are highly elaborated. In a group with a long evolutionary history, and in which environmental selection has

FIGURE 1-4
Some conspicuous features of bryophytes. *(A)* An idealized moss, showing sporophyte (above) emerging from game-tophore (below), the leaves of which are tipped by gemmae. *(B)* Detail of opening sporangium of *(A)*. *(C)* *Douinia ovata,* a junger-mannialean hepatic, showing sporophyte emerging from perianth of gametophore. *(D, E)* *Geocalyx graveolens,* jungermannialean hepatic: *(D)* Showing sporophyte emerging from marsupium of gametophore; *(E)* Showing underside of leafy stem. *(F)* *Hygrobiella laxiflora,* a jungermannialean hepatic with all leaves of similar form. (Abbreviations: *amphi.,* amphigastrium or underleaf; *apop.,* apophysis; *caly.,* calyptra; *col.,* columella; *lat. lf.,* lateral leaf; *mars.,* marsupium; *op.,* operculum; *perich.,* perichaetium; *per. t.,* peristome tooth; *rhiz.,* rhizoids; *spor.,* sporangium.)

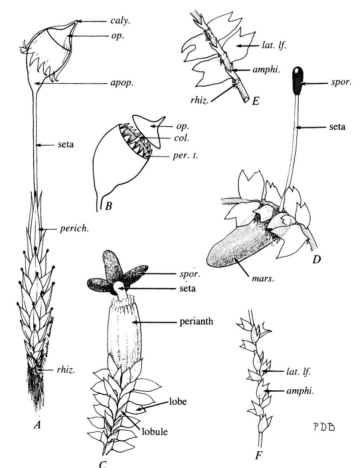

acted to produce such a diversity of forms, it is inevitable that problems will persist in evolutionary categorization of particular morphological lines.

FURTHER READING

Banbury, G. H. 1962. Geotropism of lower plants: Bryophytes. *Handb. Pflanzenphysiol.* **XVII**/2:346–350.

Bauer, L. 1967. Determination von Gametophyt und Sporophyt. *Handb. Pflanzenphysiol.* **XVIII**:235–256.

Benson-Evans, K. 1961. Environmental factors and bryophytes. *Nature* **191**:255–260.

———— 1964. Physiology of the reproduction of bryophytes. *Bryologist* **67**:431–445.

Bonnot, E. J. (ed.) 1974. Les problèmes modernes de la bryologie. *Bull. Soc. bot. France* **121**:1–360.

Bower, F. O. 1935. *Primitive Land Plants.* London: Macmillan.

Campbell, D. H. 1918. *The Structure and Development of Mosses and Ferns* (3rd ed). London: Macmillan.

Cavers, F. 1910–1911. The inter-relationships of the Bryophyta. *New Phytol.* **9**:81–112, 158–186, 193–234, 269–306, 341–353. Ibid. **10**:1–46.

Chopra, R. S. 1967. Relationship between liverworts and mosses. *Phytomorphology* **17**:70–78.

Clarke, G. C.S., and J. G. Duckett. 1979. *Bryophyte Systematics.* London: Academic Press.

Crandall-Stotler, B. 1980. Morphogenetic designs and a theory of bryophyte origins and divergence. *Bio Science* **30**:580–585.

Goebel, K. 1930. *Organographie der Pflanzen, Part 2: Bryophyten-Pteridophyten* Jena: G. Fischer.

Herzog, T. 1926. *Geographie der Moose.* Jena: G. Fischer.

Jovet-Ast, S. 1967. Bryophyta, in Boureau, E. (ed.), *Traité de Paléobotanique,* Tome 2, pp. 18–186. Paris: Masson.

Lacey, W. S. 1969. Fossil bryophytes. *Biol. Rev.* **44**:189–205.

Lewis, K. R. 1961. The genetics of bryophytes. *Trans. Br. Bryol. Soc.* **4**:111–130.

Lowery, B., D. Lee, and C. Hébant. 1980. The origin of the land plants: a new look at an old problem. *Taxon* **29**:183–197.

Miller, H. A. 1982. Bryophyte evolution and geography. *Biol. J. Linnaean Soc.* **18**:145–196.

Parihar, N. S. 1965. *An Introduction to the Embryophyta I. Bryophyta.* Allahabad, India: Central Book Depot.

Richards, P. W. 1959. Bryophyta, in Turrill, W. B. (ed.), *Vistas in Botany I,* pp. 387–420. London: Pergamon Press.

Richardson, D. H. S. 1981. *The Biology of Mosses.* New York: Wiley.

Savicz-Ljubitzkaja, L. I., and J. J. Abramov. 1959. The geological annals of Bryophyta. *Rev. Bryol. Lichénol.* **28**:330–342.

Schuster, R. M. (ed). 1983–1984. *New Manual of Bryology* (2 vols.) Nichinan, Japan: Hattori Bot. Lab.

Smith, A. J. E. 1978. Cytogenetics, biosystematics and evolution in the Bryophyta. *Adv. Bot. Res.* **6**:195–276.

Smith, G. M. 1955. *Cryptogamic Botany. Vol. 2: Bryophytes and Pterido-phytes* (2nd ed.). New York: McGraw-Hill.

Stebbins, G. L., and G. J. C. Hill. 1980. Did multicellular plants invade the land? *Am. Naturalist* **115**:342–353.

Steere, W. C. 1955. Bryology, in *A Century of Progress in the Natural Sciences, 1853–1953,* pp. 267–299. San Francisco: Calif. Acad. Sci.

—— 1964. The use of living bryophytes in the teaching of botany. *Am. Biol. Teacher* **26**:100–104.

—— 1969. A new look at evolution and phylogeny in bryophytes, in Gunckel, J. E. (ed.), *Current Topics on Plant Science,* pp. 1341–43. New York: Academic Press.

Taylor, T. N. 1982. The origin of land plants: a paleobotanical perspective. *Taxon* **31**:155–177.

Verdoorn, F. (ed.) 1932. *Manual of Bryology.* The Hague: Martinus Nijhoff.

Watson, E. V. 1971. *The Structure and Life of Bryophytes* (3rd ed). London: Hutchinson University Library.

2 The Mosses — Class Musci

Mosses are generally the most conspicuous bryophytes in the vegetation. Because other plants superficially resemble moss gametophores, this has resulted in confusion to the layman concerning the true definition of the term moss. "Moss" has been applied to many algae, to some lignified plants, including club-mosses (*Lycopodium* and *Selaginella*), Spanish moss (the pineapple relative *Tillandsia*), and lichens (*Usnea, Alectoria, Bryoria,* etc.).

Mosses show greater structural diversity than other bryophyte classes: the class Musci contains approximately 10,000 species in nearly 700 genera. In such a diverse group several clear evolutionary lines can be segregated into seven subclasses:

Subclass Andreaeidae
Subclass Sphagnidae
Subclass Tetraphidae
Subclass Polytrichidae
Subclass Buxbaumiidae
Subclass Bryidae
Subclass Archidiidae

These subclasses are founded essentially on structure of the sporophyte, especially the sporangium.

DISTINCTIVE FEATURES

Several features distinguish mosses from other bryophytes:

1. The protonema is generally an extensive, branched filamentous phase of the life cycle. In it the cross-walls are often oblique.
2. Rhizoids are always multicellular and resemble the protonema, except that they lack chlorophyll and often have brownish pigmented walls.

3. The protonema sometimes produces gemmae (= rhizoidal gemmae or "tubers").
4. The gametophore is always leafy, and the leaves are generally radially arranged in more than three rows.
5. Antheridia and archegonia usually have sterile filaments (paraphyses) intermixed among them (Fig. 2-1).
6. Leaves of the gametophore are unistratose for the most part, except at the

FIGURE 2-1
Sex organs and paraphyses of mosses (×25). (*A*) Perigonium, showing antheridia among paraphyses. (*B*) Perichaetium, showing archegonia among paraphyses. (After Scagel et al., 1982.)

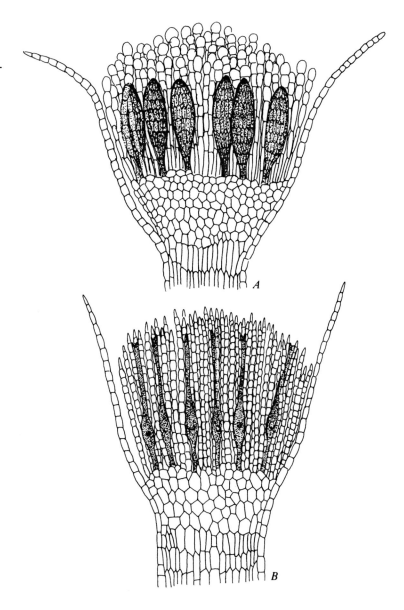

multistratose midrib (costa). The costa may be single or multiple, or even absent.

7. Leaf cells are commonly elongate and rarely possess trigones.
8. Leaf cells have simple small oil bodies, if any.
9. Leaves are rarely lobed.
10. The jacket of the sporangium generally has stomata.
11. The outer cells of the sporangium jacket lack transverse barlike thickenings or nodular thickenings.
12. The sporangium jacket is always multistratose.
13. The sporangium usually opens by means of an apical lid (operculum) (Fig. 1-4*A,B*).
14. When the operculum falls loose, it usually exposes teeth (peristome teeth) that form a ring around the opening. These teeth are often hygroscopic (Fig. 1-4*B*).
15. Within the sporangium there is usually a central mass of sterile tissue (columella) (Fig. 1-4*B*).
16. The sporangium is usually elevated on an elongate stalk (seta); the seta is usually wiry and made up mainly of thick-walled cells (Fig. 1-4*A*).
17. The calyptra usually is torn loose from the gametophore by the elongating sporophyte and protects the tip (Fig. 1-4*A*).
18. The seta elongates before the sporangium is differentiated, and is a photosynthetic organ for an extended period during sporophyte development. The presence and shape of the calyptra influence the shape and differentiation of the sporangium.
19. Generally the spores are shed from the sporangium over an extended period.

THE GAMETOPHYTE

The gametophyte shows considerable morphological variation, both in the protonematal and gametophore phases. Although it is usually a branched uniseriate filament, the protonema is sometimes thallose and/or multiseriate. The uniseriate filament is frequently heterotrichous, with a chlorophyllose creeping portion possessing transverse cross-walls (chloronema), or oblique cross-walls (caulonema), the colorless or pigmented rhizoids with oblique cross-walls, and an erect branched system with transverse cross-walls. The erect system greatly increases the photosynthetic area of the protonema.

The gametophore consists of an axis, usually termed a stem, and this axis bears leaves. Since the stem, through its lack of a lignified vascular system and other features, does not fit the strict definition of that structure, it is sometimes called a caulid (plural: caulidia). Similarly the leaves are not true leaves and are sometimes termed phyllidia. Traditionally the terms stem and leaves have been used by bryologists. The stem sometimes possesses a well-differentiated vascular system. It is never lignified, and it probably evolved independently from the vascular system of lignified plants. The stem is chlorophyllose when young, and with the leaves, is an effective photosynthetic system. Much nutrient absorption occurs in the epidermal (or cortical) cells of the stem, and since the leaves are generally unistratose, absorption of fluids is a simple procedure, and gas exchange is also direct to the cells where it is utilized.

Leaves are usually spirally arranged on the stem. The diverging leaf is generally given some support by a costa, but the nature of support of the leaf varies in different genera. Leaf orientation usually changes in response to moisture changes. These changes help to control water loss in the gametophore. The leaves often fold up against the stem when dry, a state termed imbricate, or they diverge outward when moist, an orientation termed divergent. Leaves are always sessile, with the whole base of the blade attached directly to the stem.

Cells are often diverse in size, form, and wall ornamentation in different parts of the leaf. Cells of the margin sometimes differ in size and shape from those of the remainder of the leaf, a state described as a differentiated leaf margin (Fig. 2-2 C,K). At the base of the leaf, near the point of attachment, a triangular area of cells that differ in size, shape, or color from the rest of the cells sometimes appears; these are the alar cells, and this region is termed the alar region of the leaf (Fig. 2-20,P). In cross-sectional view, the costa often shows some variety in cell structure; some cells are specialized for support, others for photosynthesis, and still others for conduction.

The gametophore frequently has many branches and bears leaves of different form on various parts of the stem. Leaves that surround the archegonia are the perichaetial leaves (Fig. 1-4 A, 2-1B). These often differ in shape and structure from the other leaves, and usually form a protective sheath around the sex organs. Leaves that surround the antheridia are the perigonial leaves (Fig. 2-1A).

Generally the leafy gametophore has rhizoids on the stem; these form a tangled mass among the leaf bases and contribute to external water conduction along the stem. The rhizoids may be colorless or reddish brown and are important for attachment of the gametophore to the substratum. Frequently these become greatly intertangled with the stems of interwoven gametophores, and result in condensed tufts or turf. The surface walls of rhizoids are usually smooth but sometimes bear tiny warts, termed papillae (Fig. 8-8A−C). Sometimes the stem possesses small chlorophyllose unistratose organs on the stem, among the leaf bases. These are paraphyllia (Fig. 8-13B−D), and appear to be important in the capillary conduction of water.

The gametophores sometimes produce gemmae on the stem or on the leaves. Occasionally the moss reproduces entirely by these vegetative organs.

THE SPOROPHYTE

The sporophyte consists of the foot that penetrates the gametophore, usually a seta that raises the sporangium well above the gametophore, and the sporangium itself, sheathed by the calyptra until it is mature. The seta sometimes has an internal conducting system, and with the outer cells often elongate

FIGURE 2-2
Diversity in leaf shapes in mosses. *(A) Rhytidium rugosum*, showing rugose surface and recurved margin (×40). *(B) Phyllogonium fulgens*, showing keeled leaf (×25). *(C) Mnium spinulosum*, showing differentiated margin with paired teeth (×25). *(D) Endotrichella elegans*, showing plications (×20). *(E) Neckeropsis undulata*, with undulations (×35). *(F) Trachypodopsis auriculata*, with marked auricles and plications (×20). *(G) Fissidens bryoides*, showing sheathing portion of leaf, attached to costa (×35). *(H) Ptilium cristcastrensis* (×40). *(I) Climacium dendroides*, stem leaf (×35). *(J) Dicranum scoparium*, a falcate leaf with differentiated alar cells (×20). *(K) Rhizomnium glabrescens*, with distinctive border (×20). *(L) Rhizofabronia persoonii*, with striking marginal teeth (×10). *(M) Polytrichum commune*, with lamellae on costa (×8). *(N) Andreaea rupestris*, an ecostate, ovate leaf (×25). *(O) Leptobryum pyriforme*, a subulate leaf (×30). *(P) Cratoneuron filicinum*, showing conspicuous alar cells (×35). *(Q) Hylocomium splendens*, stem leaf (×35), (R) *Tortula muralis*, showing costa extending as an awn (×20). *(S) Antitrichia curtipendula*, showing multiple costae (×30). *(T) Plagiomnium insigne*, showing decurrent base (×12). (After Schofield and Hébant, 1984.)

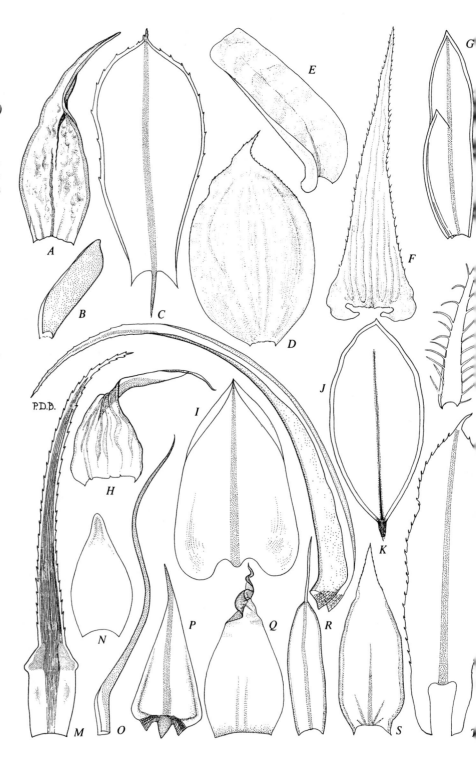

and thick-walled, making the seta relatively wiry. The sporangium jacket often has stomata, particularly in the lower portion, which lead into a limited series of intercellular spaces, making the sporangium relatively efficient in gas exchange. The stomata are exposed on the surface of the sporangium and are described as phaneropore or exposed (Fig. 2-3B); or they are embedded in the epidermal layer, with the surrounding cells overarching the stoma and forming a small external chamber; these stomata are described as cryptopore or immersed (Fig. 2-3A). An operculum and peristome teeth are usually present in the sporangium; the structure of the peristome provides important characters that distinguish the subclasses and other higher taxonomic categories (especially orders and families).

CRITERIA FOR EVOLUTIONARY INTERPRETATION

The concept of generalized versus specialized features is often used as a guide to interpret evolutionary trends. Generalized features are those that characterize organisms more closely resembling the ancestral type. The term is often used as a synonym of "primitive," which carries a strong implication for support of the primitive structure in the fossil record. The term "generalized," on the other hand, carries the assumption that the structure is relatively generalized or unspecialized, and it may or may not characterize primitive representatives. Specialization can originate through elaboration or through structural simplification. The consequence of the complexity of evolutionary pathways, assumed to be controlled through selective pressures in the environment, has led to considerable difficulty in interpretation of these structural modifications. The result is a certain amount of controversy.

Not all features can be used reliably in making interpretations, since selection acts on various aspects of elaboration, features become lost through reduction, and the plant appears generalized in one particular structure, while it is

FIGURE 2-3
Stomata of mosses. (A) *Orthotrichum ohioense,* immersed (cryptopore) stoma (×300). (B) *Funaria hygrometrica,* exposed (phaneropore) stomata (×230). (A, after Lorch, 1931. B, after Crum and Anderson, 1981.)

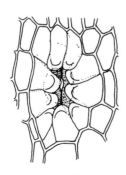

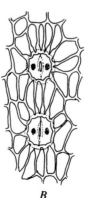

A   *B*

specialized in other features. Furthermore, in some mosses the sporophyte sometimes possesses numerous specialized details while the associated gametophore is generalized. This is inevitable, since different selective pressures act on various parts of the life cycle. Features that are considered generalized in the mosses are as follows:

1. A thallose protonematal phase instead of a heterotrichous protonematal phase. The complete absence of a protonema, where the spore is multicellular and initiates the gametophore as soon as the spore coat is ruptured, is considered the most specialized.
2. Stems of simple internal tissue organization, with outer cells thick walled and the inner cells with thin walls. Stems with a well differentiated cortex and elaborate internal tissue organization tend to be considered more specialized. Stems showing little internal differentiation are interpreted as either generalized, or specialized through reduction.
3. Stems showing little or no branching, and the branches irregular in arrangement. Stems with elaborate regular branching in a single plane would be considered more specialized.
4. Leaves radially arranged in three or more ranks. Stems with the leaves in two ranks are specialized.
5. Sex organs borne at the apex of the main shoot; sex organs borne on reduced lateral branches are specialized.
6. Smooth rhizoids are more generalized than papillose rhizoids.
7. An ornamented or fringed calyptra appears to be more specialized than a smooth simple one.
8. The production of multicellular peristome teeth composed of whole cells is more generalized than the production of peristome teeth made of remnants of cell walls.
9. In most mosses the presence of an elongate seta is more generalized than its absence.
10. The lack of a conducting system in the seta is probably more generalized than its presence, although in some mosses, where seta elongation is rapid, the absence of such a system suggests specialization by reduction.
11. Phaneropore stomata are more generalized than are cryptopore stomata.
12. A long snouted operculum is more generalized than one which is blunt.
13. Elongate cells in the epidermis of the jacket are more generalized than isodiametric cells.

THE LIFE CYCLE

The life cycle is the same in all bryophytes (see Fig. 1-3), but some details in the different stages of the cycle are unique to the mosses. The germinating spore is chlorophyllose and is usually unicellular. When the spore coat is ruptured, a tube usually emerges and becomes the first cell of the protonema; this protonematal thread cuts off cross-walls perpendicular to the long axis, and continues apical growth. This creeping protonema is richly chlorophyllose, with numerous chloroplasts in each cell, and is often termed the chloronema. As soon as the first tube ruptures the spore coat, frequently a rhizoid is also formed. This rhizoid is multicellular, with oblique cross-walls, lacks chlorophyll, and attaches the chloronema to the substratum.

As the chloronema grows it produces many branches and forms a tangled mass over the substratum. The chloronema, presumably in response to an accumulation of growth hormones, generally produces cells in which chloroplasts are fewer than in the cells of the chloronema that produces them and also differ from those of the chloronema in their response to light. These cells form a secondary protonema which is termed the caulonema. The cells of the caulonema have oblique cross-walls. The caulonema often produces perpendicular multicellular branches rich in chlorophyll.

The caulonema, generally under increased light intensity, accumulates growth hormones that cause some incipient branch cells to round up instead of elongating. This enlarging bud undergoes first a single oblique division, followed by a second oblique division at an obtuse angle to the first. Finally, a third oblique division cuts off a third cell at an oblique angle to the second, and a bulging apical cell with three distal cutting faces results. This cell is the precursor of the gametophore.

Each of the three cutting faces cuts off a single cell, and each of these ultimately forms a stem segment. From each stem segment, after further cell divisions in several planes, some of the outermost cells initiate a single leaf. In consequence, the leaves are, at first, in three ranks. The production of the stem initials from the apical cell, however, is in a spiral sequence, and the three initial cells are of unequal size. The fourth initial that is cut off, therefore, is not directly in line beneath the first. This feature, plus unequal growth of the stem, causes a three ranked arrangement of leaves to be lost in most mature moss stems. The outer cells of these stem segments also sometimes produce rhizoids; occasionally they produce paraphyllia.

When the stem branches, the branches arise just below the leaf attachment rather than in the leaf axil, thus differing from lignified plants. Branches arise from an outer cell of the stem, which becomes specialized as the apical cell of the branch. Sometimes, a small bud is initiated on the stem in the position where a branch would be predicted to arise, but this bud may develop no further. Surrounding such an incipient branch bud, reduced leaflike or filamentous structures (pseudoparaphyllia, see p. 100) often arise. Such structures also often form near the base of a new branch and persist there after the branch forms.

Ultimately the stem usually produces perichaetia (specialized clusters of leaves enclosing the archegonia) or perigonia (specialized clusters of leaves enclosing the antheridia). In more than half of the mosses these are formed on separate gametophores, and the gametophores are dioicous (i.e., with each sex on a different plant). The perichaetial and perigonial leaves tend to terminate a shoot, either the main stem, or a reduced branch. Among the sex organs there are usually sterile filaments (paraphyses). Both sex organs and paraphyses are chlorophyllose until the sex organs are mature. In most cases the sexual branch produces sex organs only once, but there are some exceptions. As the sex organs age, the cell walls tend to become red-brown.

After fertilization of the egg, the resulting zygote undergoes one transverse division. The lowermost cell undergoes further divisions to produce the foot. Sometimes the lowermost cells push in among the cells of the gametophore to become haustoria. These make a more effective means of transferring synthe-sized nutrients from the gametophore to the sporophyte since, as the sporophyte ages, its cell walls tend to become reddish-pigmented and chlorophyll is lost from the cells; thus dependence on organic nutrients produced by the gametophore becomes critical for the survival of the sporophyte. The epidermal cells of the seta also become cutinized, and water for the survival of the sporophyte must come from the gametophore. Often an internal conducting system in the seta transfers water, dissolved minerals, and metabolized material from the gametophore to the actively growing apical portion of the sporophyte.

The upper cell of this embryo is the precursor of the remainder of the sporophyte. This cell becomes the apical cell and cuts off cells that divide in several planes, producing an elongating cell mass.

While the embryo undergoes cell divisions, the cells of the venter and be-neath it (Fig. 1-1A) also divide to form a protective sheath (calyptra) around the embryo. As the embryo elongates, however, its enlargement tears the base of the calyptra from the gametophore, and this calyptra forms a protective cap at the tip of the elongating (or enlarging) embryo. Most mosses produce an elongate seta tipped by the calyptra. The calyptra prevents rapid water loss from the immature cells of the tip of the sporophyte. If the calyptra is removed too early, the sporophyte usually ceases growing, and no sporangium differen-tiates. A diversity of calyptrae are shown in Fig. 2-4.

When the seta has reached a given length (depending on the moss involved), the sporangium begins to enlarge within the calyptra. As enlargement occurs, the outer cells differentiate to become the multistratose jacket, and some outer cells may produce stomata. Within the center of the developing sporangium, a mass of cells is also sterile. These form a columella, its shape varying depen-dent on the moss subclass. Between the columella and sporangium jacket, spore mother cells are differentiated. Each of these cells undergoes meiosis and produces a group of four spores (a tetrad). These spores are initially unicellular, and generally remain unicellular until they are shed. The spore wall is sometimes elaborately ornamented with depressions, ridges, papillae, or spines.

As the sporangium enlarges it usually sheds its calyptra before the spores are released. An operculum is usually differentiated from the upper part of the sporangium by the formation of a circle of cells that cuts off a cap. Beneath the operculum, peristome teeth are usually formed. These are involved in spore

FIGURE 2-4
Calyptrae of mosses (all ×35).
(A) Atrichum selwynii. (B)
Encalypta ciliata (with some-
what eroded base). *(C) Gly-*
phomitrium humillimum. (D)
Mnium spinulosum. (E) Poly-
trichum juniperinum. (F)
Orthotrichum lyellii. (G)
Campylopus richardi. (H)
Tetraphis pellucida. (I)
Orthotrichum consimile. (J)
Ptychomitrium gardneri. (K)
Fontinalus antipyretica. (L)
Daltonia angustifolia. (After
Schofield and Hébant, 1984.)

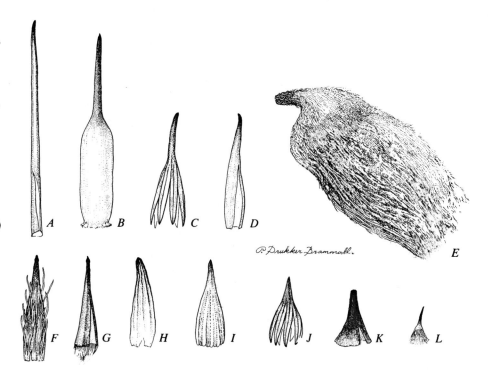

release from the sporangium. Generally the spores are air-borne to a new site. Sometimes invertebrates may inadvertently carry the spores to a suitable habitat.

FOSSIL RECORD

The fossil record for the mosses is rather poor. Fossil material that is interpreted as being a moss is given the form genus *Muscites* if its relationships are unclear. This generic designation is thus a catch-all for all problematic mosslike fossil material and is used for convenience, until adequate interpretation of the material becomes possible. In fossil material the gametophores are rarely associated with attached sporophytes, and cellular detail is often absent; thus the attributing of some fossils to the form genus *Muscites* is, at best, uncertain.

The Devonian genus *Sporogonites* (Fig. 1-2*C, D*) produced sporophytes strongly suggestive of a moss, but the nature of the gametophore is unknown. Material of late Silurian age has been interpreted as sporangia. This material, named *Torticaulis* (Fig. 1-2*E*), might be of a moss, but its state of preservation is so unsatisfactory that a clear interpretation is impossible. Furthermore, the general similarity of some moss sporangia (that lack opercula) with sporangia of ancient lignified plants is so close that they are indistinguishable, especially when in a poorly preserved state.

The earliest specimens of *Muscites* are described from upper Carboniferous sediments, and their identity is not certain, since the material consists of moss-sized plants with elongate radially arranged structures interpreted as leaves. Material from the Permian, beautifully described and interpreted by M. F. Neuberg, shows features that are commonly associated with bryophyte gametophores: radially arranged, unistratose leaves with a cellular pattern in the leaves like that in modern mosses. Neuberg described these as several genera (see p. 42). Neither sex organs nor attached sporophytes were found in this material. Modern moss genera first appear in Tertiary sediments and material from Pliocene, and interglacial deposits can be safely placed in modern species.

The frequent assertion that mosses are primitive plants is not well supported by the fossil record. It is possible that the mosses are more diverse today than they were in the ancient past, since the fossil record shows no evidence of great diversity. They illustrate the immense morphological variety that can be attained by the gametophyte. They are clearly a remarkably successful group of plants, showing as much success in the earth's vegetation as the more conspicuous lignified plants.

FURTHER READING

Anderson, L. E. 1964. Biosystematic evaluations in the Musci. *Phytomorphology* **14**:27–51.

Andrews, H. N. 1960. Notes on Belgian specimens of *Sporogonites. Paleobotanist* **7**:59–89.

Bauer, L. 1963. On the physiology of sporangium differentiation in mosses. *J. Linn. Soc. (Bot.)* **58**:343–351.

Bopp, M. 1956. Die Bedeutung der Kalyptra für die Entwicklung der Laubmoossporogone. *Ber. Deutch. Bot. Gesells.* **69**:455–468.

——— 1962. Development of the protonema and bud formation in mosses. *J. Linn. Soc. (Bot.)* **58**:305–309.

Bower, F. O. 1935. *Primitive Land Plants.* London: Macmillan.

Brotherus, V. F. 1924–1925. Musci (Laubmoose) in Engler, A., and K. Prantl (eds.), *Die Natürlichen Pflanzenfamilien* (2nd ed.), Vols. 10 and 11. Leipzig: W. Engelmann.

Campbell, D. H. 1918. *The Structure and Development of Mosses and Ferns* (3rd ed.). London: Macmillan.

Cavers, F. 1910–1911. The inter-relationships of the Bryophyta. *New Phytologist* **9**:81–112, 158–186, 193–234, 269–304, 341–353. Ibid. **10**:1–46.

Clarke, G. C. S., and J. G. Duckett (eds.) 1979. *Bryophyte Systematics.* London: Academic Press.

Correns, C. 1899. *Untersuchungen über die Vermehrung der Laubmoose durch Brutorgane und Stecklinge.* Jena: G. Fischer.

Dickson, J. H. 1973. *Bryophytes of the Pleistocene*. Cambridge: Cambridge University Press.

Edwards, D. 1979. A late Silurian flora from the Old Lower Red Sandstone of south-west Dyfed. *Palaeontology* **22**:23–52.

Goebel, K. 1930. *Organographie der Pflanzen. Part 2: Bryophyten-Pteridophyten* (3rd. ed.). Jena: G. Fischer.

Grebe, C. 1917. Studien zur Biologie und Geographie der Laubmoose. *Hedwigia* **59**:1–208.

Greene, S. W. 1960. The maturation cycle, or the stages of development of gametangia and capsules in mosses. *Trans. Br. Bryol. Soc.* **3**:736–745.

Haberlandt, G. 1886. Beiträge zur Anatomie und Physiologie der Laubmoose. *Pringsheim's Jahrb. für wissenschaftl. Bot.* **17**:359–498.

Hagerup, O. 1935. Zur Periodizität im Laubwechsel der Moose. *K. danske Vidensk. Selsk Biol. Meddels.* **11**:9–88.

Herzog, T. 1926. *Geographie der Moose*. Jena: G. Fischer.

Hughes, J. G. 1954. The physiology of reproduction in the Bryophyta. *Rapp. et Commen. Sec. 8. Congr. Int. Bot.* **16**:122–124.

Janzen, P. 1912. Die Jugendformen der Lauboose und ihre Kultur. *Ber. d. Westpreuss. Bot. Zool. Vereins:* 1–62.

Janzen, P. 1917. Die Haube der Laubmoose. *Hedwigia* **58**:157–280.

Loeske, L. 1910. *Studien zur vergleichenden Morphologie und phylogenetischen Systematik der Laubmoose*. Berlin: Max Lande.

Lorch, W. 1931. Anatomie der Laubmoose, in Linsbauer, K. (ed.), *Handbuch der Pflanzenanatomie 7 (Lieferung 28)*. Berlin: Gebrüder Borntraeger.

Neuberg, M. F. 1958. Permian true mosses of Angaraland. *J. Palaentol. Soc., India, Lucknow* **3**:22–29.

Parihar, N. S. 1965. *An Introduction to the Embryophyta I*. Allahabad, India: Central Book Depot.

Reimers, H. 1954. Bryophyta: Moose, in Engler, A. (ed.), *Syllabus der Pflanzenfamilien* (12th ed.), Vol. 1, Berlin: Gebrüder Borntraeger.

Richards, P. W. 1954. *A Book of Mosses*. London: Penguin.

——— 1959. Bryophyta, in Turrill, W. B. (ed.), *Vistas in Botany I,* pp. 387–420. London: Pergamon Press.

——— 1978. The taxonomy of the bryophytes, in Street, H. E. (ed.), *Essays in Plant Taxonomy,* pp. 177–209. London: Academic Press.

Richardson, D. H. S. 1981. *The Biology of Mosses*. New York: Wiley.

Schofield, W. B. 1981. Ecological significance of morphological characters in the moss gametophore. *Bryologist* **84**:149–165.

Schreirer, D. C. 1980. Differentiation of bryophyte conducting tissues: structure and histo-chemistry. *Bull. Torrey Bot. Club* **107**:298–307.

Schultze-Motel, W. (ed.) 1981. *Advances in Bryology,* Vol. 1. Vaduz: J. Cramer.

Schuster, R. M. (ed.). 1983–1984. *New Manual of Bryology.* (2 vols). Nichinan, Japan: Hattori Bot. Lab.

Steere, W. C. 1947. A consideration of the concept of genus in Musci. *Bryologist* **50**:247–258.

———— 1955. Bryology, in *A Century of Progress in the Natural Sciences. 1853–1953,* pp. 267–299. San Francisco: Calif. Acad. Sci.

Taylor, R. J., and A. E. Leviton (eds.) 1980. *The Mosses of North America.* San Francisco: AAAS Pacific Division.

Thieret, F. 1955. Bryophytes as economic plants. *Econ. Bot.* **10**:75–91.

Verdoorn, F. 1932. *Manual of Bryology.* The Hague: Martinus Nijhoff.

Vitt, D. H. 1981. Sphagnopsida and Bryopsida, in Parker, S. P. (ed.), *Taxonomy and Classification of Living Organisms.* New York: McGraw-Hill.

———— 1981. Adaptive modes of the moss sporophyte. *Bryologist* **84**:166–186.

Watson, E. V. 1971. *The Structure and Life of Bryophytes* (3rd ed). London: Hutchinson University Library.

3

The Lantern Mosses — Subclass Andreaeidae

The subclass Andreaeidae shows many generalized features, including the structural simplicity of both gametophyte and sporophyte. The fact that all species grow on rocks also implies generalization. Since the sporangium resembles a Chinese lantern, these mosses are called "lantern mosses."

Two genera compose the subclass Andreaeidae: *Andreaea,* with approximately 100 species, and *Andreaeobryum*, with one (Fig. 3-1). These are placed in a single family, Andreaeaceae. Although these mosses are very small in stature, usually not exceeding 2 cm, they form distinctive tight black or reddish brown tufts or turfs on bare exposed rock. *Andreaea* occurs primarily in cooler climates at temperate or frigid elevations or latitudes throughout the world. Most species grow on siliceous rocks (rocks rich in silica and thus acidic). *Andreaeobryum* is confined to northwestern Canada and adjacent Alaska; it grows on rocks rich in calcium that are thus basic in chemical reaction.

Since these genera exhibit many generalized structural features, they are discussed first. This does not imply that they are considered to be progenitors to any other subclass of mosses. They represent an entirely independent evolutionary line within the mosses.

ANDREAEA: THE GAMETOPHYTE

The spores of *Andreaea* are unicellular when they are shed, but several cell divisions occur before the spore coat is ruptured (Fig. 3-2). The protonema and rhizoids which grow from the germ tube that emerges from this multicellular mass may be uniseriate, biseriate, or multiseriate. The protonema, although branched, is a reduced thallus, attached to the substratum by rhizoids. As in most mosses, this chloronema produces a secondary protonema. In this genus the secondary protonema is composed of short unistratose or bistratose protonemal flaps oriented perpendicular to the chloronema (Fig. 3-2L). Near the

FIGURE 3-1
Diversity in the Andreaeidae.
(A) Andreaea rupestris with
opening sporangium on left
branch and unopened
sporangium on right branch.
(B) Completely dehisced
sporangium of *A. rupestris;*
note the elongate pseudopodium. *(C) Andreaeobryum
macrosporum* showing
sporangium with calyptra on
left branch, and unopened
sporangium on the right
branch; note the short seta.
(D) An opened sporangium of
*Andreaeobryum macrospo-
rum. (E) Andreaea wilsonii*
showing sporangium with
toothlike processes confined
to apex of sporangium (all
×16).

bases of these flaps, "buds" are initiated that produce the leafy gametophore
(Fig. 3-2*J*).

The stems show little internal differentiation, except that the outer cells are
somewhat thicker-walled than those of the interior (Fig. 3-3*H*). The stem bears
spirally arranged leaves that tend to be imbricated when dry and diverge when

moist. Leaves are generally unistratose through most of their width, but sometimes they have a multistratose costa; rarely some species have two costalike areas of multistratose bands that extend up from the base and, paralleling the margins, arch up toward the apex (Fig. 3-3*A–D*). The leaf surfaces are usually smooth, but sometimes a distinctly thickened papilla is on each cell; the papillae are particularly conspicuous on the abaxial surface of the leaf (Fig. 3-3*E,F*). Leaf shape varies from ovate to nearly lanceolate, and the margins vary from entire to somewhat toothed (Fig. 3-3*A–D*). The cell walls of the entire gametophore are usually darkly pigmented from nearly black to purplish-brown or red-brown. Most cells have chlorophyll when young, but this chlorophyll is obscured by the pigmented cell walls as the plant becomes older. On the lower portion of the older stems the leaves are often broken off or eroded; the whole gametophore is usually securely attached to its substratum by rhizoids of uniseriate or biseriate cells.

Branching of the stem is irregular, and the many densely compacted stems tend to form tufts or tight turfs on the substratum. The entire gametophore in many of the species is less than 2 cm tall, but several species, including

FIGURE 3-2
Spore germination of *Andreaea rupestris (A–J* ×300), showing endosporic development *(A–F)*, emergence of germ tubes *(G–I)*, and finally the appearance of a protonematal flap, rhizoids, and a bud *(J)*. *(K)* Multiseriate, somewhat thallose protonema showing both uniseriate and biseriate rhizoids as well as two gametophore "buds" (×65). *(L)* A fragment of protonema showing protonematal flaps (×65). *(A–J,* after Nishida, 1971; *K, L,* after Berggren, 1868.)

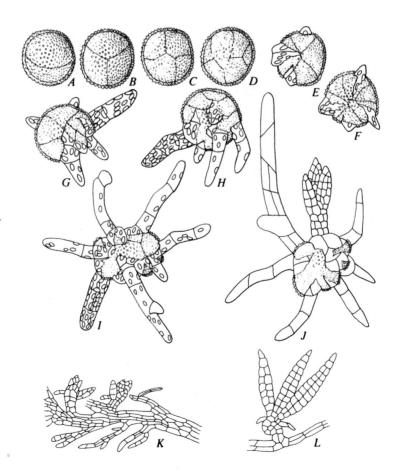

FIGURE 3-3
Features of gametophores of
Andreaeidae. *(A–G)* Leaves
(×60). *(A) Andreaea
fuegiana,* showing multistratose bands near the margins,
(B) A. rothii, showing costa,
(C) A. rupestris, lacking a
costa. *(D) A. nivalis,* showing
costa and toothed margins.
(E–H) A. rupestris. (E)
Upper part of leaf, showing
papillae on adaxial surface
(×400). *(F)* Transverse section of leaf. *(G)* Areolation of
lower portion of leaf, showing
incrassate cells (×400). *(H)*
Transverse section of stem
(×100). (*E,* after Crum and
Anderson, 1981; *F, G,* after
Ireland, 1982; *H,* after
Lorch, 1931.)

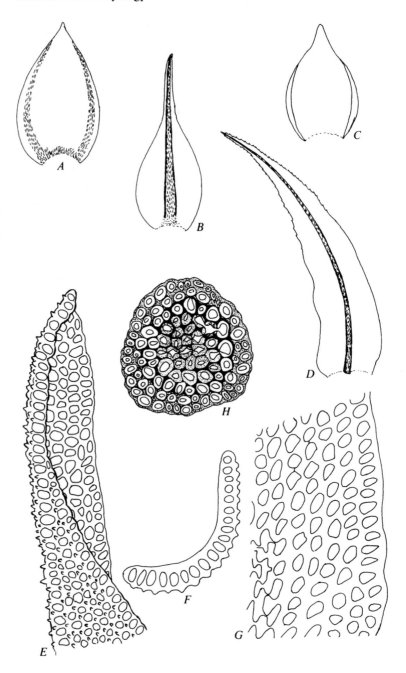

Andreaea nivalis, reach lengths of 10 cm or more. The stems are very brittle
when dry, which may result in vegetative reproduction from the fragments, but
gemmae are not produced.

Antheridia are usually on a lateral perigonial branch. The perigonial leaves are generally similar in size to the vegetative leaves and, like most perichaetial leaves, lack a costa. Since the antheridia terminate this reduced branch, they also stop the growth of that branch. The antheridia, when mature, possess a biseriate stalk several cells long, and this bears an ellipsoid antheridium. Among the several antheridia there are a few hairs that can be interpreted as paraphyses.

Perichaetia terminate the main stem, since that stem apex is used up in producing the archegonia. Near the base of the perichaetium, however, a new apical cell may be differentiated, and a branch develops. This type of branch is termed an innovation. The perichaetial leaves are usually much larger than the rest of the leaves on the stem (the vegetative leaves). Several archegonia are enclosed by the perichaetial leaves; there appear to be no paraphyses among the archegonia.

ANDREAEA: THE SPOROPHYTE

In *Andreaea,* after fertilization of the egg in one archegonium in a perichaetium, no further eggs of that perichaetium tend to be fertilized, or if more than one is fertilized, all but one embryo abort, and a single sporophyte emerges from each perichaetium. In *Andreaea* the first division of the zygote is transverse, forming two nearly equal cells. The lower cell undergoes irregular divisions to produce the foot that penetrates the gametophore, while the upper cell cuts off a small cell by an oblique division, and an apical cell is formed (Fig. 3-4*A,B*). Further cells are cut off from this apical cell until a cell mass is formed, the outer cells of which are termed amphithecium and the inner of which are termed endothecium (Fig. 3-4*E–G*). The amphithecium ultimately produces the jacket of the sporangium, while the endothecium differentiates to produce a dome-shaped mass of spore mother cells and the sterile cylinder of cells (the columella) within this mass. In this genus, as the embryo and its enveloping calyptra enlarge, further cell divisions occur in the stem beneath the foot. Elongation of these gametophytic cells pushes the sporangium up through the perichaetial leaves upon a leafless stalk termed a pseudopodium.

Development of the sporophyte differs somewhat from that in most mosses. As in other mosses, a foot is differentiated and penetrates the tissue of the stem, and the calyptra enlarges to protect the enlarging sporophyte. However, in *Andreaea* no seta is formed, but simple enlargement and elongation of the sporangium causes the calyptra to be ruptured early in sporangial development, and the fragile calyptra falls off before the sporangium is mature. The mature sporangium is elliptic; the outer cells of the jacket (exothecial cells) are thick-walled, and, as in the gametophore, the cell walls are heavily impregnated with dark brown to black pigments (Fig. 3-4*F*). There are no stomata. As the jacket matures, four or more longitudinal lines of weakness are formed, extending from the sporangium base up toward the apex, and sometimes to the apex. Within the sporangium the spore mass forms a layer like an inverted cup over and

FIGURE 3-4
Development of the sporophyte in *Andreaea. (A–E)* (×120). Early stages in development (longitudinal sections, ×120); *(E)* Shows appearance of the columella and sporogenous layer. *(F)* A nearly mature sporangium (×80), showing the apical calyptra, the central columella with overarching sporogenous layer, and the foot penetrating the apex of the pseudopodium. *(G)* Transverse section of *(F)*, showing columella (center), enlarged cells of sporogenous layer, and multistratose jacket (×100). (After Kühn, 1870.)

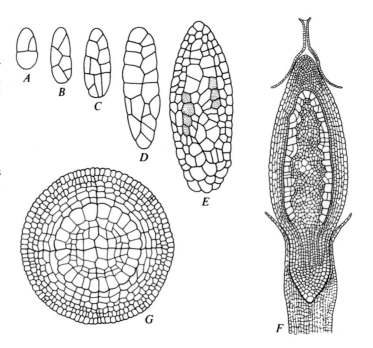

around a broadly cylindric columella. When the sporangium is mature and begins to dry out, it tends to shrink in length; the thicker-walled cells of the surface tend to shrink less than the thinner-walled cells within. This shrinkage causes the lines of weakness to gape open, and the spores are exposed and fall out. Rewetting the jacket causes the slits to close; this allows the remaining spores to be shed over a slightly extended time period (Fig. 3-1*A,B*).

ANDREAEOBRYUM: THE GAMETOPHYTE

The gametophore of *Andreaeobryum* is similar to that of *Andreaea* (Fig. 3-1*C*). The protonema is thallose, and, as in *Andreaea,* development begins within the spore coat. The leafy gametophore has spirally arranged leaves that are costate, curved, and secund (with the apices turned to one side of the stem), a feature shared with some species of *Andreaea.* The stem is structurally like that of *Andreaea,* varies from 2 to 5 cm in length, and has thick-walled epidermal cells and enlarged thin-walled inner cells. Branching is by innovations. The gametophore is attached to its rock substratum by multibranched, thick-walled, red-brown uniseriate rhizoids.

Perichaetia terminate a main shoot, and the perichaetial leaves are structurally like the vegetative leaves; the archegonia appear to be near the leaf axils. Very few (3–4) archegonia are produced in a perichaetium, and paraphyses are absent. The archegonium is very thick-stalked with a long broad neck.

Perigonia are borne in reduced lateral or terminal bulbiform branches. Perigonial leaves are smaller than vegetative leaves and lack a costa. They surround three to four elliptic, stalked antheridia that are intermixed with numerous uniseriate paraphyses, a feature shared with *Andreaea*.

ANDREAEOBRYUM: THE SPOROPHYTE

Andreaeobryum differs from *Andreaea* in many sporophytic characters. The pseudopodium is absent, the sporangium possesses a short seta, and the foot is tapered, rather than swollen. Furthermore, as in most other mosses, the calyptra persists around the sporangium until it is mature (Fig. 3-1C). Internal structure of the sporangium is as in *Andreaea*. The shape of the sporangium is very distinctive since it is turbinate (top-shaped) when mature. Four to eight dehiscence lines occupy the upper half of the sporangium, and as the sporangium dries, these toothlike divisions arch outward to release the spores within. The spores are unusually large, varying from 52 to 122 μm in diameter. In time, with alternate wetting and drying, and flexing inward and outward, these "teeth" break off and leave a small exposed pedestal consisting of the remainder of the sporangium.

RELATIONSHIPS AND FOSSIL RECORD

The two genera are very closely related. *Andreaeobryum* appears to be more closely related to costate species of *Andreaea*, with which it is almost identical in gametophytic features. It is more specialized in its ecology: on limestone instead of siliceous rock. The nature of the sporophyte is also more specialized than in *Andreaea* (presence of seta, nature of dehiscence).

Andreaea appears in the fossil record in glacial, interglacial, and postglacial deposits. In view of the habitat of the genus on exposed rock surfaces, the possibility that it would appear in the fossil record is limited. Only vegetative material is known, and modern species are represented. Since many modern species grow on volcanic rock, fossils should be sought on such substrata that were buried by volcanic ash in earlier geologic time.

The Andreaeidae appear to be closely related to no other group of mosses, although they share some basic gametophytic features with the subclass Tetraphidae, as is discussed later (see Chapter 5). The presence of a pseudopodium and the nature of the sporogenous layer are very similar to the subclass Sphagnidae, but the mode of formation of the sporogenous layer is strikingly different. Within the Andreaeidae the costate species of *Andreaea* appear to be the more generalized, especially those in which the perichaetial leaves are essentially like the vegetative leaves.

In sporophytic features there are several species that show an apparent reduction series, with an increase in specialization of the sporangium for spore dispersal (Fig. 3-5). In *A. nivalis* the dehiscence lines extend the entire length of the sporangium, while in *A. morrisonensis* they are confined to the upper half. In *A. wilsonii* the dehiscence lines are more numerous and are confined to near the apex of a cylindric sporangium; and when the sporangium opens, there are toothlike processes, as in *Andreaeobryum*. In *Andreaeobryum* the lines of dehiscence form toothlike processes, and spores are efficiently shed. The very enlarged spores in *Andreaeobryum* also suggest specialization.

ECOLOGY

The Andreaeidae often invade bare exposed rock surfaces, a feature they share with some representatives of the subclass Bryidae. They do so by production of rhizoids that penetrate irregularities of the rock surface and appear to cement themselves to the surface in a manner not clearly understood. The tolerance of the gametophore to complete desiccation also makes them effective rock colonists. Often other bryophytes and some algae invade colonies of *Andreaea*, but most clones tend to be relatively pure. Sporophytes are initiated late in the autumn in most species of temperate latitudes, and spores are shed in the early spring or when the snow cover disappears to expose populations.

FIGURE 3-5
Sporangia of Andreaeidae.
(A) Andreaea morrisonensis
(×35). *(B) A. rupestris* (×25).
(C) A. wilsonii (×50). *(D) A.
nivalis* (×30). *(E) Andreaeobryum macrosporum* (×35).

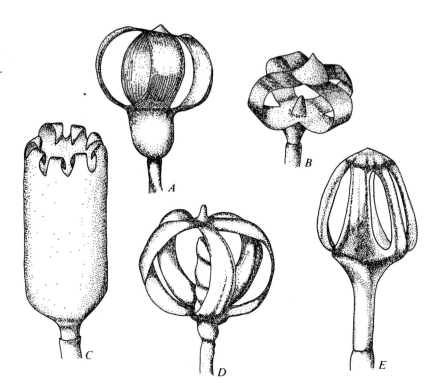

GENERALIZATIONS

Feature that distinguish the subclass from the other mosses are rather few:

1. Biseriate rhizoids
2. Multiseriate protonemata
3. Dehiscence of the sporangium by longitudinal lines

Except for the subclass Sphagnidae, the dome-shaped sporogenous layer is found only in the Andreaeidae. A pseudopodium also characterizes these two subclasses, but a similar structure appears rarely in the Bryidae.

FURTHER READING

Berggren, S. 1868. Studier öfver mossornas byggnad och utveckling. I: Andreaeaceae. *Lund Univ. Års-Skr.* **4**:1–31.

Bruch, P. H., W. Ph. Schimper, and T. Gümbel. 1855. Bryologia Europea seu genera muscorum europaeorum monographice illustrata. **6**:131–156 (fasc. 62-64.).

Chen, P-C., and T-L. Wan. 1958. A preliminary study of the Chinese *Andreaea. Acta Phytotax. Sin.* **7**:91–104.

Clifford, H. T. 1953. Victorian Musci I: Introduction and Andreaeaceae. *Proc. Roy. Soc. Victoria (N.S.)* **64**:4–9.

Greene, S. W. 1969. Studies in Antarctic bryology II. *Andreaea, Neuroloma. Rev. Bryol. Lichénol.* **36**:139–146.

Kühn, E. 1870. Zur Entwicklungsgeschichte der Andreaeaceen. *Mitth. Gesamt. Bot.* **1**:1–56.

Murray, B. 1982. Andreaeopsida, in Parker, S. P. (ed.) *Synopsis and Classification of Living Organisms.* New York: McGraw-Hill.

Nishida, Y. 1971. Studies on the formation of the protonema and the leafy shoot in *Andreaea rupestris* var. *fauriei. Bot. Mag. (Tokyo)* **84**:187–192.

Pottier, J. 1920. La parenté des Andréaeacées et des Hépatiques et un cas tératologigue qui la confirme. *Bull. Mus. Hist. Nat.* **1920**:337–344.

Redhead, S. A. 1973. Observations on the rhizoids of *Andreaea. Bryologist* **76**:185–187.

Schultze-Motel, W. 1970. Monographie der Laubmoosgattung *Andreaea.* I Die costaten Arten. *Willdenowia* **6**:25–110.

Steere, W. C., and B. Murray. 1976. *Andreaeobryum macrosporum,* a new genus and species of Musci from northern Canada and Alaska. *Phytologia* **33**:407–410.

Takaki, N. 1953. On the genus *Andreaea* of Japan. *J. Hattori Bot. Lab.* **10**:30–34.

4 The Peat Mosses — Subclass Sphagnidae

The peat mosses include a single genus *Sphagnum* in the family Sphagnaceae of the order Sphagnales. There are probably approximately 150 recognizable species, although more than 300 have been described. Since the form of the gametophore has a strong tendency to be structurally altered when growing under different environmental extremes, these environmental modifications have been given formal names. This has resulted in a multiplication in the number of species described.

Sphagnum is distributed throughout the world, and reaches its greatest abundance in the cooler temperate portion of the Northern Hemisphere where it dominates the vegetation of wetlands (Fig. 4-1). All species grow in wet areas, either in water or in sites where seepage water is available. Since *Sphagnum* is an important peat-former, it is commercially the most valuable of the bryophytes.

The genus *Sphagnum* differs from all other mosses in detail of both gametophyte and sporophyte. It shows no strong relationships with other moss subclasses, although it has many features in common with them, and presumably shared a common ancestor.

THE GAMETOPHYTE

The spores of *Sphagnum* are unicellular when shed (Fig. 4-2*D*). They have a brief period of viability (up to six months in some species), but can germinate soon after release. In a germinating spore a germ tube ruptures the spore coat and forms a short uniseriate filament; soon its terminal cell produces a unistratose multicellular thallus, and the cells near the base of this thallus initiate uniseriate rhizoids with oblique crosswalls (Fig. 4-2*E*). Ultimately a bud originates opposite the rhizoids and initiates an apical cell which produces one leafy gametophore. Sometimes filaments form on the thallus margin; the terminal cells

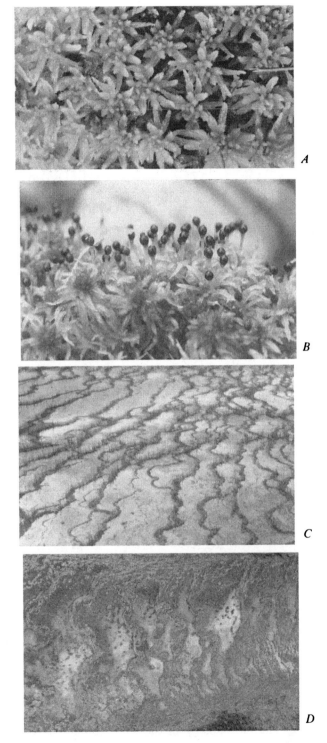

FIGURE 4-1
Sphagnidae and *Sphagnum*-dominated peatland. *(A)* *Sphagnum pulchrum* viewed from above (×1.5). *(B) S. jensenii,* showing sporophytes (×1.5). *(C)* Patterned rich fen, S. E. Yukon, Canada. *(D)* Patterned rich fen, Fort McMurray area, Alberta, Canada. (Provided by Dale H. Vitt.)

A

B

C

D

FIGURE 4-2

Gametophytic features of *Sphagnum. (A–C) S. palustre* stems. *(A)* Transverse section of stem of divergent branch, showing cortical cells with fibril thickenings and pores (×90). *(B)* Transverse section of main stem, showing multistratose cortex with pores and fibrils (×40). *(C)* Schematic longitudinal and transverse section of main stem, showing the emergency of one fascicle of branches on right; note also the central cylinder (×60). *(D)* germinating spore (×1000). *(E)* Thalloid protonematal phase with rhizoids (×100). *(F)* filamentous protonema (×330). (*C*, after Schimper, 1858; *D, F*, after Nawaschin, 1897.

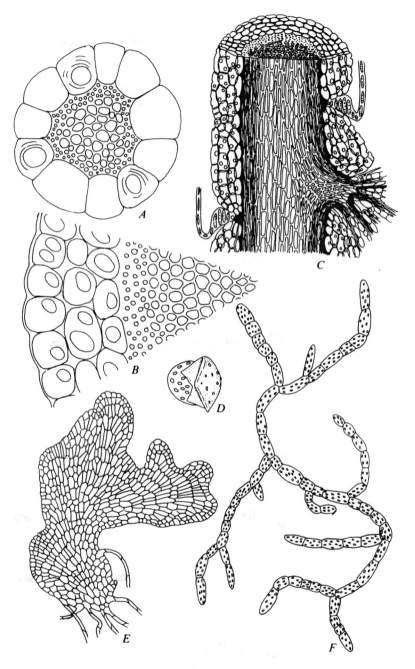

of these filaments initiate extra prothallia, and thus vegetatively reproduce the proteonematal phase from which further gametophores can arise.

The morphology of the leafy gametophore is unique (Fig. 4-3). The main stem is usually erect with the stem leaves spirally arranged and widely spaced; the mature stem leaves have little or no chlorophyll. All leaves are unistratose

FIGURE 4-3
Habit sketches of several species of *Sphagnum* (×4). *(A) S.
papillosum. (B) S. pylaesii.
(C) S. squarrosum. (D) S.
rubellum,* antheridium-
producing gametophore; note
tips of divergent branches that
are more deeply pigmented;
these hold the antheridia. *(E)
S. macrophyllum.*

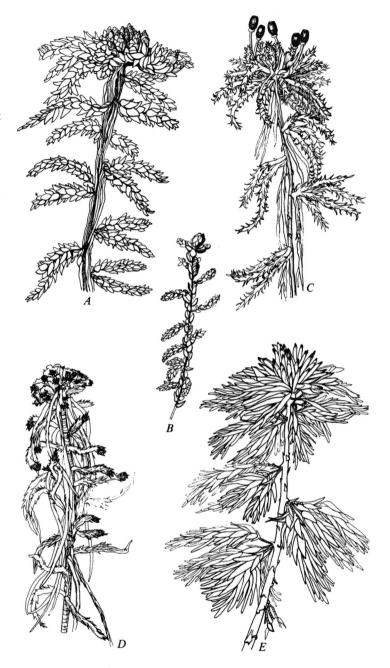

and have no costa (Fig. 4-4). The stem tissue is usually differentiated into three
distinct layers (Fig. 4-2*A*–*C*). The outer cortical cells are empty, colorless, and
often larger and thinner-walled than the cells of the central axis. These outer
cortical cells often have pores in the exposed surface, and sometimes have delicate fibril thickenings forming bands on the inner surface of the cell wall, as

FIGURE 4-4

Gametophytic features of *Sphagnum.* *(A, B) S. rubel-lum.* *(A)* Transverse section of main stem, showing multistratose cortex and central cylinder of outer cells of narrower diameter and inner cells of wider diameter (×120). *(B)* Transverse section of divergent branch, showing unistratose cortex with one retort cell apparent (×120). *(C) S. cuspidatum* stem of divergent branch, showing retort cells (×40). *(D–G)* Transverse sections of leaves from divergent branches (×100). *(D) S. rubellum;* *(E) S. magellanicum;* *(F) S. fuscum;* *(G) S. cuspidatum.* *(H)* Antheridium of *S. rubellum* (×40). *I–R)* Leaves of divergent branches (left of pair) and main stem (right of pair) (×20). *(I–J) S. tenellum;* *(K,L) S. squarrosum;* *(M, N) S. girgensohnii;* *(O, P) S. obtusum;* *(Q, R) S. teres.* (*A, B,* redrawn from Nyholm, 1969; *C,* after Eddy, 1977.)

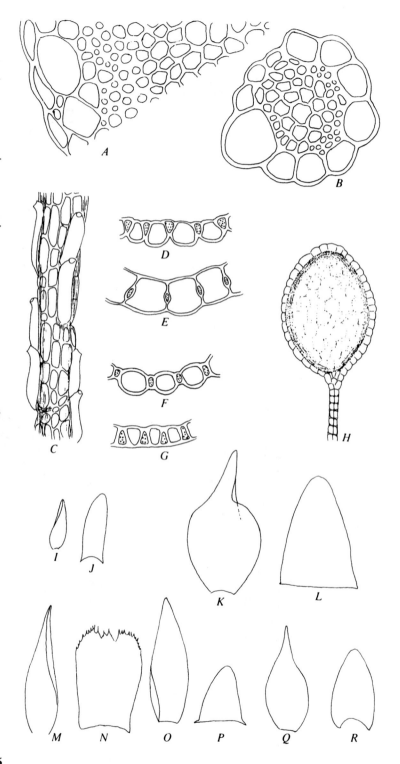

well as pores on the walls of the adjacent inner cells of the cortex. The axial cylinder is formed of a multistratose outer layer of small, pigmented thick-walled cells and a central mass of somewhat enlarged cells.

On the stem, in the position where a fourth leaf would be expected to arise, a fascicle of branches emerges, and thus these fascicles are also spirally arranged. Each fascicle consists of two or more divergent branches and two or more very slender pendent branches which droop downward, closely investing the stem (Fig. 4-3*A,C,D*). At the stem apex a crowded group of young branches forms a head or capitulum. The branches also have spirally arranged leaves, and these usually closely overlap. The branch leaves are made up of a network of elongate chlorophyllose cells, 5 or 6 of which surround each hyaline swollen, much larger, porose dead cell (Fig. 4-5). These porose cells usually also have annular fibril wall thickenings. The main stem leaves also have hyaline cells, and these cells possess or lack transverse fibrils. The chlorophyllose cells vary from triangular to trapezoidal or elliptic in cross-sectional view and are a small fraction of the diameter of the adjacent hyaline cells, which are nearly circular in cross-section. The main stem leaves usually differ in shape and structure from those of the branches.

In the stems of divergent branches, the outer cortical layer is usually unistratose, and sometimes these cells are of two types. Some of the enlarged cells are longitudinally elongate, and the pore is confined to a protuberance at the upper end of the cell; this cell is termed a retort cell, the function of which is not clear

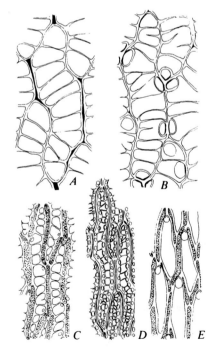

FIGURE 4-5
Details of cells of leaves from divergent branches of *Sphagnum* (×100). Note especially the fibril bands and pores of the enlarged hyaline cells. *(A, B) S. magellanicum. (A)* adaxial surface; *(B)* abaxial surface. *(C) S. cuspidatum,* abaxial surface. *(D) S. subsecundum,* abaxial surface. *(E) S. sericeum,* abaxial surface. (After Eddy, 1977.)

(Fig. 4-4C). The remainder of the outer cortical cells are elongate or quadrate, and lack pores. Sometimes the cortical cells are fibrillose and porose, and occasionally they are not clearly different from the remainder of the stem cells.

The innumerable enlarged porose dead cells in *Sphagnum* gametophores make them capable of absorbing considerable water. In some species the dried gametophores are said to absorb up to 20 times their weight in fluid. The combination of porose cells and pendent branches makes each gametophore an efficient capillary wick, and colonies expanding radially outward from a water source can carry water and dissolved nutrients to the perimeter of the colony.

Mature gametophores lack rhizoids. The gametophores vary in length from 5 to 10 cm of living stem, but undecomposed stems sometimes reach lengths of several decimeters.

Vegetative reproduction is also important in *Sphagnum*, although gemmae appear not to be produced. In culture, the protonema appears to be able to produce gemmalike structures. Young branches of the gametophore, when broken loose, are capable of forming new gametophores. The stems fragment readily, and fragmentation is an important means of asexual reproduction. Injured leaves sometimes initiate a thallus at the point of injury; these can produce new gametophores.

Most species of *Sphagnum* are dioicous, but some are monoicous and some may be either monoicous or dioicous. Initially the sexual branches are in the capitulum, but elongation of the main stem isolates them in lateral fascicles. The antheridial branches are specialized divergent branches in which a stalked spherical antheridium (Fig. 4-4H) occurs at the base of each of the leaves in the upper third of the branch. Antheridial branches are often deeply pigmented with red or brown near the branch tip (Fig. 4-3D). Many antheridial branches are borne on each gametophore.

Perichaetia are also short modified divergent branches. The perichaetial leaves are often considerably larger than vegetative leaves; they surround several archegonia that terminate the perichaetial branch, and the apical cell is used up in producing the archegonia. No paraphyses are present. Several perichaetia are usually produced on each gametophore.

THE SPOROPHYTE

When fertilization of the egg in one archegonium occurs, no others in the same perichaetium can be fertilized, or if others are fertilized, only one tends to form a mature sporangium. The first division of the zygote is transverse and forms two approximately equal cells (Fig. 4-6A). Further divisions of the upper cell result in a linear series of 6–8 cells (Fig. 4-6B,C). The lower cell undergoes irregular division to produce the foot that penetrates the upper part of the gametophore, whereas the linear series will form the remainder of the sporophyte (Fig. 4-6D,E,G). Longitudinal divisions of the linear series of cells

FIGURE 4-6
Sporophytic features of
Sphagnum. *(A–E,G)* Longitudinal sections showing development of the embryo. The
differentiation of the amphithecium is shown in *(F)*,
which is a transverse section
of a stage earlier than *(G)*.
The differentiation of the
columella (central and un-
stippled cells) and sporogenous layers (stippled cells) is
shown in *(G)* and *(H)*, with
(H) representing the transverse section of *(G)*. *(I)* A
mature sporangium
(×45), showing the oper-
culum differentiated, the
dome-shaped sporogenous
area, the dome-shaped columella, and the expanded foot.
Archegonia are apparent at
the apex of the pseudopo-
dium, and shreds of the calyptra persist at the base of the
sporangium. *(J)* A view of a
pseudostoma in the epidermis
of the sporangium jacket.
(A–E, ×500, after Bryan,
1920; *F–H,* ×100, after
Waldner, 1887.)

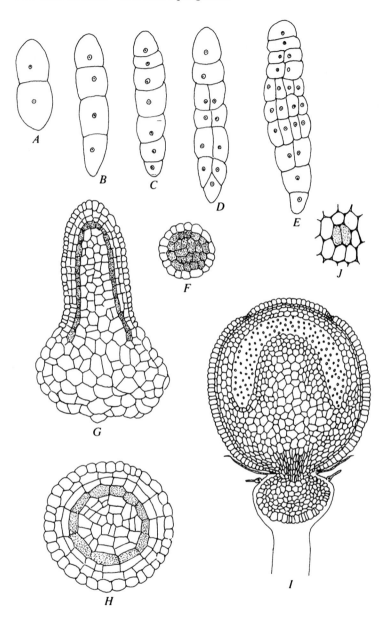

produce a multiseriate structure in which the outer cells are termed amphithecium and the inner are termed endothecium (Fig. 4-6*F–H*).

Further divisions of the endothecium result in the formation of the columella.
The amphithecium differentiates into outer cells that undergo wall thickening
and form the sterile jacket of the sporangium, while the inner cells form a mass
of spore mother cells that overarch the dome-shaped columella beneath (Fig.

4-6*l*). Each of these cells divides meiotically to form a tetrad of spores. As the embryo develops, a gametophytic stalklike pseudopodium elongates at its base; as the sporangium approaches maturity, the pseudopodium elevates the sporangium beyond the perichaetium. The calyptra sheathes the subspherical sporangium until it is nearly mature. The exceedingly delicate calyptra is torn irregularly as the sporangium enlarges, leaving a tattered fringe at the base of the sporangium.

The sporangium has no seta, and the expanded foot burrows into the upper portion of the pseudopodium. The sporangium differentiates an operculum near its apex. The jacket has numerous pairs of cells in the epidermis that resemble stomata. These may represent rudimentary stomata, but they are not functional since no hole is present between the two guard cells.

Spore dispersal in *Sphagnum* is unlike that of any other moss. As the mature subspherical sporangium dries out, it shrinks in diameter, and the columella collapses and is replaced by gaseous material. The shrinking sporangium places the gas under considerable pressure (up to five atmospheres, or equivalent to the air pressure in large tires of trucks of intermediate trailer-truck size). This pressure against the operculum forces it to be thrown off violently, and the spores are shot into the air (Fig. 4-7). Sporangia break loose from the pseudopodia soon after spore dispersal.

In many species, annual production of sporangia is very abundant, while in others it is infrequent, often as a result of the absence of both sexes of the same species near enough to allow sexual reproduction. Sporangium production is strictly seasonal, at least in the Northern Hemisphere, where a few species mature sporophytes in spring while most mature them in summer. Antheridia are fully formed by late summer, but fertilization takes place in late winter, often

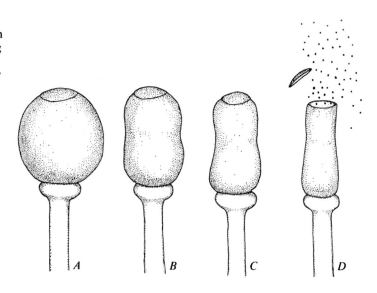

FIGURE 4-7
Dehiscence is the sporangium of *Sphagnum* (×25), showing "air-gun" mechanism. *(A)* Mature sporangium with fully differentiated operculum, surmounting pseudopodium. *(B, C)* Sporangium, showing shrinkage near equator, placing sporangium contents under considerable pressure. *(D)* When the pressure is extreme, the operculum and the spores beneath it are thrown off. (Modified after Ingold, 1939.)

during early thaws. It takes about four months from fertilization until spore maturity.

Many species of *Sphagnum* have brightly pigmented cell walls. Sometimes this color becomes very vivid, when reds, red-purple, orange, and brown betray the extensive mats of peat mosses in moorland and on subalpine slopes and ledges.

RELATIONSHIPS AND FOSSIL RECORD

Some authors would place *Sphagnum* in a separate Class Sphagnopsida because it shows unique features such as the following:

1. Fascicles of divergent and pendent branches
2. Leaf cells in a network of elongate chlorophyllose cells enclosing enlarged, empty, hyaline, porose cells with transverse fibril bands and pores on the surface
3. A thallose protonematal phase
4. Explosive dehiscence of the sporangium
5. Remarkable adsorptive capability

The position of the antheridia differs from other mosses, but is similar to some leafy hepatics. The nature of the pseudopodium and the position of the sporogenous layer are much as in *Andreaea*, although the origin of the sporogenous layer differs in these two genera.

Permian deposits of fossils in the Soviet Union yielded material of the genera *Junjagia*, *Vorcutannularia,* and *Protosphagnum* with a leaf cell pattern similar to that of *Sphagnum* (Fig. 4-8). However, the leaves are costate. These genera were treated by M. F. Neuberg as belonging to a fossil order, Protosphagnales. These data suggest a group of mosses from which both the Sphagnidae and Bryidae might have evolved, but since only gametophytic material is available, it is hazardous to speculate.

ECOLOGY

In the Northern Hemisphere, where glaciation has left numerous water bodies, *Sphagnum* species are important pioneers in vegetation establishment, and they control the dynamics of the vegetation of extensive areas.

Several species of *Sphagnum* grow submerged, at least at high water, often colonizing pond and lake margins. It is probable that much of the expansion of these pioneer species is by vegetative means when the colony forms a spongy mat in the shallows of the water body. The cell walls of *Sphagnum* have a remarkable adsorptive capacity, being able to selectively adsorb basic ions and release hydrogen ions, thus increasing the acidity of their aquatic medium.

FIGURE 4-8
Fossil mosses that show some similarities to Sphagnidae. (*A, B*) *Vorcutannularia plicata.* (*A*) Reconstruction of gametophore, showing whorls of costate leaves (×2). (*B*) Cellular detail, showing elongate cells (stippled) surrounding clearer cells, similar to chlorophyllose and hyaline cells in *Sphagnum* (×100). (*C–E*) *Protosphagnum nervatum.* (*C*) Single leaf (×6). (*D*) Detail of leaf margin, showing differentiated cells (×150). (*E*) Detail of areolation of middle of leaf showing elongate cells surrounding enlarged cells (×150). (After Neuberg, 1960.)

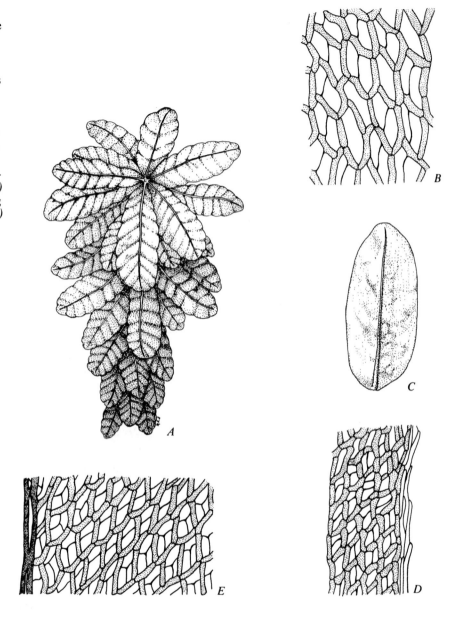

Acidity of the water can sometimes reach a pH of near 3. This acidity (combined with low oxygen availability), also inhibits the survival of many organisms, including decomposers, so that organic material accumulates in the lake marginal shallows, and peat is formed. Other plants including herbs and woody plants, colonize this peaty substratum. With the passage of time, the *Sphagnum* mat and the undecomposed material beneath it expand toward the center of the water body and finally fill in the whole lake, replacing it by the colonizing forest that invades the peaty substratum (Fig. 4-9).

Sphagnum growing around lakes surrounded by establed forest can also invade that forest. Since *Sphagnum* plants act as a sponge to fluids, they can hold a reservoir of water. These water-saturated plants can cause the forest floor to become so wet that the forest can be drowned out and be replaced by an expansion of the peatland.

In boreal regions at high latitudes the presence of permanently frozen ground not far below the soil surface impedes drainage of the water downward. This results in conditions that favor *Sphagnum* growth, especially in open areas. The situation is somewhat complicated by the invasion of the open sites by conifer forests. In open sites the summer temperatures can cause melting of the upper layers of the frozen soil. After many seasons of a freeze-thaw cycle,

FIGURE 4-9
Diagram illustrating the origin of peat deposits from open water in a depression. (*A*) With marginal aquatics, through encroachment of a *Sphagnum*-dominated peat mat; (*B,C*) To a complete filling of the basin by peat and its encroachment by forest vegetation. (After Dachnowski, 1912.)

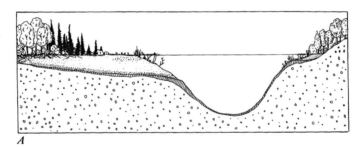

A

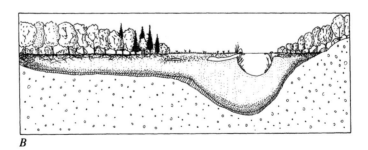

B

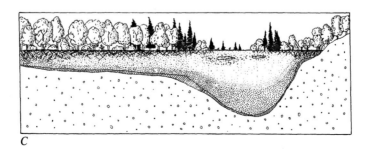

C

the permafrost layer is further from the soil surface, and the soil becomes less water-saturated. Forest is able to invade the better drained sites where *Sphagnum*, other mosses, and smaller lignified plants originally grew and produced a peat-rich substratum. When the coniferous forest establishes itself on the site, however, the deep shade of the forest lowers the summer temperature of the forest floor and prevents the summer melting of the frozen ground to the greater depths. The frozen layer then moves closer to the soil surface and surface water accumulates. This causes the trees to be killed, and *Sphagnum* is able to reinvade the newly opened site. In consequence, such areas show cyclic successions of forest and peatland vegetation.

The peat production, acidification of the aquatic site, and inhibition of decomposition lead to preservation of wind-blown spores and pollen that settle on the surface of the lake, become waterlogged, and sink to the bottom. If the lake is deep enough, these annual deposits of spores (and other material) supply an undisturbed record, on the lake bottom, of the local vegetation and regional climatic change through time. This subfossil material provides a record of the sequence of vegetational succession from the time soon after the origin of the lake following glaciation up to the present. Cores of these lake bottom deposits have been sampled in the Northern Hemisphere, and the sequence of varying vegetational patterns and changes in regional climate have been reconstructed.

ECONOMIC SIGNIFICANCE

Peatlands have an immense capacity to retain large quantities of water. This feature makes peatlands a significant influence in the hydrology and climate of the area where they are extensive.

The economic significance of *Sphagnum* is a product of both its structure and ecology. *Sphagnum* mats control and impede drainage in the terrain of the high north and on mountain slopes. This buildup of peat may enhance or inhibit forest growth, dependent upon circumstances, the economic importance of which is apparent. *Sphagnum* peat has been cut into blocks and dried, and burned as fuel. Pulverized dried *Sphagnum* peat and dried gametophores are valuable as insulation and packing material since the empty cells enclose considerable "dead" air space, and the product is light in weight.

During the Franco-Prussian, Russo-Japanese, and First World wars, dried *Sphagnum* was used in surgical dressings. Its high absorptive capacity for fluids plus its antibiotic characteristics made it especially important. During World War I it was considered more effective than cotton in the manufacture of surgical dressings, and the cotton, then in scarce supply, was released for use as wadding in gun cartridge manufacture.

Pulverized dried *Sphagnum* has been used as bedding material for both livestock and people. Its use by aboriginal people as baby diapers is of particular interest, since the moss is readily available and is biodegradable. Aboriginal people have used *Sphagnum* mixed with tallow or grease as a salve for wounds, and "sphagnol" has been extracted as a curative for skin diseases and as relief from itching, particularly from insect bites. In China, herbalists used a decoction of *Sphagnum* gametophores as a medicine for acute hemorrhage and also for eye diseases. Pulverized *Sphagnum* peat can be used as an absorbing material for oil spills. The combination of its buoyancy and remarkable absorptive capacity, plus the fact that it can be easily burned or the oil extracted, make it especially valuable. Its use as packing material and in horticulture make peat harvesting and processing a multimillion dollar industry. Since growth rates of peat may vary from 1 to 9 and up to 12 metric tons per hectare per year, it could become an important renewable resource, if intelligently harvested.

FURTHER READING

Andrus, R. E. 1980. Sphagnaceae (peat moss family) of New York State. *N. Y. State Museum Bull.* **442**:i–vi, 1–89.

Bell, P. R. 1959. The ability of *Sphagnum* to absorb cations preferentially from dilute solutions resembling natural waters. *J. Ecol.* **47**:351–355.

Bendz, G., Martensson, O., and Nilsson, E. 1966. Moss pigments 6. On the pigmentation of *Sphagnum* species. *Bot. Not.* **120**:345–354.

Blomquist, H. L. 1938. Peatmosses of the southeastern states. *J. Elisha Mitchell Sci. Soc.* **54**:1–21.

Boatman, D. J. 1977. Observations on the growth of *Sphagnum cuspidatum* in a bog pool on the Silver Flowe National Nature Reserve. *J. Ecol.* **65**:119–126.

Braithwaite, R. 1880. *The Sphagnaceae or Peat-Mosses of Europe and North America.* London: David Bogue.

Clymo, R. S. 1963. Ion exchange in *Sphagnum* and its relation to bog ecology. *Ann. Bot.* **27**:309–324.

—— 1964. The origin of acidity in *Sphagnum* bogs. *Bryologist* **67**:427–431.

—— 1970. The growth of *Sphagnum*: methods of measurement. *J. Ecol.* **58**:13–49.

Conway, V. 1949. The bogs of central Minnesota. *Ecol. Monogr.* **19**:173–206.

Coupal B., and J. M. Lalancette. 1976. The treatment of waste water with peat moss. *Water Res.* **10**:1071.

Craigie, J. S., and W. S. G. Maass. 1966. The cation exchanger in *Sphagnum* spp. *Ann. Bot. (N.S.)* **30**:153–154.

Crum, H. A. 1984. Sphagnopsida. North American Flora (Ser. II) Part 11:1–180.

Damman, A. W. H. 1978. Distribution and movement of elements in ombrotrophic peat bogs. *Oikos* **30**:480–495.

—— 1979. Geographic patterns in peatland development in eastern North America. *Proc. Int. Symp. Classif. Peat and Peatlands, Int. Peat Soc.,* pp. 42–57.

Dansereau, P., and F. Segadas-Vianna. 1952. Ecological study of the peat bogs of eastern North America I. Structure and evolution of vegetation. *Can. J. Bot.* **30**:490–520.

Drury, W. H. 1956. Bog flats and physiographic processes in the Upper Kuskoquim River region, Alaska. *Contrib. Gray Herb., Harvard Univ.* **178**:1–30.

Eddy, A. 1977. *Sphagnales of Tropical Asia. Bull. British Museum (Nat. Hist.)*, Vol. 5.

——— 1979. Taxonomy and evolution of *Sphagnum*, in Clarke, G. C. S., and Duckett, J. G. (eds.), *Bryophyte Systematics*, pp. 109–121. London: Academic Press.

Eurola, S. 1962. Über die regionale Einteilung der südfinnischen Moore. *Ann. Bot. Soc. Bot. Fenn. 'Vanamo'* **33**:1–243.

Gorham, E. 1953. Some early ideas concerning the nature, origin and development of peat lands. *J. Ecol.* **41**:257–274.

——— 1957. The development of peat lands. *Q. Rev. Biol.* **32**:145–166.

Heinselman, H. L. 1963. Forest sites, bog processes, and peatland types in the Glacial Lake Agassiz Region, Minnesota. *Ecol. Monogr.* **33**:327–374.

——— 1970. Landscape evolution, peatland types, and the environment in the Lake Agassiz Peatlands Natural Area, Minnesota. *Ecol. Monogr.* **40**:235–261.

Hennezel, F., and B. Coupal. 1972. Peatmoss—a natural absorbent for oil spills. *Can. Min. Met. Bull.* **65**:51–53.

Hill, M. O. 1978. Sphagnopsida, in Smith, A. J. E. (ed.), *The Moss Flora of Britain and Ireland*, pp. 30–78. Cambridge: Cambridge University Press.

Horton, D. G., D. H. Vitt, and N. G. Slack. 1979. Habitats of circumboreal-subarctic *Sphagna*. I. *Can. J. Bot.* **57**:2285–2317.

Isoviita, P. 1966. Studies in *Sphagnum* L. I. Nomenclatural revision of the European taxa. *Acta Bot. Fenn.* **3**:199–264.

Katz, N. J. 1926. *Sphagnum* bogs in central Russia: phytosociology, ecology and succession. *J. Ecol.* **14**:177–202.

Kivinen, E., and P. Pakarinen. 1981. Geographical distribution of peat resources and major peatland complex types in the world. *Ann. Acad. Sci. Fenn. Series A. III Geologica-Geographica*, No. 132.

Leverin, H. A. 1943. Peat moss or *Sphagnum* moss. Its use in agriculture industry and in the home. Canada Dept. of Mines & Res. Publ. 809, 1–10.

Lane, D. M. 1977. Extent of vegetative reproduction in eleven species of *Sphagnum* from northern Michigan. *Michigan Bot.* **16**:83–89.

Lange, B. 1982. Key to northern boreal and arctic species of *Sphagnum*, based on characters of the stem and leaves. *Lindbergia* **8**:1–29.

Miller, N. G. 1981. Bogs, bales and BTU's: a primer on peat. *Horticulture* **59**:38–45.

Moore, P. D., and D. J. Bellamy. 1973. *Peatlands*. New York: Springer-Verlag.

Mörnsjö, T. 1969. Studies on vegetation and development of a peatland in Scania, South Sweden. *Opera Bot.* **24**:1–186.

Nichols, G. E. 1918. The *Sphagnum* moss and its use in surgical dressings. *J. N.Y. Bot. Garden* **19**:203–220.

Noguchi, A. 1958. Germination of spores in two species of *Sphagnum. J. Hattori Bot. Lab.* **19**:71–75.

Pakarinen, P. 1978. Production and nutrient ecology of three *Sphagnum* species in southern Finnish raised bogs. *Ann. Bot. Fenn.* **15**:15–26.

Paton, J. A., and P. J. Goodman. 1955. The conditions promoting anthocyanin formation in *Sphagnum nemoreum* Scop. *Trans. Br. Bryol. Soc.* **2**:261–267.

Pedersen, A. 1975. Growth measurements of five *Sphagnum* species in South Norway. *Norwegian J. Bot.* **22**:277–284.

Puustjarvi, V. 1956. On the cation exchange capacity of peats and on the factors of influence upon its formation. *Acta Agr. Scand.* **6**:410–449.

Rose, F. 1953. A survey of the ecology of the British lowland bogs. *Proc. Linn. Soc. London* **164**:186–211.

Savicz-Ljubitzkaja, L., and Z. N. Smirnova. 1968. *Handbook of Sphagnaceae of the U.S.S.R.* Leningrad: Akad. Nauk. SSSR Komarov Bot. Inst. (in Russian).

Sjörs, H. 1961. Surface patterns in boreal peatland. *Endeavour* **20**:217–224.

Suzuki, H. 1972. Distribution of *Sphagnum* species in Japan and an attempt to classify the moors basing on their contribution. *J. Hattori Bot. Lab* **35**:2–24.

Tallis. J. H. 1962. The identification of *Sphagnum* spores. *Trans. Br. Bryol. Soc.* **4**:209–213.

Vidal, H. 1963 (1968). Importance of northwestern and central European bogs as regulators of water balance and climate. in *Trans. Second Int. Peat. Congr., Leningrad* **I**:167–168.

Vitt, D. H., P. Achuff, and R. E. Andrus. 1975. The vegetation and chemical properties of patterned fens in the Swan Hills, north-central Alberta. *Can. J. Bot.* **53**:2776–2795.

———, and R. E. Andrus. 1977. The genus *Sphagnum* in Alberta. *Can. J. Bot.* **55**:331–357.

———, H. Crum, and J. A. Snider. 1975. The vertical distribution of *Sphagnum* species in hummock-hollow complexes in northern Michigan. *Michigan Bot.* **14**:190–200.

———, and N. G. Slack. 1975. An analysis of the vegetation of *Sphagnum*-dominated kettle-hole bogs in relation to environmental gradients. *Can. J. Bot.* **53**:332–359.

——— 1984. Niche diversification of *Sphagnum* relative to environmental factors in northern Minnesota peatlands. *Can. J. Bot.* **62**:1409–1430.

Warén, H. 1976. Untersuchungen über Sphagnumreiche Pflanzengesellschaften der Moore Finnlands. *Acta Soc. Fauna Flora Fenn.* **55**(8):1–33.

Warnstorf, C. 1881. *Die Europäischen Torfmose. Eine Kritik und Beschreibung derselben.* Berlin.

——— 1911. Sphagnales-Sphagnaceae (Sphagnologia Universalis), in Engler, A. (ed.), *Das Pflanzenreich.,* Vol. 51, 546 pp. Leipzig.

Watson, W. 1918. Sphagna, their habits, adaptations and associates. *Ann. Bot.* **32**:536–551.

Wells, E. D. 1981. Peatlands of eastern Newfoundland: distribution, morphology, vegetation, and nutrient status. *Can. J. Bot.* **59**:1977–1997.

Worley, I. A. 1981. Maine peatlands. Maine State Planning Office Report No. 73. Augusta, Maine.

5

The Four-Toothed Mosses — Subclass Tetraphidae

This subclass is often considered to be the most generalized of the "true" mosses. The gametophore shares many features with the Bryidae; the sporophyte, although superficially resembling that of the Bryidae, has such an unusual peristome that the subclass is treated as an independent evolutionary line.

The subclass Tetraphidae contains a single order, Tetraphidales, with one family, Tetraphidaceae, and two genera: *Tetraphis* (two species) and *Tetrodontium* (three species). The sporophyte has the most structurally generalized peristome teeth known in the mosses. The gametophyte also shows numerous generalized features in the nature of the protonema and the gametophore (Fig. 5-1). The family Calomniaceae, with the single genus *Calomnion* (three species) of Australasian distribution, is sometimes placed in the subclass Tetraphidae. It shows many gametophytic features extremely similar to *Tetraphis*. No peristome teeth are produced in *Calomnion*, and since its gametophytic features are the same as those of many genera of the subclass Bryidae, it is placed conservatively in that subclass. It is possible that careful study of the development of the potential peristomial layers of *Calomnion* may contribute to a better understanding of the relationships of this genus. Details of the protonemal stage have not been reported and would be valuable in making interpretations.

The genus *Tetraphis* exhibits a circumpolar distribution in coniferous and mixed forests of the Northern Hemisphere. It grows mainly on decomposing wood in somewhat shaded sites, but sometimes appears on peaty banks and occasionally on sandstone. The genus *Tetrodontium* is distributed in widely scattered localities throughout the Northern Hemisphere and is also in New Zealand, where it is rare. It is confined to siliceous rock, and grows attached to shaded faces of grottoes and crannies.

FIGURE 5-1

Tetraphidae. (*A*–*C*, *E*–*F*)
Tetraphis pellucida. (*A*)
Sporangium with calyptra
(×14). (*B*) Gametophore bearing sporophyte with calyptra
removed and operculum in
place (×14). (*C*) Sporangium
with operculum removed,
showing peristome teeth
(×14). (*D*) *Tetraphis geniculata* sporophyte (×14). (*E*)
Gametophore with perigon-
ium on left branch (×45). (*F*)
Gemma cup with gemmae
(×45). (*G*) *Tetrodontium
brownianum* on rock substratum; note the protonematal
flaps affixed to the substratum (×22).

50

THE GAMETOPHYTE

Spores of both genera are unicellular when shed. The germinating spore produces a uniseriate much-branched filamentous chlorophyll-rich protonema that is attached to the substratum by rhizoids. This protonema produces perpendicular protonemal flaps. These flaps are chlorophyllose and unistratose and persist for long periods in *Tetrodontium*, but in *Tetraphis* they disappear soon after the leafy gametophores begin to form turfs (Fig. 5-2*A,B*).

Buds that produce leafy gametophores originate near the base of the protonemal flaps or from the caulonema that arises in the same position. The gametophores bear radially arranged costate leaves in which all cells have chlorophyll. The gametophores are greatly reduced in *Tetrodontium* (often 0.5–1 mm tall), sometimes forming only perichaetia and perigonia, with the protonemal phase long persistent and apparently carrying on most of the photosynthesis (Fig. 5-1*G*). In *Tetraphis* the gametophores are usually less than 1 cm tall. The first gametophores produced in *Tetraphis* tend to form an apical cuplike cluster of reduced blunt leaves that enclose gemmae (Fig. 5-1*F*). The cuplike cluster probably enhances raindrop splash dispersal of the gemmae. The gemmae are lens-shaped and long-stalked (Fig. 5-2*D*). These vegetative diaspores, when shed, germinate in much the same way as spores and produce the protonemal phase with protonemal flaps before the production of more gametophores. In *Tetraphis* the flaps are broadly obovate and taper from the filamentous protonema (Fig. 5-2*B*). The flaps are sometimes produced on the leafy stem of the gametophore, and can serve as vegetative diaspores.

Later, from the same protonema that produces gemmiferous shoots, leafy shoots form that are terminated by a cluster of elongate perichaetial leaves, which surround 8–10 archegonia, among which are a few filamentous paraphyses. On all stems the leaves are in 5–9 ranks: the lowermost leaves are small, but those on the upper portion of the stem are bright green, ovate and costate, and the hexagonal cells are filled with numerous chloroplasts. Uniseriate persistent rhizoids attach the base of the gametophore to the substratum. The stem anatomy shows some cellular differentiation, with thick-walled cortical cells, cells of wider diameter forming most of the inner part of the stem, and with a narrow central strand of small cells composed of several hydroids, cells specialized for water conduction (Fig. 5-2*E*).

After archegonia are produced, the apical cell of the stem ceases further growth, and any growth from that stem must be accomplished by the initiation of lateral buds, producing innovations. If fertilization of the eggs in a perichaetium does not take place, in the following year one or two innovations form within the perichaetium. Each is terminated by a bulbiform perigonium that encloses numerous antheridia intermixed with many filamentous paraphyses, or one branch may bear a perichaetium and the other a perigonium.

FIGURE 5-2
Gametophytic features of
Tetraphidae. *(A, B)* Protonematal flaps. *(A) Tetrodontium
brownianum* (×140). *(B–E)
Tetraphis pellucida* (×70).
(C) Young sporophyte developing directly from diploid
protonema (×60). *(D)*
Gemma (×60). *(E)* Transverse
section of stem showing central strand with hydroids
(×80). (*A*, after Bruch et al.,
1843. *B, C*, after Bauer,
1956).

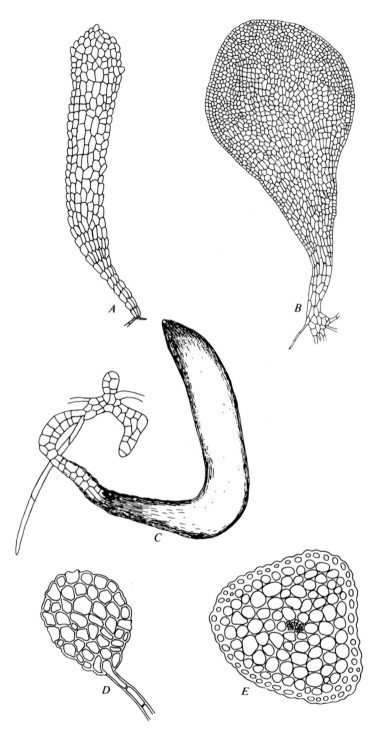

THE SPOROPHYTE

Development of the zygote in *Tetraphis* proceeds much as in *Andreaea* except that the sporogenous layer surrounds the cylindrical columella, a feature that characterizes most other subclasses of mosses. The jacket of the sporangium is formed from the amphithecium, whereas all other cells in the sporangium originate from the endothecium. Growth of the embryonic sporophyte is accompanied by growth of tissue below the venter. In this way a large, pleated, campanulate calyptra is formed that sheathes the embryo. As the sporophyte elongates, the base of the calyptra is torn loose and carried upward on the tip of the elongating seta (Fig. 5-1*A*). The young sporophyte is chlorophyllose and the tapered foot penetrates the apex of the gametophore. This allows for transfer of synthesized material, water, and dissolved minerals from the gametophore to the developing sporophyte. The seta has a central conducting strand. Most of the cells external to the central strand are thick-walled and give the seta rigidity. Such thick-walled cells are termed stereids.

As the sporangium enlarges, the calyptra usually sheathes only the upper half, but it persists until the sporangium is mature. The exothecial cells in the sporangium of *Tetraphis* are predominantly elongate and thick-walled. There are no stomata in the sporangial wall of *Tetraphis,* but these are a few in *Tetrodontium.*

Within the sporangium of both *Tetraphis* and *Tetrodontium,* the columella forms a multicellular central cylinder, and the sporogenous layer surrounds it. An apical cone of the sporangium jacket is cut off as an operculum and, beneath this, all of the cells immediately above the sporogenous cylinder and columella divide more or less equally to form four wedge-shaped peristome teeth consisting of whole cells (Fig. 5-3). When the operculum is shed, the peristome teeth are exposed. Upon drying, they gape open somewhat, and some of the spores are released with an abrupt movement of the sporophyte. Some movement results when the seta twists as it dries out and untwists as it becomes moistened again. Spores are protected from excessive rewetting since the peristome teeth move inward when wet. Dehydration causes the teeth to gape open again, and more spores are released. This relatively slow process increases the chances that some spores will be released under conditions that favor their effective dispersal and germination.

The persistent protonemal flaps in *Tetrodontium* are narrowly lanceolate and biseriate; occasionally they are forked (Fig. 5-2*A*). Both perichaetial and perigonial plants are produced from the same protonema. The part of the gametophore that actually bears the sex organs is essentially stemless and consists of a mass of cells to which the reduced perichaetial or perigonial leaves are attached. The perichaetia enclose a few archegonia and intermixed short paraphyses; the perigonia enclose a few antheridia and filiform paraphyses. Short, slender leafy branches sometimes arise at the base of perichaetia. These vegetative branches are rarely more than 2 mm long and bear narrowly ovate leaves. Perhaps, since the gametophore is so greatly reduced, it is important

FIGURE 5-3
Tetraphis pellucida peristome
teeth (×100). *(A)* Longitudinal
view of a single multicellular
tooth. *(B)* Transverse section
of two peristome teeth.

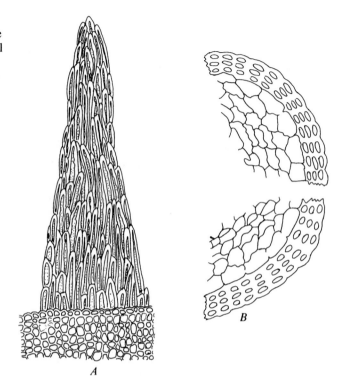

that the sporophyte be a more efficient photosynthetic structure, and thus it
possesses stomata that might improve its efficiency in gas exchange and reduce
its dependence on photosynthate derived from the very diminutive gametophore. The reduced gametophore may also strongly influence the restriction
of this genus to constantly humid shaded sites where water loss is likely to be
impeded, and where moisture is more readily available for absorption by the
sporophyte.

RELATIONSHIPS

The subclass shows some features in common with the subclass Polytrichidae, especially in its possession of multicellular peristome teeth, and some
with the Andreaeidae, especially the protonematal flaps, but the greatest
number of gametophytic and sporophytic features are shared with the subclass
Bryidae. These features include the general form and structure of the
sporophyte (the structure and presence of the seta, the mode of formation and
basic structure of the sporangium), and the gametophyte (nature of protonema
and spores, general form of the gametophore, structure and position of the
rhizoids). The subclass appears to represent a very generalized type of peristome and, to a large degree, a very generalized gametophore and protonema.

EXPERIMENTAL RESEARCH

Leopold Bauer was able to culture fragments of young sporophyte segments of *Tetraphis pellucida* that gave rise to diploid protonemata. These protonemata, upon forming buds, produced sporophytes directly without a leafy gametophore (Fig. 5-2C). As will be shown later, in the Bryidae, diploid gametophores have been induced from sporophyte fragments, and from these leafy gametophores sporangia arose without a sexual stage. This remarkable plasticity in the morphology of the diploid phase suggests a possible procedure in which polyploid races of some species could originate in nature. It also emphasizes the very generalized condition of the cells that have a potentiality to produce variant morphology dependent on the external (and presumably internal) conditions.

GENERALIZATIONS

The subclass Tetraphidae is unique in the mode of formation of the peristome of four multicellular teeth. A protonema with protonematal flaps is not of general occurrence in the mosses, nor is the formation of the gemma-cups with complex lenticular gemmae borne on slender stalks. The change in sexuality of a gametophore from one year to the next appears also to be extremely rare. In general morphology the gametophore resembles the subclass Bryidae, and, except for the nature and formation of the peristome, the sporophyte is also very similar to that subclass.

Within the subclass, the genus *Tetraphis* is the more generalized in gametophore morphology while *Tetrodontium* is remarkably reduced and specialized. The sporophyte of *Tetrodontium* appears to be more generalized than that of *Tetraphis*. The relationships of the subclass are unclear. The multicellular peristome suggests closest relationships with the Polytrichidae, yet the structure of the gametophore is very close to that of some genera of the subclass Bryidae.

FURTHER READING

Bauer, L. 1956. Über vegetative Sporogonbildung bei einer diploiden Sippe von *Georgia pellucida*. *Planta* **46**:604–618.

Bruch, P., W. P. Schimper, and T. Guembel. 1843. *Tetraphis*, in *Bryologia Europaea*, fasc. 17, pp. 1–7 and plate. Ibid., *Tetrodontium*, pp. 1–5 and plate.

Forman, R. T. T. 1962. The family Tetraphidaceae in North America: Continental distribution and ecology. *Bryologist* **65**:280–285.

—— 1964. Growth under controlled conditions to explain the hierarchical distribution of a moss, *Tetraphis pellucida*. *Ecol. Monogr.* **34**:1–25.

Hodgetts, W. J. 1915. Vegetative production of flattened protonema in *Tetraphis pellucida*. *New Phytol.* **14**:43–49.

Muraoka, S., and A. Noguchi. 1961. The formation of gemmae and their germination in *Tetraphis pellucida* and *T. geniculata*. *Miscell. Bryol. Lichenol.* **2**:83–85.

Nishida, Y. 1970. Studies on the formation of the protonemata of *Tetraphis pellucida*. *Bot. Mag. Tokyo* **83**:423–427.

Schneider, M. J., and A. J. Sharp. 1962. Observations on the reproduction and development of the gametophyte of *Tetraphis pellucida* in culture. *Bryologist* **65**:154–166.

Wallner, J. 1932. Zur Klarung der Frage nach dem morphologischen Wert der Protonemabäumchen von *Georgia pellucida*. *Hedwigia* **72**:175–183.

Weber, W. A. 1977. *Tetraphis pellucida* and *T. geniculata*: Scindulae as diagnostic features in bryophytes. *Bryologist* **80**:164–167.

6 The Hair-Cap Mosses — Subclass Polytrichidae

The hair-cap mosses have received considerable study. In most elementary botany textbooks, if mosses are treated at all, the genus *Polytrichum* is used as the representative moss. This is probably a result of the broad distribution and easy availability of *P. commune* and *P. juniperinum,* especially in temperate climates. Furthermore, these mosses are relatively large, so that most features can be seen readily.

The subclass Polytrichidae contains a single order, Polytrichales, with two families, Polytrichaceae and Dawsoniaceae. There are at least 19 genera and approximately 370 species (Fig. 6-1).

The subclass is distributed throughout the world from the high Arctic to Antartica, from sea level to high alpine elevations. The Polytrichidae grow most frequently on siliceous or acidic substrata; most occur directly on mineral soil, but many grow on peat, and a few are on rock. Many species are found in exposed sites while others thrive best in somewhat shaded areas.

THE GAMETOPHYTE

The spores of Polytrichidae are extremely small (5–8 μm in diameter in *Dawsonia*). Indeed, a single sporangium in some species of *Dawsonia* can produce more than 65 million spores. The spores are unicellular and, upon germination, produce a uniseriate germ tube. This tube produces branching rhizoids and green creeping branches of the chloronema, from which erect branches often arise. The chloronema (including the erect portion) has perpendicular transverse walls. The erect portion is usually richly chlorophyllose. The caulonema and rhizoids have oblique cross-walls. The entire protonema, since it consists of different kinds of hairlike branches, is described as heterotrichous (much as in Fig. 8-4C). In time, the caulonematal portion produces buds, and these buds initiate leafy gametophores. In some genera the protonemata form a green algalike coating over mineral soil. In some species of *Pogonatum* (includ-

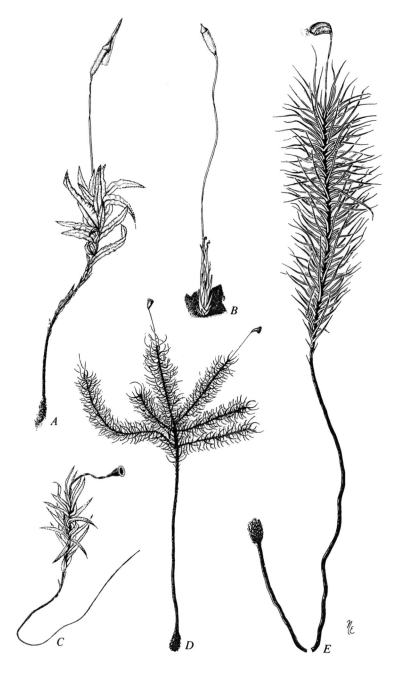

ing *P. contortum* and *P. tahitiense*), this coating disappears when the leafy
gametophores form a continuous turf. In other species of *Pogonatum* (e.g., *P.
pensilvanicum*, *P. spinulosum*), as well as species of *Racelopus* and
Racelopodopsis, the protonemata persist for several years.

The gametophore reaches its greatest structural complexity in the Polytrichidae, and because of this tissue differentiation, free-standing erect stems of *Dawsonia superba* sometimes reach heights of up to 60 cm or more (Fig. 6-1*E*). Other moss gametophores are frequently larger, but generally they must either be supported by an aquatic medium or recline or are epiphytic. In some subantarctic sites, stems of *Polytrichum strictum* reach lengths of several meters, but these are in dense undecomposed turfs, of which only the upper few centimeters appear to be living.

The stem shows considerable tissue differentiation. In both *Dawsonia* and *Polytrichum* there is a well-defined conducting system in both the stem and leaves (Figs. 6-2 and 6-3). Numerous rhizoids emerge from the epidermal cells of the subterranean portion of the stem in some species of *Polytrichum*. Beneath the epidermis is a multistratose cortex of thick-walled cells (stereids), separated from the broad central strand by a single layer of enlarged cells, sometimes interpreted as an endodermis. The central cylinder of cells of this subterranean portion is made up mainly of thick-walled supportive stereids, among which are water conducting cells, the hydroids. These are analogous to the tracheid elements of lignified plants. There are also three clusters of phloemlike metabolite conducting cells in the outer part of the central strand,

FIGURE 6-2
Transverse section of central portion of a stem of *Polytrichum commune* (×400), showing central hydrome and encircling cells. The leptoids are not easily distinguishable from the matrix of cells that make up the cortical cells; only a few layers are shown. (After Janzen, 1918.)

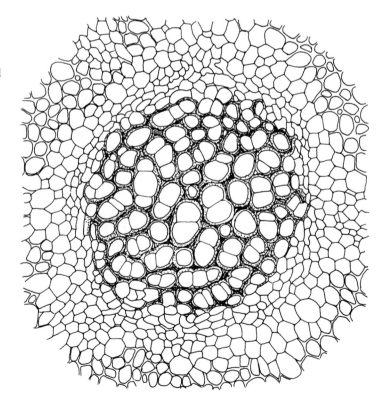

FIGURE 6-3
Diagrammatic representation
of the conducting system of
Polytrichum commune, showing hydroids (h) and leptoids
(l) arranged much as are
vascular tissues in lignified
plants (×450). Note the perfo-
rations of the end-walls and
the presence of nuclei in the
leptoids. (Redrawn after Hébant, 1977.)

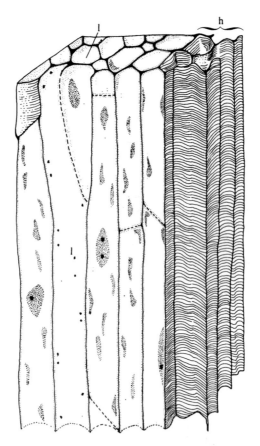

which interrupt the "endodermis." These conducting cells are termed leptoids
and are analogous to sieve elements. In the aerial portion of the stem of the
same plant, there are also similar layers of cells, except that the central portion
is made up mainly of hydroids (and is termed the hydrome), and it is encircled
by a discontinuous layer of leptoids (that make up the leptome). There is a
broad cortex, but there is no layer that resembles the "endodermis" of the subterranean portion. This hydrome-leptome system forms a relatively efficient internal conducting system in the gametophore.

   The leaves are usually in six to eight spirally arranged ranks. Leaves tend to
be long-lanceolate, costate, and often possess an abruptly expanded colorless
base that sheaths the stem (Fig. 6-4*A,D*). The sheathing bases, sometimes
accompanied by a felt of rhizoids that close the stem, form an effective external capillary conducting system. The sheathing base usually lacks chlorophyll.
Its cells are generally elongate, while cells of the blade are mainly quadrate.
The transition from blade to sheath cells is often abrupt, and a "hinge tissue"
is frequently present. This tissue, by swelling when wet and shrinking when
dry, controls inward and outward flexing of the leaf blade.

FIGURE 6-4
Leaves of Polytrichidae, showing diversity in structure. (*A*) *Polytrichum juniperinum*, showing lamellae on costa and involute margins forming a chamber enclosing lamellae (×20). (*B*) *Bartramiopsis lescurii*, showing toothed lamellae and ciliate margins (×30). (*C*) *Pseudoracelopus philippinensis* (×25), showing absence of lamellae. (*D*) *Oligotrichum hercynicum*, showing sinuose lamellae (×30). (*E*) *Atrichum selwynii* showing lamellae on costa, and differentiated leaf margin (×25).

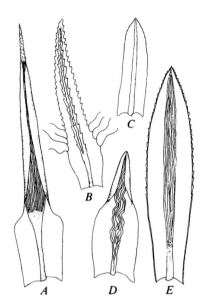

On the adaxial surface of the leaf, especially on the costa, but not always confined to it, are few to many unistratose parallel flaps of tissue that extend the length of the leaf (Figs. 6-4, 6-5, 6-6, and 6-7). These lamellae are richly filled with chlorophyll and considerably increase the photosynthetic area of the leaf. The blade that extends laterally from the costa tends to be unistratose and is sometimes chlorophyllose, but it may be colorless, and the blade margins may be folded over the lamellae, enclosing them in a longitudinal chamber and preventing rapid water loss in them. The sheathing base lacks lamellae. The interior of the costa is composed of thick-walled supportive cells, with scattered hydroids and leptoids (Fig. 6-6*B* and 6-7*C*), sometimes leading via leaf traces to the hydrome-leptome system of the stem.

The abaxial surface of the leaf and the upper cells of the lamellae are often cutinized. In most genera, the leaves are markedly divergent when moist, exposing their upper surfaces to maximum illumination. As the leaves dry out they may become contorted and much twisted or become imbricated against the stem, slowing down further water loss. Leaf margins may be toothed or toothless (entire). In some genera in which the gametophores consist of only tiny sexual shoots, the leaves lack lamellae and are not markedly chlorophyllose. In most such genera (*Pseudoracelopus, Racelopus, Racelopodopsis*), the heterotrichous protonemata are perennial and form a cobweblike coating over the substratum (usually mineral soil). The stems of some genera show very little internal tissue differentiation, and sometimes lack hydroids or leptoids, especially when grown under extremely humid conditions. This is true even in species that normally have an elaborate conducting system.

Only *Alophosia* produces gemmae; these gemmae are disclike, as in *Tet-*

FIGURE 6-5
Detail of lamellae of *Polytrichum commune* as recorded by scanning electron microscopy; scales given with each figure. *(A)* View of upper surface of leaf about halfway between apex and base, showing toothed margins of lamina and the numerous parallel lamellae as viewed from above. *(B)* Diagonal view of lamellae, showing the grooved apical cell of the lamellae. (Provided by Dale H. Vitt.)

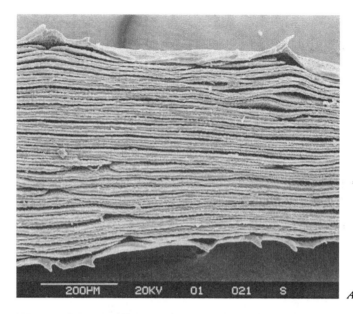

A

B

raphis, but are much larger. Fragments of the subterranean stem can also act as vegetative diaspores in many species of Polytrichidae. In artificial culture almost any portion of a gametophore can vegetatively regenerate a gametophore. It is possible also to induce callus tissue in culture, and these masses of tissue, when isolated and given suitable growth conditions, can be induced to produce "buds" that grow into leafy gametophores.

Within the Polytrichidae a certain variety of diversification in gametophore structure suggests trends in structural reduction. The genera that show these features are not necessarily derived from each other in a neat linear fashion; the extant plants simply show persistent examples of a pattern of structural simplification that may have occurred through the passage of time. In *Dawsonia* and *Polytrichum,* for example, there is the remarkable differentiation of a conducting strand with associated complexity in the nature of the costa, the presence of a sheathing leaf base, and elaboration of the lamellae (which tend to be numerous and many cells high). In both genera there are some species in which the lamellae contain all of the chlorophyll of the leaf, and the hyaline leaf lamina overlaps the lamellae, resulting in an enclosed "chlorenchyma."

Considerable reduction from this elaborate leaf and stem structure is represented in *Atrichum,* in which the lamellae, restricted to a narrow costa, are reduced to a few cells in height. There is no sheathing leaf base, and the entire

FIGURE 6-6
Transverse sections through leaves of Polytrichidae. (*A*) *Oligotrichum aligerum,* showing lamellae on both surfaces (×300). (*B*) *Polytrichum commune,* showing chlorophyll-rich lamellae with notched apical cells (×450); note also the stereid cells above and below the guide cells. Hydroids are in small groups (usually pairs), especially below the guide cells. (*A,* after Lorch, 1931; *B,* after Hébant, 1977.)

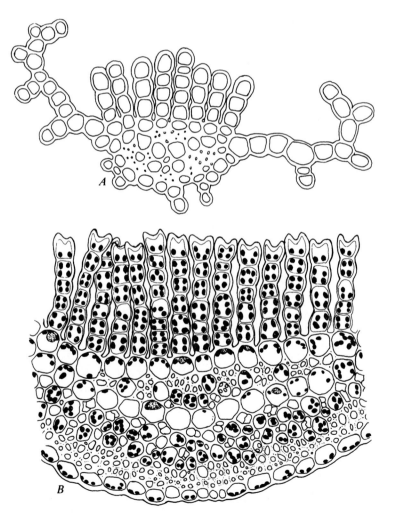

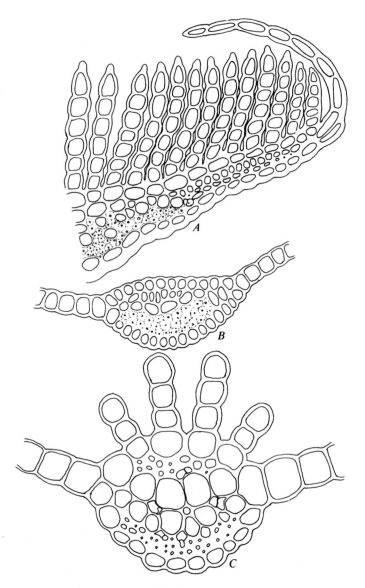

leaf is chlorophyllose; there is a bistratose margin of elongate cells. In
Alophosia the entire blade is bistratose, the costa lacks lamellae, and there is no
tissue differentiation in the costa. The most extreme reduction is in genera in
which the gametophyte has a persistent protonema which bears only reduced
sexual shoots.

Gametophores are dioicous in most Polytrichidae (Fig. 6-8*A*,*C*). Sometimes
the perigonial stem of the male gametophore bears normal vegetative leaves in
its early stages of growth, but then produces an apical perigonium; this is composed of expanded, often reddish or brownish perigonial leaves that form the

rim of a cuplike structure. At the axils of each of these leaves are many antheridia and associated filamentous paraphyses. These paraphyses are multiseriate and club-shaped in some genera, while in others they are uniseriate. Antheridia are elongate, with a tapering short stalk. In the center of the perigonium, the apical cell renews growth and continues to produce further

FIGURE 6-8

Polytrichum juniperinum. (*A*) Habit sketch, showing sporophyte with calyptra (×4). (*B*) Sporangium with calyptra removed, showing operculum, angled sporangium, and basal apophysis (×8). (*C*) Antheridium-producing gametophore, habit sketch, showing two rosettes of perigonial leaves (×4). (*D*) Longitudinal section through a sporangium showing peristome teeth and epiphragm (×120), (*E*) Transverse section through sporangium (×120), showing multicellular teeth. (*D, E,* after Flowers, 1973.)

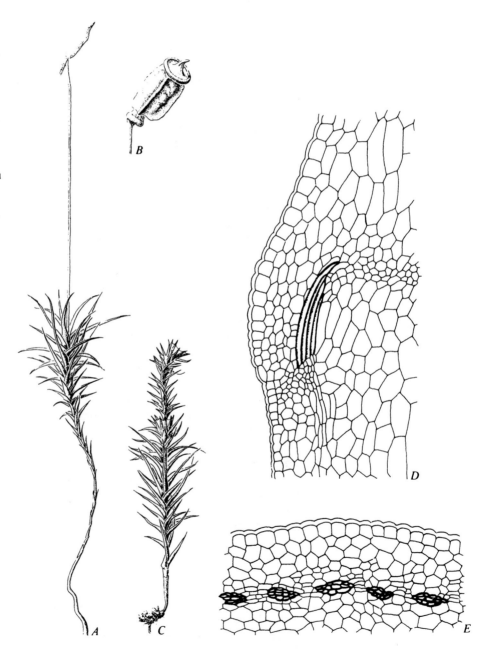

length of stem with associated vegetative leaves; and within the next year this extension of the stem is terminated by a new perigonium. In consequence, more than 5-year-old perigonial stems can possess a series of perigonial buds aligned like beads along the stem length. In genera like *Racelopus* where the antheridial gametophore is reduced to a perigonium arising from a protonematal "bud," the perigonium is annual.

Sometimes sperm dispersal is enhanced by the "splash cup" structure of the perigonium. Raindrops falling into the "cup" cause the sperms (or even antheridia) to be splashed some distance from the gametophore, and possibly closer to a perichaetium. Invertebrates are also sometimes attracted to mucilage produced in both the perigonium and perichaetium; moving from the perigonium to the perichaetium, they carry bryophyte sperms on their bodies.

The perichaetial leaves are at the apex of a main stem or, if the stem is branched, terminate the branches. Perichaetial leaves tend to be narrower than vegetative leaves, sometimes lack lamellae and chlorophyll, and are imbricated around the archegonia and paraphyses that are structurally like those in the perigonium. In the perichaetium the apical cell of the stem is used up in producing the archegonia, and consequently production of a perichaetium completes the growth of that stem tip. Generally, however, innovations are formed just below the perichaetium, and growth of the gametophore continues.

Generally, if branching occurs, it is irregular and by innovations; these innovations often appear after the production of perichaetia. In some genera, however, the branching forms a dendroid gametophore, in *Dendroligotrichum* (Fig. 6-1*D*), and *Microdendron,* for example, in which the main stem forms an apical cluster of banches, and the gametophore consequently resembles a tiny tree. Some species of *Pogonatum* and *Polytrichum* produce several branches that flare outward from a short erect shoot. In these genera the stems frequently form dense turfs, while dendroid gametophores are more widely spaced.

The development of the sporophyte and calyptra is usually similar to that in *Tetraphis*, but an apical mass of hairs is often produced by the calyptra, and these hairs grow downward to invest the calyptra and the sporangium beneath (Figs. 2-3*E* and 6-8*A*). Hairs on the calyptra appear to influence the shape and orientation of the apex of the sporangium and thus the shape of the sporangium. If, as the sporangium matures, much of it is not covered by the hairy calyptra, the cells of the face least exposed elongate more rapidly, while the cells of the other face are arrested in growth; this results in an asymmetric curved sporangium.

THE SPOROPHYTE

The sporophyte always has a seta and produces a tapered foot that penetrates the gametophore. The seta, generally smooth, is made up mainly of thick-walled cells, and generally has a central conducting strand of hydroids and leptoids. The young sporophyte is photosynthetic, but as it ages, the cell walls become deeply pigmented reddish or brown (Fig. 6-9*B,C*).

The sporangium is generally elongate, but the cylinder is sometimes flattened, often conspicuously angled (sometimes four angled, like a box), broadly bell-shaped, with the mouth flaring outward, ovoid or simply a straight or curved cylinder. The epidermal cells are usually smooth, but are sometimes

FIGURE 6-9
Sporophytic features of Poly-trichidae. (*A*) Longitudinal section of sporangium of *Pogonatum aloides,* showing hairy calyptra (*caly*), the well-differentiated apical operculum (*op.*), the epiphragm (*epi*), a central columella (*col.*), and the sporogenous layer bounded by air spaces (*a.s.*), traversed by trabeculae (×50). (*B*) Transverse section of seta of *Polytrichium juniperinum,* showing the stereids that make up the outer cortex, an inner region of parenchymatous cells, the central cylinder of hydroids (very center), and leptoids (bounding the hydroids) (×180). (*C*) Longitudinal section of seta of *P. juniperinum* (×180). (*A*, after Janzen, 1918; *B, C,* after Vaisey, 1888.)

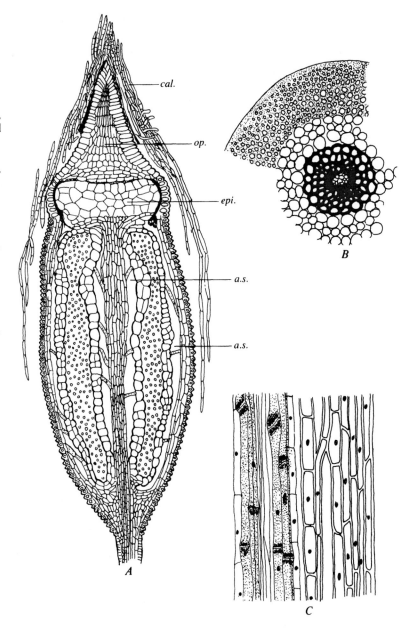

papillose. The operculum also varies from conic to long-snouted rostrate (Fig. 6-10*B*,*E*) or apiculate (Fig. 6-10*I*).

The sporangium always has a multistratose jacket, the epidermis of which sometimes has stomata. In several genera the base of the sporangium has a well-differentiated bulge (the apophysis, Fig. 6-8*B*) above the site where the sporangium joins the seta, and the stomata are often confined to a groove between the apophysis and the body of the sporangium. In others the stomata (Fig. 6-11*C*) are scattered over the sporangium, and in still others stomata are absent. The stomata often have guard cells that are incompletely separated, so that the stomatal apparatus consists of a single binucleate cell with a central slit. Occasionally more than two cells form the stomatal apparatus. An operculum is always differentiated. This falls loose when the sporangium matures and dries out. When mature, the sporangium and the operculum both shrink in diameter, but to differing degrees, and thus the operculum comes loose and falls away.

The sporogenous layer is in the same position as in *Tetraphis*, but in some Polytrichidae (including *Polytrichum*), between the columella and sporogenous

FIGURE 6-10
Sporangial diversity in Polytrichidae (×12). (*A*) *Psilopilum cavifolium. (B) Atrichum selwynii. (C) Dawsonia superba. (D) Polytrichum sexangulare. (E) Lyellia crispa. (F) Polytrichadelphus magellanicus. (G) Alophosia azorica. (H) Bartramiopsis lescurii. (I) Polytrichum commune.*

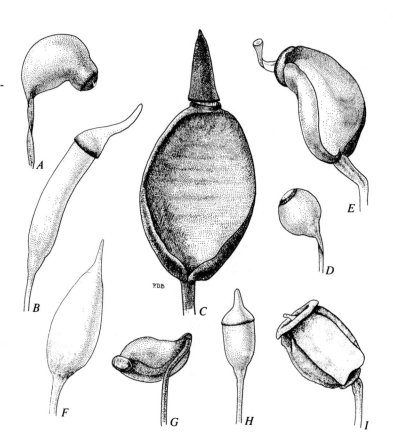

FIGURE 6-11
Sporophytic details of Polytrichidae. *(A)* View of apex of
sporangium of *Atrichum undulatum* after operculum has
fallen, showing peristome
teeth curving over epiphragm
(×40). *(B)* Detail of two peristome teeth of *A. undulatum*
viewed from outer surface
(×280). *(C)* Single stoma of
Polytrichum alpinum, showing guard cells (×400). *(B,*
after Janzen, 1918.)

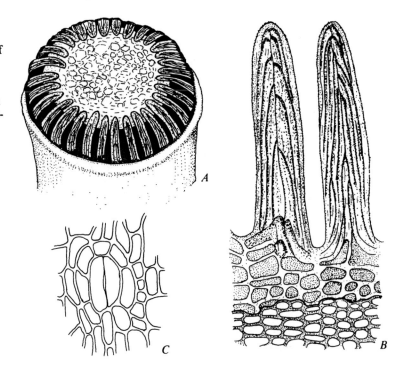

layer is a cylinder of air spaces, and yet another air space cylinder is between
the sporogenous layer and the jacket. These air space cylinders are traversed
by supportive filaments. Any pressure on these air cavities can cause ejection
of the spore through the spaces between the peristome teeth (fig. 6-9*A*).

Generally multicellular peristome teeth are present. These teeth are derived
from a restricted number of concentric layers of cells located beneath the
operculum and from the inner layers of the jacket of the sporangium (Fig.
6-8*D,E*). In *Dawsonia* the many teeth are long and filiform. They are derived
from seven or more concentric layers of the jacket and from the outer cells of
the apex of the columella (Fig. 6-12). When the operculum falls, the peristome
teeth twist about each other and thus close the mouth to rapid shedding of the
spores. Spore dispersal occurs over an extended period of time and results from
any jarring of the sporangium wall that will cause the spores to be puffed out
through the gaps among the peristome teeth. In peristomial structure and development, the situation in *Dawsonia* is more specialized than in Tetraphidae,
since the peristome teeth are derived from a restricted series of concentric
layers.

In most other genera the peristome teeth are formed from four concentric
layers; they are blunt and form a single concentric ring of roughly 32 to 64 teeth
(Fig. 6-11*A,B*). The apex of the columella is expanded to form a membranelike
epiphragm that nearly closes the sporangium mouth. The margin of the

FIGURE 6-12
Transverse section through
upper part of sporangium of
Dawsonia (×400), showing
the many layers of peristome
teeth. (After Goebel, 1930.)

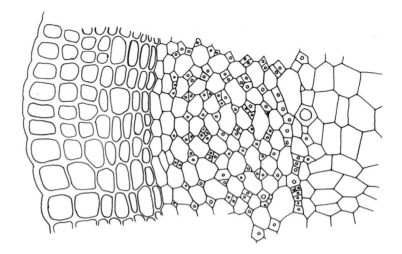

epiphragm is attached to the inner faces of the peristome teeth that arch over the epiphragm. Between each of the teeth and the point of attachment of the margin of the epiphragm are tiny gaps, and it is through these holes that the spores must be ejected. As in *Dawsonia,* this ensures that the spores are disseminated over an extended period of time.

Finally, some genera have no peristome teeth; in all of them (including *Alophosia, Bartramiopsis,* and *Lyellia*), an epiphragm is present. In *Alophosia* and *Lyellia* the teeth are replaced by a flattened ring that is perforated by the columella and closed by the epiphragm. In all three of these genera the spores are shed when the sporangium shrinks in length upon drying, leaving a gap between the columella and the edge of the ring or the mouth of the sporangium. Raindrops can strike the capsule and discharge the spores. In *Bartramiopsis* the mouth of the sporangium is relatively wide, and spore dispersal is aided by the wide gap between the columella and the expanded opening.

RELATIONSHIPS

The subclass Polytrichidae forms an isolated evolutionary line, both in sporophytic and gametophytic features. It shares many general features with other mosses, but several features are confined essentially to the group:

1. The gametophore has both hydroids and leptoids.
2. Leaves have longitudinal lamellae.
3. The antheridial stem has indeterminate growth and axillary antheridia.
4. The multicellular peristome teeth are formed of a restricted number of concentric cell layers.
5. The columella apex often forms an expanded epiphragm.

ECOLOGY AND USES

The presence of particular species of *Polytrichum* in specific conifer forests makes them valuable indicators of site types. *Polytrichum* is often used to produce a rich dark green turf in Japanese gardens. Since the gametophores can be transplanted, and vegetative expansion of the colonies is relatively rapid, these make extensive and attractive perennial tall turfs. From excavations of Roman Britain, small neatly woven mats have been found that were made from *Polytrichum* stems. The wiry stems, stripped of leaves, are also used in making brooms. In early times turf of living plants of *Polytrichum* was cut out to manufacture sleeping mattresses.

FURTHER READING

Allorge, V. 1949. Quelques observations sur *Alophosia azorica* (Ren et Card.). Card. *Revue Bryol. Lichénol.* **18**:172–174.

Blaikley, N. M. 1932. Absorption and conduction of water and transpiration in *Polytrichum commune. Ann. Bot.* **46**:289–300.

Bowen, E. J. 1931. Water conduction, in *Polytrichum commune. Ann. Bot.* **45**:175–200.

Chopra, R. S., and N. N. Bhandari. 1959. Cyto-morphological studies of the genus *Atrichum* Palis. *Res. Bull. (N.S.) Panjab Univ.* **10**:221–231.

―――― and P. A. Sharma. 1958. Cyto-morphology of the genus *Pogonatum* Palis. *Phytomorphology* **8**:41–60.

―――― 1958. Cyto-morphological studies of *Oligotrichum* Lam. & DeCand. *J. Indian Bot. Soc.* **38**:400–414.

Eschrisch, W. and M. Steiner. 1967. Autoradiographische Untersuchungen zum Stofftransport bei *Polytrichum commune. Planta* **74**:330–349.

―――― 1968. Die Struktur des Leitgewebesystems von *Polytrichum commune. Planta* **82**:33–49.

Finnochio, A. F. 1967. Pitting of cells in moss gametophores. *Bull. Torrey Bot. Club* **94**:18–20.

Hagen, I. 1914. Florarbejder til en norsk løvmosflora XIX Polytrichaceae. *Norsk Vidensk. Selsk. Skr.* **1913**(1):1–77.

Harvey-Gibson, R. J., and D. Miller-Brown. 1927. Fertilization of Bryophyta. *Polytrichum commune* (Preliminary note). *Ann. Bot.* **41**:190–191.

Hébant, C. 1977. *The Conducting Tissues of Bryophytes.* Vaduz: J. Cramer.

Ireland, R. R. 1969. Taxonomic studies on the genus *Atrichum* in North America. *Can. J. Bot.* **47**:357–368.

Kawai, I. 1969. Studies on the affinity regarding the conducting tissue of the sporophyte in the Musci (2). On the seta in some species of the family Polytrichidae. *Sci. Rep. Kanazawa Univ.* **14**:39–51.

Lindberg, S. O. 1868. Observationes de formis praesertim europaeis Polytrichoidearum (Bryacearum nematodontearum). *Not. Sällsk. Faun. Fl. Fenn.* **9**:91–158.

Longton, R. E. 1970. Growth and reproduction of *Polytrichum alpestre* Hoppe in Antarctic regions, in Holdgate, M. W. (ed.), *Antarctic Ecology.* Vol. 2, pp. 818–837. London: Academic Press.

———— 1974. Genecological differentiation in bryophytes. *J. Hattori. Bot. Lab.* **38**:49–65.

Lorch, W. 1908. Die Polytrichaceen. Eine biologische Monographie. *Afhandl. K. bayer. Akad. Wiss. math.-phys. Kl.* (Munchen) **23**:447–546.

Osada, T. 1965. Japanese Polytrichaceae I. Introduction and the genus *Pogonatum. J. Hattori Bot. Lab.* **28**:171–201.

Paolillo, D. J. 1968. The effect of the calyptra on capsule symmetry in *Polytrichum juniperinum* Hedw. *Bryologist* **71**:327–334.

Sarafis, V. 1971. A biological account of *Polytrichum commune. N. Z. J. Bot.* **9**:711–724.

Scheirer, D. C. 1972. Anatomical studies in the Polytrichaceae I. The gametophore of *Dendroligotrichum dendroides* (Hedw.) Broth. *Bryologist* **75**:305–314.

———— 1973. Hydrolyzed walls in the water-conducting cells of *Dendroligotrichum* (Bryophyta): histochemistry and ultrastructure. *Planta* **115**:37–46.

———— 1975. Anatomical studies in the Polytrichaceae II. Histochemical observations on thickened lateral walls of hydroids of *Dendroligotrichum. Bryologist* **78**:113–123.

———— 1976. Some fine structural observations on the rhizome of *Dendroligotrichum* (Bryophyta). *Protoplasma* **89**:323–337.

Sharma, P. D., and R. S. Chopra. 1964. The life history of *Lyellia crispa* R. Br. *Bryologist* **67**:329–343.

Smith, G. L. 1971. A conspectus of the genera of Polytrichaceae. *Mem. N.Y. Bot. Gardens* **21**:1–83.

———— 1974. New developments in the taxonomy of Polytrichaceae. *J. Hattori Bot. Lab.* **38**:143–150.

Stevenson, D. W. 1974. Ultrastructure of the nacreous leptoids (sieve elements) in the polytrichaceous moss *Atrichum undulatum. Am. J. Bot.* **61**:414–421.

Tansley, A. G., and E. Chick. 1901. Notes on the conducting tissue-system in Bryophyta. *Ann. Bot.* **15**:1–42.

Thrower, S. L. 1964. An investigation of translocation in *Dawsonia superba* C.M. *Trans. Br. Bryol. Soc.* **4**:664–667.

Ward, M. 1960. Callus tissues from the mosses *Polytrichum* and *Atrichum. Science* **132**:1401–1402.

Wiggleworth, G. 1956. Further notes on *Polytrichum commune* L. *Trans. Br. Bryol. Soc.* **3**:115–120.

Wijk, R. van der. 1920. Über den Bau und die Entwicklung der Peristomzähne bei *Polytrichum. Rec. Trav. Bot. Néerl.* **26**:289–394.

Zanten, B. O. van. 1973. A taxonomic revision of the genus *Dawsonia* R. Br. *Lindbergia* **2**:1–48.

——— 1974. The hygroscopic movement of the leaves of *Dawsonia* and some other Polytrichaceae. *Bull. Soc. Bot. France, Colloque Bryol.* **121**:63–66.

7 The "Bug" Mosses — Subclass Buxbaumiidae

Since the gametophyte of *Buxbaumia* is microscopic, this moss is never noticed until sporophytes appear. The sporangium of *B. aphylla* is roughly bug-shaped and, like a bug, is often glossy and dorsiventrally flattened.

The subclass contains a single order, Buxbaumiales, with two families: Buxbaumiaceae, with *Buxbaumia* (14 species); and Diphysciaceae, with *Diphyscium* (22 species), *Muscoflorshuetzia* (1 species), and *Theriotia* (2 species). *Buxbaumia* occurs primarily in the temperate portion of the Northern Hemisphere, but some species are tropical, and several are widely scattered in the Southern Hemisphere. *Buxbaumia* grows on nutrient-poor soil with some organic material or on rotten logs. *Diphyscium* is widely distributed in the Northern Hemisphere, but is present also at middle latitudes in both hemispheres. Most species grow on soil, but several are on siliceous rock. *Theriotia* grows in rock crevices along streams in East Asia, and *Muscoflorshuetzia* is terrestial in Chile. Many species occur in relatively well-illuminated sites, but some grow in shade (Fig. 7-1).

Many authors would include this subclass within the Bryidae since, in most gametophytic features, it matches that subclass. In sporophytic features, especially the nature and development of the peristome teeth, it is unique. Since the gametophore is extremely reduced in the Buxbaumiaceae and is also reduced to a great degree in the Diphysciaceae, this adds to the difficulties in interpreting the relationships of the subclass. Gametophytic simplicity limits the number of features available for comparison. Here *Buxbaumia* and the family Diphysciaceae are treated individually, since they show conspicuous differences.

BUXBAUMIA: THE GAMETOPHYTE

The spore germinates to produce a short germ tube of cylindrical cells, and this initiates both rhizoids and chloronematal filaments, both of which bear branches. Short branches of the protonema give rise to "buds" at their tips, and

FIGURE 7-1
Habit sketches of Bux-
baumiidae. (*A*) *Musco-
florschuetzia* (×15). (*B*) *Diphyscium foliosum* (×8). (*C*)
Theriotia lorifolia (×8). (*D*)
Buxbaumia aphylla (×10).
(*E*) External view of peristome of *B. aphylla,* showing
pleated endostome, external
exostomial layer, and torn
margin where operculum was
attached; note also the columella emerging through the
endostome (×50.)

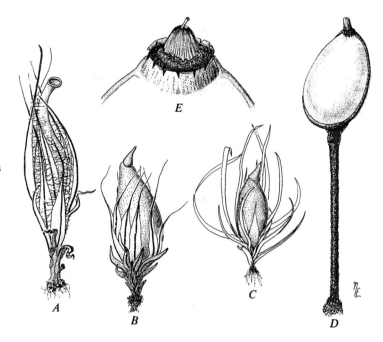

these develop into either perichaetia or perigonia. There is no caulonema. The
perigonium is the most reduced gametophore known in the bryophytes. When
mature, it consists of a single unistratose flap of chlorophyllose tissue that
sheathes a single short-stalked, spherical antheridium (Figs. 7-2*B* and 7-3*F*).
The perichaetium is also very small, consisting of a short budlike stem and
three to four unistratose nonchlorophyllose perichaetial leaves that enclose one
or two archegonia and a few reduced paraphyses (Fig. 7-2*A*). The marginal
cells of the perichaetial leaves often become elongate and ciliate, and these cilia
sometimes form protonematal threads, which possibly serve to expand the population vegetatively (Fig. 7-3*A*). In culture, at least, the chloronema often
produces chains of rounded cells that can break off to form vegetative
propagules.

BUXBAUMIA: THE SPOROPHYTE

The young sporophyte is green and a small conic calyptra caps the tip of the
seta as it elongates (Fig. 7-3*A*). The extremely abbreviated stem of the
perichaetium enlarges into a bulbiform structure in which the foot of the
sporophyte is embedded. The seta is rough-papillose and is usually reddish,
even when young (Fig. 7-1*D*); much of this pigment is in the thick-walled
cells that give rigidity to the seta; there is a small central conducting strand.

When mature, the sporangium is usually ovoid and tapered to a short conic
operculum (Figs. 7-1*D* and 7-3*B*). There is always an expanded apophysis. The

FIGURE 7-2
Buxbaumiidae: *Buxbaumia aphylla*. (*A*) Archegonium-producing gametophore (essentially a perichaetium; ×300); (*B*) Protonematal threads bearing several antheridium-producing "buds" (×300). (Scanning electron micrographs provided by Dale M. J. Mueller.)

A

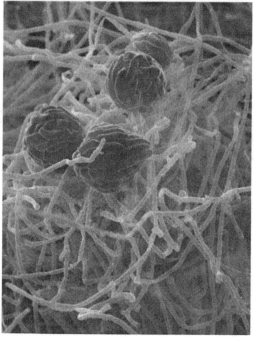

B

FIGURE 7-3

Buxbaumiidae: *Buxbaumia.*
(*A*) Female gametophore of *B. aphylla* with immature sporophyte enclosed in calyptra and sheathed by ciliate leaves of the gametophore (×200). (*B*) Longitudinal median section through sporangium of *B. viridis* (somewhat diagrammatic) showing the large columella, the extensive air spaces traversed by trabecular filaments, and the basal apophysis (×20). (*C*) Detail of transverse section of jacket of *B. viridis,* showing its multistratose nature and the filaments (×360). (*D,E*) Views of stomata (×300). (*D*) Transverse section through stoma, showing guard cells and intercellular space. (*E*) Surface view of two stomata. (*F*) Habit sketch of single antheridial "branch" (×225). (After Janzen, 1918.)

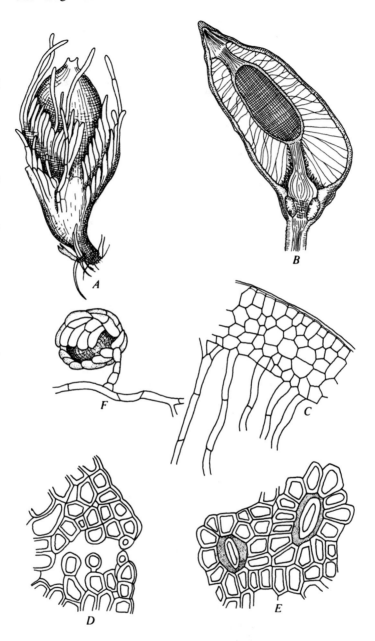

calyptra usually falls from the developing sporangium before the operculum has differentiated and before much sporangium enlargement has occurred. The sporangium is clearly asymmetric and is oriented at an oblique angle at the apex of the seta so that one face forms a slanted, and often flattened, surface.

Stomata are occasional in the apophysis and lower part of the sporangium (Fig. 7-3*D,E*). The jacket is multistratose. As in the Polytrichidae, a large cyl-

FIGURE 7-4
Transverse section through
peristome teeth of *Buxbaumia
viridis* (×500) showing pleated
endostome (*e*.), two rows of
exostomial teeth (*ex. t.*), and
columella (*col*.). (After Goebel, 1930.)

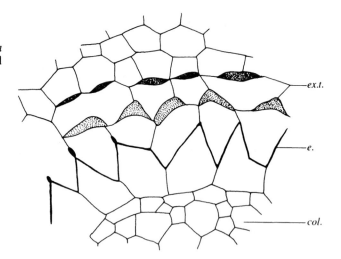

inder of intercellular spaces is between the jacket and sporogenous layer (Fig.
7-3*B,C*). Since the upper surface of the sporangium is flattened, any pressure
on the surface, such as pelting raindrops, can compress the air spaces within
and cause the spores to be ejected through the narrow peristomial opening. In
temperate regions, at least, the sporophytes begin to develop in late summer,
continue to develop over the winter, and mature in the next spring.

In some species the peristome is the most complex found in the mosses
(Figs. 7-1*E* and 7-4). The peristome teeth are derived from many concentric
cell layers, sometimes as many as 10, which is similar to the situation in *Dawsonia*. The most complex peristome type, as in *Buxbaumia viridis,* has an outer
multistratose, multicellular ring, similar to the cells that form the basal ring to
which the teeth of the Polytrichidae are attached. (In *B. aphylla* this ring has
jagged teeth on the upper edge.) Inward from this ring there can be as many as
four concentric series of articulated teeth. The outer teeth are relatively short
and, proceeding inward, become longer in each row; approximately 32 opaque
teeth are in each circle. These articulated teeth are formed of cell wall fragments and are derived from parts of cell walls of two concentric cell layers.
Each tooth forms around the radial wall of the outer cell layer by deposition of
additional cellulose to the corners of two adjacent cells and on the transverse
face of the concentric cell layer directly associated with the two adjacent cell
corners. Disintegration of the rest of the cell walls, as the peristome matures,
leaves each tooth isolated.

These teeth are said to be diplolepidous in structure, since the outer face of
each tooth is formed by two "scales" derived from parts of two cells while the
inner face of the tooth is formed of one "scale" derived from part of one cell.
The teeth are not hygroscopic, and appear to serve no active function in spore
dispersal. The articulated outer teeth are termed the exostome. There is also a
central truncated cone of fused peristome teeth. This cone is formed also from

the inner and outer faces of two concentric layers of cells and is isolated as a hyaline cone when the residue of the radial cell walls of those two concentric cell layers disintegrate. This is the endostome, and is often somewhat twisted when dry. Such an endostome has 32 pleats and characterizes all species of *Buxbaumia,* while the number of rows of exostomial teeth varies among the species.

The entire sporophyte may be up to 2 cm tall with the sporangium up to 0.5 cm long. The mature sporangium is chestnut-colored and shiny or sometimes yellowish-brown. In *Buxbaumia viridis* the epidermis of the upper surface of the sporangium splits and curls back. The function of this feature to the moss is not apparent, but it serves as an easily observed character useful in identification.

DIPHYSCIACEAE: THE GAMETOPHYTE

In the Diphysciaceae the gametophyte differs conspicuously from that of *Buxbaumia* (Fig. 7-1*A–C*). Only in *Diphyscium* and *Theriotia* are the details adequately described, and so most of the information that follows is based on these genera. The spore produces a germ tube that gives rise to a short-branched protonema initially similar to that of *Buxbaumia,* with a reclining branched chloronema affixed to the substratum by rhizoids. Ultimately, either the terminal cell of this chloronema and/or a lateral bud produces an erect rich-green trumpet-shaped protonemal flap (Fig. 7-5). This caulonematal flap has a multistratose stalk, and the apex forms a concave disc. At the base of the flap a "bud" produces a leafy gametophore.

Most Diphysciaceae are dioicous, but some species of *Diphyscium* are monoicous. In dioicous species of *Diphyscium* the vegetative stems are always short and bear many strap-shaped, blunt, dark green, costate leaves in which the blade is bistratose. Rhizoids attach the gametophores to the substratum. The perichaetium is made up of mainly nonchlorophyllose leaves, each with a

FIGURE 7-5
Trumpet-shaped protonematal flap of *Diphyscium* (×220), showing external view in center and transverse sectional views from various levels. (After Nishida and Saito, 1961.)

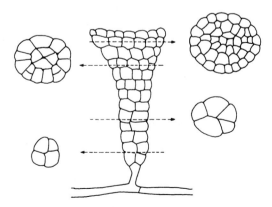

long costa extending beyond the apex as an awn (Fig. 7-6*A,B*). The blade is unistratose, and the upper margins have numerous cilia or are jaggedly toothed. These leaves form a conspicuous tuft at the apex of the short perichaetial stem. The perichaetial leaves surround several archegonia among which are uniseriate filamentous paraphyses (Fig. 7-6*B*). The archegonia terminate the perichaetium, and any further growth from the short stem must be by innovations. The entire leafy shoot is usually less than 1 cm tall. The perigonial branches superficially resemble vegetative stems, but the apex of the shoot produces numerous elongate antheridia and associated filamentous paraphyses (Fig. 7-6*C,D,E*).

Theriotia differs from *Diphyscium* in the nature of its vegetative leaves, which are always multistratose and long and tapered (Fig. 7-1*C*). Of particular interest is the internal structure of the leaves in which some cells are specialized for photosynthesis (chlorocysts) and others are hyaline (leucocysts). In the basal part of the leaf most of the body is composed of leucocysts, and there is a median layer of chlorocysts. The unistratose median band of chlorocysts continues in the apical part of the leaf, but the epidermal cells are also chlorocysts, and layers of leucocysts are both above and below the median band of chlorocysts. In *Muscoflorschuetzia* the gametophore is extremely similar to *Diphyscium*, but the sporangium differs, especially in its lack of peristome teeth (Fig. 7-1*A*).

DIPHYSCIACEAE: THE SPOROPHYTE

In *Diphyscium* the sporangium is similar to that of *Buxbaumia*, but the seta is extremely short, and the mature golden brown sporangium is immersed among the perichaetial leaves (Fig. 7-1*B*). The endostome is similar to that of *Buxbaumia* but has only 16 pleats. The exostome is never in more than one layer and consists of approximately 16 teeth that are often rudimentary.

EVOLUTION AND RELATIONSHIPS

In the Buxbaumiidae several examples of sporangial structure persist that suggest the stages in a reduction series. In *Buxbaumia* the long seta and complex peristome of several layers of teeth represents the most generalized type. Within *Buxbaumia* are species that have a single row of exostomial articulated teeth plus the endostome and an irregular multicellular peristomial ring outside the articulated teeth. Some species also have a very short seta. The seta in *Diphyscium* and *Theriotia* is even further shortened; the endostome is reduced to 16 pleats, and exostomial teeth may be absent. *Muscoflorshuetzia* has no peristome teeth; this presumably represents a reduced specialized sporophyte.

The reduction series in the gametophores is a complete reversal in structural complexity compared to that of the associated sporophytes. It is possible that

FIGURE 7-6
Diphyscium foliosum, gametophytic details. (*A*) Habit
sketch of perichaetial plant
(×8). (*B*) Detail of archegonia
with paraphyses and imma-
ture perichaetial leaves
(×100). (*C*) Antheridium-
producing plant (×15). (*D*)
Mature (left) and immature
(right) antheridia, with immature paraphyses (×225). (*E*)
Mature paraphysis (×300),
showing unusual double cell
walls. (After Janzen, 1918.)

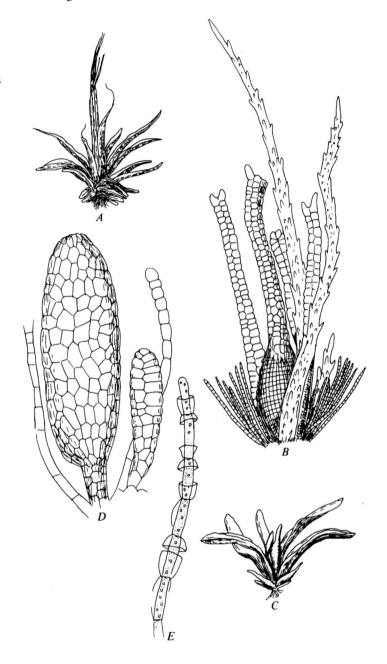

an elaborate enlarged sporophyte is necessary if the gametophore is greatly
reduced, otherwise the sporophyte might produce too little photosynthate to
survive to maturity. Thus, in the Diphysciaceae where the sporophyte is
greatly reduced, the gametophore is short, leafy, and an effective photosynthetic plant. In *Buxbaumia* extreme reduction in the gametophore has occurred,

and the sexual branches are reduced to few sex organs with abbreviated protective sheaths.

The Buxbaumiidae show some affinities to the Polytrichidae, where the leaf, though lacking lamellae, is structurally similar. The ring of multicellular "teeth" in the outer peristome of the Buxbaumiidae is simillar to the basal part of the peristome in the Polytrichidae. The origin of the peristome from numerous concentric layers also resembes the Polytrichidae. The internal "bladder" of air spaces in the sporangium is also similar to many Polytrichidae, as is the sporangium orientation (like that in *Dawsonia*), the position of the small operculum (like that in *Dawsonia* and *Notoligotrichum*), and the bellowslike method of spore dispersal. Paraphyses of some species of *Atrichum* (Polytrichidae) are also similar to those of the Diphysciaceae.

There are also many similarities with the subclass Bryidae, where the exostomial teeth are always articulated and an endostome is often present, and these peristomial layers are of cell-wall fragments rather than whole cells. The vegetative gametophore is indistinguishable from that of some Bryidae.

The subclass seems transitional and might represent part of the stock that retains features of the progenitors that could have been derived from Polytrichidae-like ancestors and given rise to both the Buxbaumiidae and Bryide. Features that characterize the subclass include:

1. The perigonium of *Buxbaumia*, reduced to a single antheridium and its surrounding unistratose sheath
2. The unusual trumpet-shaped protonematal flaps in *Diphyscium* and *Theriotia*
3. The pleated truncated endostome
4. The multiple rows of exostomial teeth in *Buxbaumia*
5. The peristome composed of both multicellular "teeth" and teeth formed of cell wall fragments

FURTHER READING

Britton, E. G. 1901. Some further observations on *Buxbaumia*. *Bryologist* **4**:33–34.

Castaldo, R. 1972–73. Contributo alla cuonoscenza dei muschi saprofiti: le Buxbaumiaceae (Bryales), con particolare riguardo a *Buxbaumia indusiata* Brid. *Delpinoa,* **14-15**(n.s.):35–48.

———, S. Giordano, and R. Ligrone. 1981. Scanning electron microscope characterization of spores of European Buxbaumiaceae. *J. Bryol.* **11**:743–746.

Chen, P.-C., and S.-C. Lee. 1964. Two new species of *Buxbaumia* in China. *Acta Phytotax. Sinica* **9**:277–280.

Crosby, M. R. 1977. *Florschuetzia,* a new genus of Buxbaumiaceae (Musci) from Southern Chile. *Bryologist* **80**:149–152.

Dening, K. 1928–29. Entwicklungsgeschichtliche Untersuchungen an Gametophyten von *Buxbaumia aphylla* (L.). *Verh. Naturhist. Vereines Preuss. Rheinl. Westphalens* **85**:306–444.

Eastwood, S. K. 1936. Notes on *Buxbaumia aphylla* (Linnaeus) Hedwig. *Bryologist* **39**:127–129.

Hancock, J.A., and G. R. Brassard. 1974. Phenology, sporophyte production, and life history of *Buxbaumia aphylla* in Newfoundland, Canada. *Bryologist* **77**:501–513.

———— 1974. Element content of moss sporophytes: *Buxbaumia aphylla. Can. J. Bot.* **52**:1861–1865.

Ligrone, R., R. Gambardella, R. Castaldo, S. Giordano, and M. Sposito. 1982. Gametophyte and sporophyte ultrastructure in *Buxbaumia piperi* Best (Buxbaumiales, Musci). *J. Hattori Bot. Lab.* **52**:465–499.

McClymont, J. W. 1950. Notes on *Buxbaumia. Bryologist* **53**:267–277.

Mueller, D. M. J. 1972. Observations in the ultrastructure of *Buxbaumia* protonema. Plasmodesmata in the cross-walls. *Bryologist* **75**:63–68.

Nishida, Y. 1971. On the development of the gametophytes of *Buxbaumia aphylla. J. Japn. Bot.* **46**:199–206.

———— and S. Saito. 1961. Studies of the germination of the spore in some mosses II. *Diphyscium fulvifolium* Mitt. and *Sphagnum cuspidatum* Ehrh. *Bot. Mag. Tokyo* **74**:91–97.

Noguchi, A. 1935. Icones of the Japanese Buxbaumiales. *J. Japn. Bot.* **11**:267–275 and Plate II.

Stone, I. G. 1983. *Buxbaumia* in Australia, including one new species, *B. thornsborneae. J. Bryol.* **12**:541–552.

Taylor, J. 1972. The habitat and orientation of capsules of *Buxbaumia aphylla* in the Douglas Lake region of Michigan. *Michigan Bot.* **2**:70–73.

8 The Jointed-Toothed Mosses — Subclass Bryidae

More than 95% of the mosses belong to the subclass Bryidae. Usually 11 orders are recognized, and these contain approximately 90 families with at least 650 genera and more than 9,000 species. Even more than among the other subclasses, the number of species recognized is undoubtedly greatly exaggerated. Unquestionably many species names will disappear as genera are more carefully studied, because it is probable that many species have been recognized several times under different names.

The subclass is widely distributed on the earth's surface, from the highest latitudes on exposed mineral soil or rock in both hemispheres, to the lowest latitudes on many substrata, and nearly to the highest elevations where plants can survive. Indeed, living Bryidae have been found on earth transported on "ice islands" in the Arctic Sea, and these islands have approached the North Polar region. Some Bryidae grow near hot springs, but none can tolerate extremely high temperatures. Others grow in arid climates, but are always restricted to sites where moisture is available at some period during the year. The Bryidae occupy most habitats: submerged in water (to depths of 20 m in clear lakes); as floating mats; on the soil surface (even under small quartz pebbles, e.g., *Aschisma kansanum*); on rock surfaces; on humus; and on tree trunks and branches from the base to the canopy.

The Bryidae reach their greatest abundance and diversity in the Northern Hemisphere and at tropical and subtropical latitudes, especially in humid temperate climates. In such areas they often form thick soft carpets or turfs, sheathe tree trunks, and festoon branches with pendulous banners. Since the Bryidae exhibit a diversity of colors (from whitish-green through pale-greens, golden-greens, browns, reds, and purples to dark-green and nearly black), they often provide a spectacular display, especially when the gametophores are moist. Most Bryidae thrive best in somewhat shaded sites where evaporation is slow and nutrient is enriched, especially by rainwash from the tree canopy or through soil water seepage. Many Bryidae are very specific in their substratum requirements. Some are restricted to lime-rich substrata of a particular texture, while others grow only on lime-poor substrata. Some are confined to shaded cavern mouths and crevices, while others grow only in areas of high illumination. Some inhabit only organic substrata. Several Bryidae are restricted to animal excrement or decomposed animal remains (many genera of the family Splachnaceae). Some Bryidae are found only on thick, evergreen leaves of shrubs.

Since such a diversity of genera is involved, it is difficult to recognize consistent features common to all Bryidae. Some of this diversity is shown in Figs. 8-1 and 8-2. A number of features separate the Bryidae from most other subclasses of mosses:

1. Rhizoidal gemmae are produced in many genera.
2. The gametophore sometimes has bilateral symmetry.
3. Leaf cells occasionally possess remarkably complex ornamentation on the exposed cell surfaces.
4. The gametophore never has leptoids, and often lacks hydroids.
5. The gametophore sometimes shows regular or pinnate branching.
6. Paraphyllia and/or pseudoparaphyllia sometimes occur on the stems.
7. Gemmae are often produced on the gametophore.
8. Often the sexual branches are reduced in size and are lateral, with the main stem having indeterminate growth.
9. Peristome teeth are composed primarily of remnants of cell walls rather than of whole cells.
10. Peristome teeth are never constructed from more than three concentric layers of cells.
11. Peristome teeth are distinctly articulated and are usually hygroscopic.
12. Peristome teeth usually possess elaborate ornamentation on the surfaces of cell plates.
13. Hygroscopic peristome teeth assist in gradual dissemination of the spores.

With other subclasses of mosses they share the following features:

1. Spores are usually unicellular.
2. The protonema is usually heterotrichous.
3. Most gametophores have spirally arranged leaves.
4. Gametophores usually have rhizoids.
5. Leaves are usually unistratose, except at the costae and margins, where they are sometimes multistratose.
6. Paraphyses usually are among the sex organs.
7. The calyptra is usually smooth, but occasionally has a much ornamented surface (with hairs or papillae); sometimes the lower margin of the calyptra has an elaborate fringe.
8. The sporophyte usually has a seta in which both hydroids and leptoids may be present.
9. The sporangium usually has an operculum.
10. The columella and archesporium are cylindric.
11. The sporangium often has stomata and an apophysis.

FIGURE 8-1
Habit sketches of acrocarpous Bryidae. (*A*) *Dicranum polysetum*, showing falcate-secund leaves and multiple sporophytes from a single perichaetium (×2). (*B*) *Bryum argenteum*, showing julaceous shoots (×7). (*C*) *Encalypta rhaptocarpa*, showing large mitrate calyptra (×6). (*D*) *Paludella squarrosa*, showing neatly arranged, squarrose leaves (×5). (*E*) *Plagiopus oederi*, showing spherical sporangium (×7). (*F*) *Pleuridium bolanderi*, showing immersed sporangium (×25). (*G*) *Funaria hygrometrica* (×10). (*H*) *Distichium capillaceum*, showing leaves arranged in two neat rows (distichous) (×6). (*I*) *Schistidium apocarpum*, showing immersed sporangium (×8).

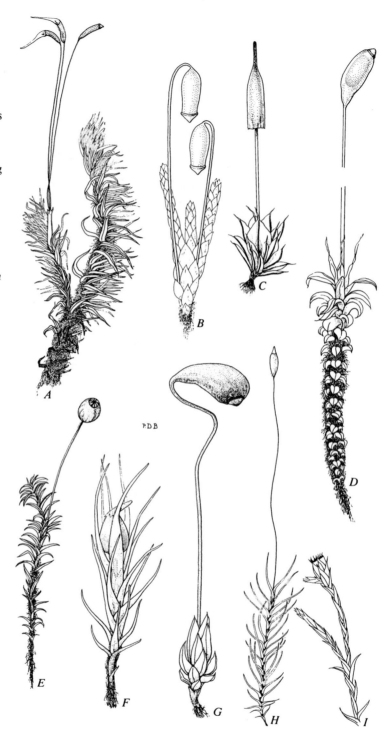

FIGURE 8-2
Habit sketches of pleurocarpous Bryidae. *(A) Neckera douglasii*, a pendulous moss, usually epiphytic on trees (×1). *(B) Climacium dendroides*, showing the miniature treelike habit (dendroid) (×1), *(C) Dendroalsia abietina* (×1), showing the numerous sporophytes produced (ventral view). *(D) Hylocomium splendens*, showing the multipinnate branching and the stepwise separation of each year's gametophytic growth (×1). *(E) Hypnum circinale* (×2). *(F) Ptilium cristacastrensis*, showing the neat pinnate arrangement of branches (×1).

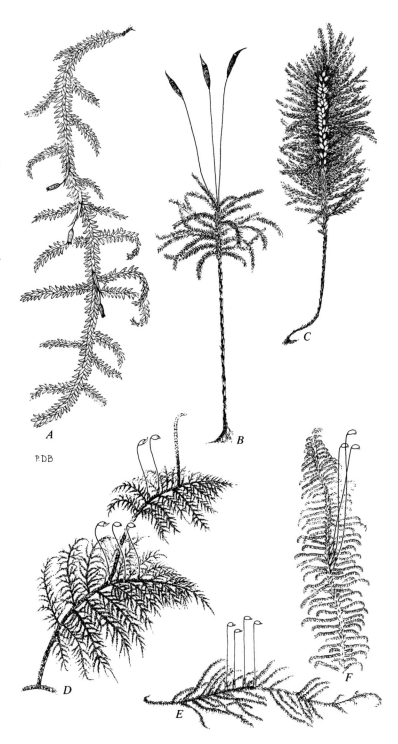

P. DB

The following discussion treats some of the diversity within the Bryidae at each stage of the life cycle. The discussion does not attempt to be comprehensive, but indicates certain trends that allow the Bryidae to occupy such a wide range of habitats.

SPORES AND THEIR SURFACE ORNAMENTATION

The use of the scanning electron microscope has contributed considerable information on spore structure in mosses (Fig. 8-3), but very few species have been studied in detail. Most Bryidae have unicellular spores 10–15 μm in diameter, but in a few genera, the spores are multicellular (e.g., *Dicnemon*). More than 50,000 spores are usually produced in each sporangium.

Even when examined with the light microscope, ornamentation of the spore surface is shown to differ markedly among species. Indeed, in some genera, including *Bruchia* and *Encalypta,* such ornamentation is useful in determining some species. Many Bryidae have spores with many circular depressions embedded in a network so that the spore looks like a diminutive golf ball. Such ornamentation is called foveolate. Other spores have spiny, warty, smooth, or reticulate surfaces.

Most spores are spherical, but some show a pyramidal surface that betrays the earlier position of the spore in a tetrad. In others the spores are kidney-shaped (reniform), and for the same reason. Since the spore shape, size, and ornamentation is so similar in many Bryidae, it is difficult to decipher their identity in the subfossil record.

SPORE GERMINATION PATTERNS

K. Nehira (1976) recognized eight spore germination patterns in the Bryidae. The following discussion treats four basic patterns. The spore is unicellular in most Bryidae, and rupture of the spore coat produces a chlorophyllous germ tube that represents the first cell of the chloronema. Later, either by rupturing the spore coat opposite the germ tube or by branching from the germ tube itself, a colorless rhizoid is formed that attaches the developing protonema to the substratum. Subsequent branching of the chloronema results in a much branched reclining uniseriate system with rhizoids emerging at intervals. Generally, this chloronema produces erect, richly chlorophyllous uniseriate branches, and usually upon these branches or associated with them, multistratose buds form that differentiate an apical cell which is the precursor to the leafy gametophore (Fig. 8-4C). There are modifications of this general pattern, and they may be significant in influencing where the protonema can establish itself. This protonema type is heterotrichous, a common type in the subclass, and present also in the Polytrichidae.

Occasionally when the spore is shed, it undergoes several cell divisions

FIGURE 8-3
Diversity of spores in Bryidae as recorded by scanning electron microscopy: Encalyptaceae (scales given under each spore). *(A) Encalyptra flowersiana*, distal face. *(B) E. affinis*, distal face. *(C) Bryobrittonia longipes*, distal face. *(D) E. asperifolia*, distal face. *(E) E. microstoma*, distal face. *(F) E. asperifolia*, proximal face. (Provided by Diana G. Horton.)

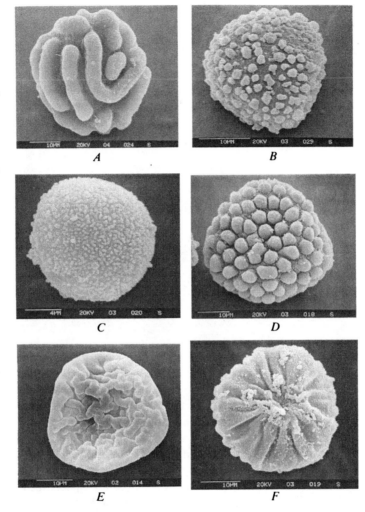

before the spore coat is ruptured and this multicellular mass produces chloronematal branches. From the chloronemata rhizoids emerge, and "buds" are differentiated, but no caulonema is formed. Sometimes the germinating spore, instead of producing a germ tube, produces a cell mass after it ruptures the spore coat; in this mass an apical cell differentiates and rhizoids emerge (Fig. 8-4*A*). Finally, in some Bryidae (as in *Dicnemon*), the spore is multicellular when it is shed. This multicellular mass differentiates an apical cell and produces rhizoids, thus bypassing the protonematal stage entirely. The Bryidae therefore exhibit diverse stages in reduction of the complexity of the protonema.

The protonemata of *Mittenia* and *Schistostega* produce chains of swollen

FIGURE 8-4
Protonemata of Bryidae. *(A)*
Drummondia sinensis (×190).
(B) Anomodon giraldii
(×100). *(C) Funaria hygro-*
metrica (×150). (After Nishida, 1978.)

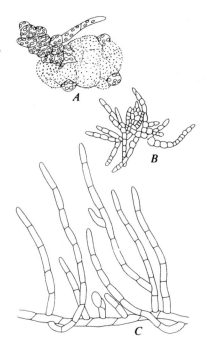

lenslike cells. The convex faces of these cells appear to focus light upon the
chloroplasts within and thus allow the protonema to live in deeply shaded sites.
These mosses are the "luminous mosses," since the lenslike cells also reflect
the light outward making the protonema glow yellow-green from the reflected
light.

Protonemata can reproduce vegetatively from the chloronematal or caulonematal branches by cells that break from the brittle filaments (Fig. 8-4*B*).
Rhizoids also sometimes produce specialized vegetative propagules. These
rhizoidal gemmae form at the end of a filament as a cluster of two or more
enlarged, somewhat thicker-walled cells. They take on the same reddish-brown
pigmentation of older rhizoids and cannot germinate to produce new protonemata until they are broken from the rhizoid (Fig. 8-11*D,E*).

In a few Bryidae the protonema persists indefinitely, and the gametophore is
reduced to perigonial and/or perichaetial "buds" (e.g., *Discelium*).

APICAL CELL AND LEAF ARRANGEMENT OF THE GAMETOPHORE

The apical cell in most Bryidae possesses three cutting faces, and leaf arrangement is, in consequence, initially in three ranks (Fig. 8-5*J,L*). This three-
ranked arrangement is altered since the three initial segments are unequal in
size, and growth of the stem generally produces a spiral arrangement in more
than three ranks (Fig. 8-5). In some Bryidae, although radially arranged on the

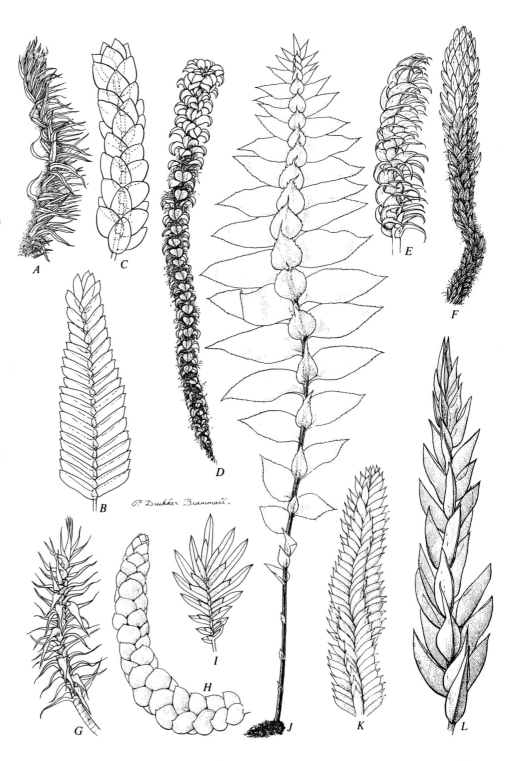

FIGURE 8-5

Leafy gametophores of Bryidae, showing diverse leaf arrangements *(A) Dicranum polysetum*, with leaves falcate and secund (×1). *(B) Phyllogonium fulgens*, showing leaves in two ranks (distichous) (×6). *(C) Hookeria lucens* showing complanate (flattened) leafy shoot (×8). *(D) Paludella squarrosa*, with squarrose leaves in five ranks (×3). *(E) Hypnum revolutum*, with falcate and secund leaves (×10). *(F) Aulacomnium turgidum*, with leaves evenly spirally arranged (×3). *(G) Campylium stellatum*, with leaves squarrose (×15). *(H) Myuroclada maximowiczii*, showing julaceous shoot with imbricated leaves (×5). *(I) Fissidens bryoides*, showing leaves in two ranks, and with sheathing flaps (×10). *(J) Cyathophorum bulbosum*, with distinctive underleaves, viewed from under surface (×3). *(K) Phyllodrepanium fulvum*, with ranked leaves (×7). *(L) Fontinalis antipyretica*, showing three-ranked arrangement (×5). (After Schofield and Hébant, 1984.)

P. Duikker Brammall.

91

stem, the leaf orientation gives the leafy shoot a dorsiventrally flattened or complanate appearance (Figs. 8-2*A* and 8-5*C*). In still others the three cell rows give rise to leaves of differing size: two lateral rows are of large leaves, and a third row is of smaller leaves (Fig. 8-5*J*). Such a leaf arrangement is common in the family Hypopterygiaceae. In a few mosses, including the common genus *Fissidens,* the apical cell at first has three but later has two cutting faces, and the leaves are in two lateral rows (Fig. 8-5*I*).

STRUCTURE OF VEGETATIVE LEAVES

There is considerable diversity in morphology and anatomy of vegetative leaves in this subclass (Fig. 2-2*A*–*L,O*–*T*). Leaves are predominantly unistratose and all cells are chlorophyllose, at least when the leaves are immature. The cell shape and arrangement within the leaf, however, differ markedly among genera and species and provide some of the most reliable characters that are used to distinguish among them. Differences in cell size, ornamentation of surfaces, shape, and color in various parts of the leaf are characteristic for each species (Figs. 8-6 and 8-7). The leaf margins are often composed of cell shapes distinctly different than those of the rest of the leaf (Fig. 8-7*B*). The margin is either toothed (Fig. 8-6*A,C,D*) or without teeth (entire) (Figs. 8-6*B* and 8-7*F*); it is usually unistratose, but it can be bistratose or even multistratose. The cells near the leaf apex may be longer or shorter than the rest of the leaf cells. At the leaf base the alar cells are often conspicuously different from the rest of the leaf cells (Fig. 8-6*B,E*). Ornamentation of cell surfaces is often very complex (e.g., papillae), especially in some acrocarpous Bryidae (Figs. 8-7*C,F,* 8-8, and 8-9).

The nature of the costa is relatively constant within a given species, but within the subclass it exhibits considerable variety (Figs. 2-2 and 8-10). Most genera of acrocarpous Bryidae have a single costa. The internal anatomy of the costa often shows, in cross-section, both enlarged cells continuous with those of the lamina (guide cells) and small thick-walled cells (stereids) that help to hold the leaf rigid as it diverges when moistened (Fig. 8-10*B*–*I*). Lamellae and surface filaments are infrequent in the Bryidae and are restricted to a few acrocarpous genera, including *Pterygoneurum* and *Aligrimmia* (Fig. 8-10*A*).

Sometimes, as in the family Leucobryaceae, the leaf shows specialization into chlorophyllose and hyaline cells (Fig. 8-10*C*). In genera of this family the outer cells of the multistratose leaf are hyaline (and these often have internal pores), while there is a unistratose internal layer of chlorophyllose cells.

In a few acrocarpous, and many pleurocarpous, Bryidae the costa is double or forked (Fig. 2-2*Q*). A single costa sometimes extends beyond the tip of the blade as a hair point (Fig. 2-2*R*), or may stop at or below the leaf apex. Sometimes the costa is confined to the leaf base; then it is usually double. In some Bryidae the costa has internal hydroids.

Some leaves show conspicuous longitudinal pleats (plications, Fig. 2-2*D,F*), regular transverse waves (undulations, Fig. 2-2*E*), or the surface may be sim-

FIGURE 8-6
Detail of areolation of leaves of Bryidae. *(A) Rhacomitrium lanuginosum,* portion of margin near apex, showing strongly papillose cells, and cells with nodulose walls (×190). *(B) Leucodontopsis geniculata,* basal cells showing well-differentiated alar region (×150). *(C) Syrrhopodon texanus,* cells near leaf base, showing distinctive area of hyaline cells (×100). *(D) Dicranum scoparium,* cells from near leaf apex, showing pronounced pits in cell walls (×140). *(E) Hygrohypnum ochraceum,* showing somewhat enlarged alar cells (×210). (After Crum and Anderson, 1981.)

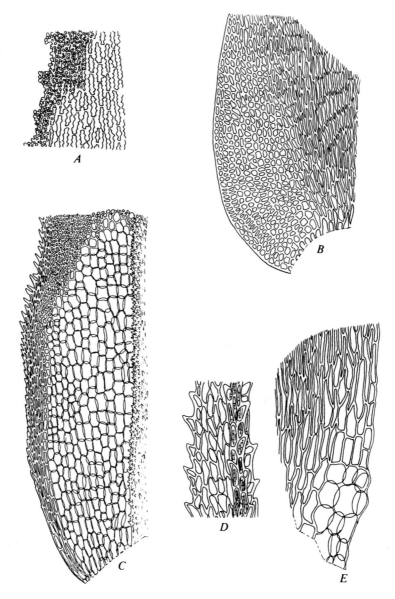

ply flat. Margins are either plane, erect, curved under (recurved or revolute), or incurved over the upper surface (involute).

Upon drying, the leaf often shows a characteristic morphology that differs strikingly from the moist leaf. It may become greatly corkscrew twisted or contorted, or it may simply fold inward and become imbricated on the stem. Such changes between wet and dry conditions, besides making the gametophore strikingly different in appearance in the different states of dryness, also appear to be significant in the drought tolerance of the plant.

FIGURE 8-7
Leaf apices of Bryidae, showing areolation. *(A) Tayloria
serrata*, showing thin-walled
cells (×60). *(B) Mnium spinulosum*, showing margin of differentiated elongate cells
(×90). *(C) Aulacomnium acu-
minatum*, showing single papilla on each cell, and cells
with pronounced trigones
(×85). *(D) Tortula fragilis*
(×150). *(E) Pterigynandrum
filiforme*, showing projecting ends of cells (×170). *(F)
Orthotrichum ohioense*,
showing cells with many
papillae (×200). (After
Crum and Anderson, 1981.)

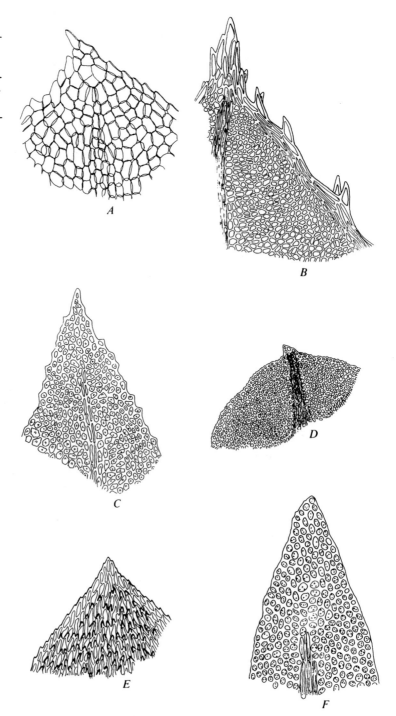

FIGURE 8-8

Leaf surface ornamentation in Bryidae as revealed by scanning electron microscopy. Scale provided with each figure. *(A) Macromitrium submucronifolium,* view of back of leaf, showing smooth costa and papillose cells of lamina. *(B) Thelia asprella,* showing unusual papillae. *(C) Papillaria* sp. leaf surface. (Provided by Dale H. Vitt.)

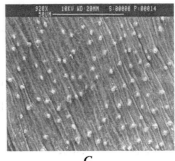

A

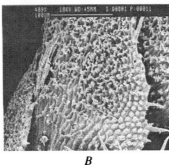

B

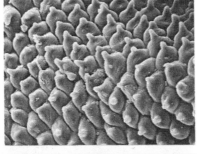

C

FIGURE 8-9

Leaf surface ornamentation in Bryidae as revealed by scanning electron microscopy. *(A) Cardotiella subappendiculata* (×1000). *(B) Encalypta procera,* transverse section (×2000). *(C) Macromitrium subfragile* (×1200). *(A, C,* provided by Dale H. Vitt; *B,* provided by Diana G. Horton.)

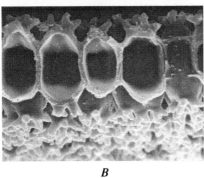

A

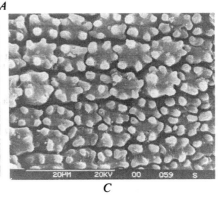

B

C

95

FIGURE 8-10
Transverse sections of leaves of Bryidae. *(A) Aligrimmia peruviana*, showing adaxial lamellae (×60). *(B) Campylopus pilifer*, showing strong ridges on abaxial surface (×330). *(C) Leucobryum glaucum*, showing single inner row of chlorocysts completely embedded in porose leucocysts (×300). *(D) Grimmia arizonae* (×750). *(E) Trichodon tenuifolius* (×750). *(F) Amblystegium montanae* (×600). *(G) Symblepharis tenuis* (×600). *(H) Grimmia elatior* (×750). *(I) Plagiomnium medium* (×450). *(A, D–I,* after Kawai, 1968; *B,* after Crum and Anderson, 1981; *C,* after Lorch, 1931.)

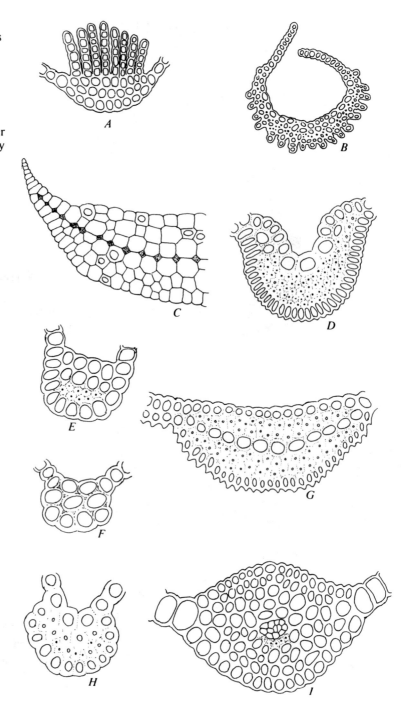

VEGETATIVE REPRODUCTIVE STRUCTURES

As in all mosses, vegetative reproduction is important for expansion of local populations and for wider dispersal of the species. Since gametophores are frequently brittle when dry, terrestrial species, when trampled, become fragmented, and the fragments can be dispersed to new sites. These fragments can produce protonemata which then proceed through bud production to form new gametophores. If a differentiated apical cell is already present in the fragment, or the fragment is a living stem (in which a new apical cell can be differentiated as an innovation), a new stem is produced directly from the fragment.

The gametophores in some Bryidae produce brittle specialized branches that serve as vegetative diaspores (Fig. 8-11*A,B*); or the leaves are brittle, and the fragments serve as diaspores. Sometimes diminutive branches form near the leaf axils, especially near stem apices. Such diaspores are produced commonly by some Bryidae in early stages in the growth of the gametophore, before sexual organs appear.

Gemmae occur frequently in acrocarpous Bryidae (Fig. 8-11). They form sometimes at the leaf apex, on the surfaces of the blade or costa, or on the stem (Fig. 8-11*F–H*). Rarely the stem produces an elongate almost leafless extension that is terminated by a mass of gemmae, as in *Aulacomnium androgynum* (Fig. 8-11*C*). Since gemmae are small and easily carried by wind or water, they serve as effective diaspores for the moss.

Rhizoidal gemmae are produced mainly by acrocarpous Bryidae (Fig. 8-11*D,E*). Sometimes they appear (as in *Leptobryum*) before the leafy gametophores. Generally they are produced abundantly only when the gametophore is vegetative. As the gametophore begins to bear sporophytes, these rhizoidal gemmae tend to become uncommon. They are important in expansion of populations in open sites. Their morphology is often distinctive for given species.

STEM STRUCTURE AND BRANCHING PATTERNS

The epidermal cells of the stem are often thicker walled than those of other cortical cells beneath (Fig. 8-12). These outer cells are chlorophyllose when young, but as the stem ages, cell walls often become red, brown, or yellow. Rarely the epidermal cells are papillose. The epidermal cells in some mosses are enlarged and thin-walled, but generally are the same size as the cortical cells. The outer cells are usually all termed cortical cells and the term "epidermal" is avoided. These outer cells give rise to all emergences from the stem, including rhizoids, paraphyllia (Fig. 8-13*B–D*), pseudoparaphyllia (Fig. 8-13*A*), leaves, and branches. Paraphyllia are confined to pleurocarpous Bryidae and are restricted to a few genera. Paraphyllia are extremely small green leaflike structures or merely small uniseriate or multiseriate determinate emergences which are scattered along the stem surface among the leaves. Such paraphyllia considerably increase the photosynthetic area, but their main sig-

FIGURE 8-11
Vegetative diaspores of Bryidae. *(A) Dicranum flagellare*, with brittle flagelliferous branches (×8). *(B) Pohlia erecta*, showing axillary "bulbils" (×20). *(C) Aulacomnium androgynum* with gemmae surmounting flagelliferous shoot (×6); detail of gemma at left. *(D)* Rhizoidal gemma of *Leptobryum pyriforme* (×100). *(E)* rhizoidal gemmae of *Tayloria serrata* (×50). *(F) Calymperes richardii*, with gemmae on leaf apex (×40). *(G) Encalypta procera*, with gemmae near the base of leaf (×16). *(H) Syrrhopodon parasiticus* with gemmae on leaf costa (×27). *(A, D,* after Scagel et al., 1982; *B, C, E,* after Correns, 1899; *F–H,* after Crum and Anderson, 1981.)

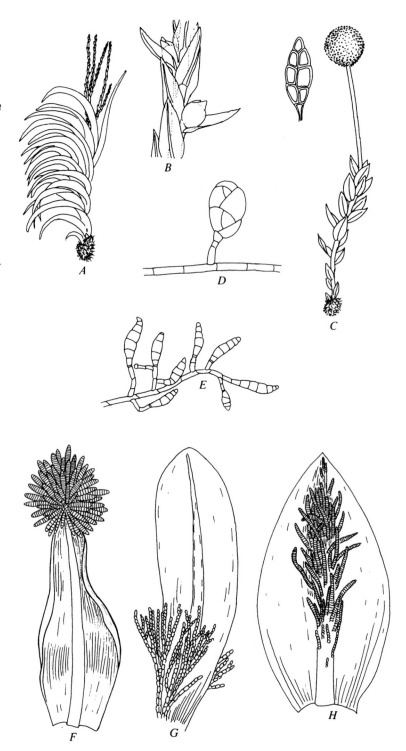

FIGURE 8-12
Transverse sections of stems of Bryidae. *(A) Pseudobryum cinclidiodes,* showing the hyaline cortical cells, the large region of parenchymatous cells, the central strand, and the smaller strands in the cortex leading toward leaf bases (×55). *(B) Pilotrichidium antillarum,* showing thin-walled outer cortical cells, an extensive area of stereid cells, and a central cylinder of enlarged cells (×50). *(C) Brachythecium* sp., showing both transverse and longitudinal views; note the stereid cells of the cortex and the distinctive central strand (×165). *(A, B,* after Lorch, 1931; *C,* after Hébant, 1977.)

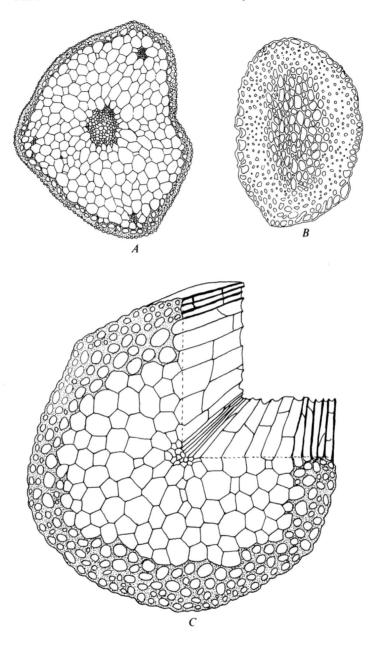

nificance is probably in external conduction of water and dissolved nutrients. Pseudoparaphyllia are also found only in pleurocarpous mosses, and are of relatively frequent occurrence. They are small uniseriate unbranched filaments or irregularly shaped, narrowly deltoid structures confined to branch buds that never produce branches, or to bases of branches. The main function of

FIGURE 8-13
Pseudoparaphyllia and paraphyllia of Bryidae. *(A)* Pseudoparaphyllia of *Porotrichum
bigelovii* around "bud" of incipient branch (×340). *(B, C)*
Hylocomium splendens. (B)
Short length of stem, showing
paraphyllia on stem surface
(×12). *(C)* Single paraphyllium (×340). *(D)* Paraphyllium of *Thuidium deli-
catulum* (×340). *(A*, after
Schofield and Thomson,
1966; *B*, after Crum and Anderson, 1981; *C, D*, after
Scagel et al., 1982.)

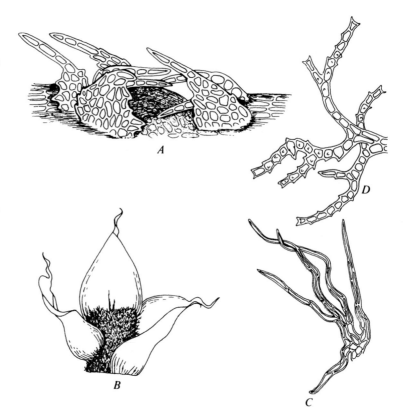

pseudoparaphyllia appears to be protection of the apical cell of a branch bud
(whether or not the bud ultimately produces a branch). Rhizoids arise from the
stem or from the leaf bases. Those that arise from large cells surrounding
branch buds produce a much branched uniseriate system with a main axis and
numerous small branches. These are macronemata. Others arise from scattered
cells of the stem, have fewer branches, and are usually smaller; these are
micronemata. Macronemata appear to be similar to pseudoparaphyllia, while
micronemata appear to be functionally important in capillary water conduction. Rhizoids are smooth or sometimes papillose (Fig. 8-14).

Beneath the cortex, the stem is often composed of enlarged cells that form a
wide multistratose central layer; sometimes this layer occupies the rest of the
stem's diameter, or the center of the stem may have a central strand. This
strand may be composed of thick-walled cells that include some hydroids, or it
may be a cluster of very small cells with no hydroids.

Two main trends in the position of the sex organs are shown in the subclass
Bryidae. Sometimes, as noted in several subclasses discussed earlier, the
perigonia and perichaetia terminate the apex of the main stem; any further
growth of that stem must be by production of lateral innovations. These innovations usually form just below the perichaetium or perigonium. This results

in erect, often tufted, and usually sparsely branched plants. In such mosses, the sporophyte emerges from the terminal perichaetium, although growth of an innovation after sporophyte production may make the sporophyte appear to be lateral. This kind of sporophyte position is termed acrocarpous (Fig. 8-1).

The sexual branches are greatly reduced in many species and are lateral on the main stem. Continued growth from the apex of the main stem makes growth of that stem indeterminate. This usually results in prostrate, freely branched plants that form interwoven mats. In such stems the sporophytes are lateral, and several perichaetia, each with its sporophyte, may be on one stem; this is termed pleurocarpous (Fig. 8-2). It is generally quite easy to distinguish acrocarpous and pleurocarpous mosses.

FIGURE 8-14
Surface ornamentation of rhizoids of Bryidae as revealed by electron microscopy: Bartramiaceae. *(A, B) Plagiopus oederi. (C, D) Leiomela bartramioides. (E, F) Conostomum tetragonum* (figures on right ×1000, on left ×2000). (Provided by Tooru Hirohama and Zennoske Iwatsuki.)

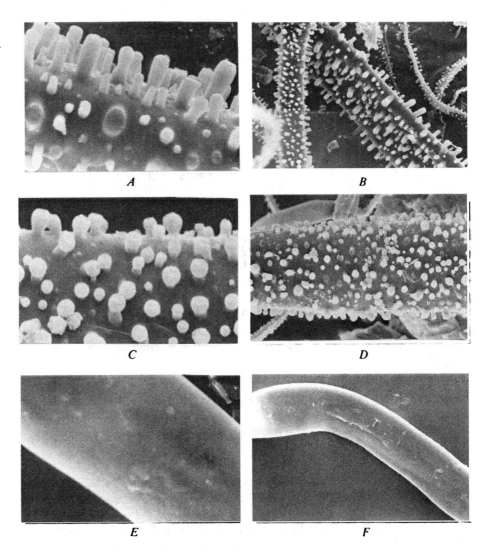

More than 60% of the Bryidae are dioicous. Those that are monoicous have antheridia that tend to mature at a different time than the archegonia on the same stem, and thus self-fertilization is generally prevented. Some dioicous Bryidae have tiny perigonial plants that are epiphytic on archegonium-producing plants. These consist of only perigonia with a few rhizoids. The perichaetial plant produces a substance that inhibits the growth of this male gametophore; if the spore germinates independently from the perichaetium-bearing plant, it produces perigonium-bearing gametophores morphologically similar to the perichaetial gametophores. Such diminutive gametophores are called dwarf males, and the sexuality of the plant is often described as phyllodioicous: the gametophore appears to be dioicous because the perigonium is attached to leaves of the perichaetium-bearing plant. The presence of such dwarf males assures sexual reproduction in the clone, but after the first production of spores in that clone, most succeeding dwarf males will be from the spores of that clone, and self-fertilization will occur. This allows short-term survival of the population but slows down any potential evolutionary change. In some Bryidae dwarf males are always produced, whether epiphytic or separate from the perichaetial plant.

THE CALYPTRA

Janzen (1916) made a detailed study of calyptra morphology in the mosses. Among the Bryidae smooth hood-shaped (cucullate) calyptrae are most common. Structure is relatively uniform within a species among the various other shapes of calyptrae and serves as a useful taxonomic feature (Fig. 2-4). Furthermore, the size and shape of the calyptra appears to strongly affect the shape and orientation of the capsule. Some calyptrae have papillae or bristles (Fig. 2-4*F*); others are smooth, but are conspicuously pleated (Fig. 2-4*I*). In most calyptrae the lower margin is entire, but sometimes a distinct fringe develops after the calyptra is torn loose from the gametophore as the seta elongates (Fig. 2-4*G,L*). The calyptra is sometimes an elongated cone that completely encloses the sporangium, as in *Encalypta* (Figs. 2-4*B* and 8-1*C*).

PERIGONIA AND PERICHAETIA

The paraphyses of Bryidae, whether among antheridia or archegonia, tend to be uniseriate and chlorophyllose, at least when young. The apical cell may be very swollen and blunt or not swollen and pointed. Occasionally the paraphyses have a biseriate upper portion.

Perigonia are sometimes composed of leaves similar to the vegetative leaves, but sometimes perigonial leaves are greatly enlarged and form a petallike rosette around the anteridia and paraphyses (Fig. 8-15*C,D*). This is especially true in many of the genera in the family Mniaceae (Fig. 8-15*D*). The

FIGURE 8-15
Perichaetia and perigonia of Bryidae. *(A–D)* Perigonia. *(E–H)* Perichaetia. *(A) Pohlia wahlenbergii* (×6). *(B) Dendroalsia abietina* (×15). *(C) Breutelia affinis* (×5). *(D) Rhizomnium glabrescens* (×2). *(E) Hedwigia ciliata* (×8). *(F) Brachythecium frigidum* (×20). *(G) Dicnemon calycinum* (×8). *(H) Fontinalis antipyretica* (×8). (After Schofield & Hébant, 1984.)

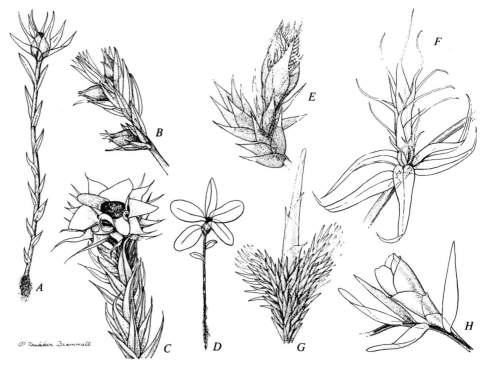

rosettelike perigonium acts as an effective splash-cup from which raindrops can disperse the sperms some distance from the antheridial gametophore and nearer to the perichaetium (Fig. 8-15*C,D*). Budlike perigonia are common among both acrocarpous and pleurocarpous Bryidae. The budlike perigonia and perichaetia enclose the sex organs so that water is retained longer, and they protect the antheridia from drying out before sperms are released (Fig. 8-15*A,B*).

The perichaetial leaves in acrocarpous Bryidae are often structurally similar to the vegetative leaves. Sometimes they are larger and have little or no chlorophyll. The perichaetial leaves often lack a costa when the vegetative leaves possess one. Differentiated perichaetial leaves often have an expanded base, taper abruptly to the apex, and form a budlike perichaetium (Fig. 8-15*E,F*). When the perichaetial leaves are much elongated, they sheathe each other to enclose the archegonia and paraphyses in a protective sleeve.

The perichaetia in pleurocarpous Bryidae tend to be bulbiform, whether the leaves are short or elongated. Sometimes, as in *Hedwigia,* the perichaetial leaf margins have elaborately branched cilia (Fig. 8-15*E*); occasionally, as in *Fontinalis,* the perichaetial leaves are blunt (Fig. 8-15*H*). Archegnoia vary in the length of the neck but are otherwise of relatively uniform structure among the Bryidae. They are chlorophyllose up to fertilization of the egg, after which those in which fertilization does not take place turn dark reddish-brown (Fig. 1-1*A*).

SPOROPHYTE DIVERSITY

The sporophyte in most Bryidae has a pointed foot that penetrates the gametophore, an elongate, smooth, wiry seta, and an enlarged sporangium that is sheathed by the calyptra, at least until the sporangium is nearly mature. The sporangium is chlorophyllose until mature and varies in shape from cylindric through ovoid (shaped like an egg) to obovoid (like an egg with the broad end upward), ellipsoid, or spherical. Occasionally the sporangium is clearly pear-shaped (pyriform, Fig. 8-16*B*) or shaped like a top (turbinate, Fig. 8-16*I*). The epidermal wall of the jacket generally possesses stomata, and these are either cryptopore or phanerophore (Fig. 2-3). The exothecial cells are commonly square (quadrate), and they are usually smooth. Sometimes the surface is longitudinally grooved (Fig. 8-16*C,H*).

The sporangium often has a neck; usually this is very abbreviated, but sometimes it is clearly differentiated and expanded as an apophysis (or hypophysis, Fig. 8-16*E*). In *Splachnum,* the apophysis sometimes forms an umbrellalike expansion at the sporangium base. This is bright opalescent red-purple in *Splachnum rubrum* and exudes a foul odour that attracts flies. The flies accidentally get spores on their bodies and when they move to dung or other decomposing animal material, they scatter the spores to a suitable substratum for the *Splachnum.*

The anatomy of the sporangium of Bryidae is similar to that of many Polytrichidae. Generally it has a multistratose jacket, a cylindric sporogenous layer, and a cylindric columella. Sometimes there are intercellular spaces in a cylinder, traversed by supportive filaments, between the jacket and the sporogenous layer (Fig. 8-17).

An operculum is usually differentiated, and its shedding sometimes is aided by enlarged elastic cells in a ring (annulus) that, when ruptured, coils outward to release the operculum. The operculum varies from conic to convex. It is usually constructed of cells that are smaller and thicker-walled than those of the rest of the sporangium jacket. The operculum therefore shrinks less when drying than the rest of the jacket. The operculum usually breaks loose from the top of the columella and falls away, but sometimes much of the columella falls away with it. Rarely the operculum remains attached to the columella even when the spores are shed. Such sporangia usually shrink as they dry, forcing the mouth open and releasing the spores.

The seta anatomy is much like that of the Polytrichidae. Sometimes a central strand with hydroids and leptoids is present; often there is no differentiation of a central strand (Fig. 8-18). The seta, usually smooth, is occasionally papillose, and in rare cases (as in *Eriopus,* Fig. 8-18*A*), has many uniseriate chlorophyllose filaments over its surface, which greatly increase its photosynthetic area. Often the linear cells of the seta are longitudinally arranged in somewhat helical rows, and the cell wall thickness varies on different sides of the seta. This appears to influence the twisting and untwisting of the seta as it dries and becomes rehydrated. This twisting of the seta results in twitching which shakes the

FIGURE 8-16
The diversity of sporangia in the Bryidae. *(A) Trematodon longicollis* (×25). *(B) Leptobryum pyriforme* (×12). *(C) Philonotis fontana* (×10). *(D) Barbula cylindrica* (×15). *(E) Splachnum ampullaceum* (×15). *(F) Oncophorus wahlenbergii,* showing the struma where the sporangium joins the seta (×20). *(G) Bryum argenteum* (×12). *(H) Orthotrichum lyellii* (×10). *(I) Voitia hyperborea* (×12), showing the absence of an operculum. *(J) Amblystegium serpens* (×15). *(K) Scouleria aquatica,* showing the columella with the persistent operculum; the peristome teeth have fallen away (×25.)

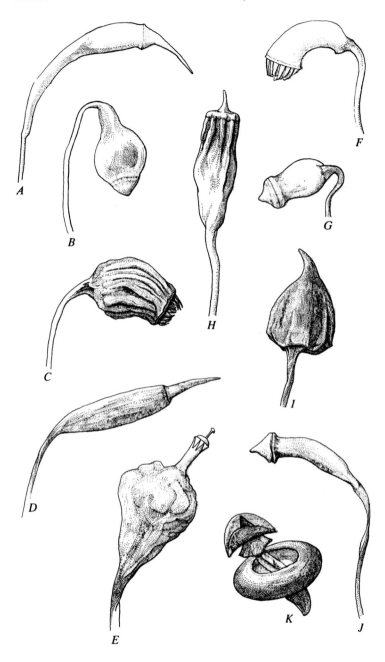

sporangium and aids in spore release. The seta is usually rigid, but in some species of *Splachnum* it elongates very rapidly and holds the sporangium aloft only when turgor permits it.

FIGURE 8-17
Longitudinal sections through
sporangia of Bryidae. *(A)*
Funaria hygrometrica, showing the enlarged columella
(col.), bounded by the
sporogenous layer; the large
area of air spaces *(a.s.)*,
traversed by trabeculae; the
enlarged cells of the annulus
(ann.), at the base of the
operculum *(op.)*; and the stomata *(sto.)*, near the base of
the sporangium (×70). *(B)*
Splachnum ampullaceum,
showing the enlarged hypophysis (×20). *(C) Micromi-*
trium tenerum, showing the
absence of a columella when
the sporangium is mature, and
the small foot penetrating the
apex of the gametophore
(×75). *(A*, after *Den danske*
mosflora, 1976; *B*, after Janzen, 1918; *C*, after Goebel,
1930.)

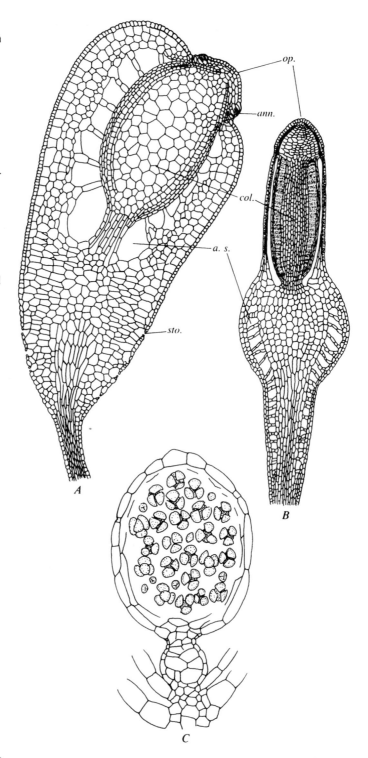

FIGURE 8-18

Detail of the seta in Bryidae.
(A) Eriopus uleanus, transverse section, showing filaments on surface and central strand of hydroids (×70). *(B) Funaria hygrometrica,* showing transverse and longitudinal views; the stereid layer is conspicuous in the cortex, and the central strand is of internal hydroids and a cylinder of leptoid-like cells external to them (×300). *(C) Bryum caespiticium,* transverse section of upper portion of stem of gametophore, showing the seta with haustorium-like cells penetrating the stem cells; note also the central strand of hydroids in the foot of the seta (×100). *(D) Dicranum majus,* transverse section, showing wide region of stereid cells and the well-differentiated central strand with hydroids (×100). *(A, C, D,* after Lorch, 1931; *B,* after Hébant, 1977.)

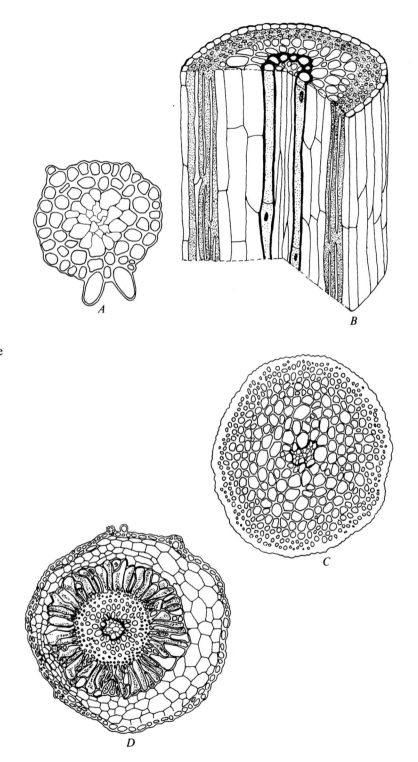

PERISTOMES AND SPORE DISPERSAL

Peristome teeth are composed of fragments of numerous cells, but no whole cells remain as in the Polytrichidae and Tetraphidae. In the Bryidae, never more than three concentric cell layers are involved in peristome tooth production. The cell layers that produce the peristome form a multicellular dome or cone beneath the operculum, and the broad part of this cone is attached to the inner boundary wall of the jacket, just beneath the dehiscence line of the operculum. Peristome teeth form in two basic patterns. Many acrocarpous Bryidae have a single circle of teeth. Each tooth is formed from the abutting tangential walls of two concentric cell layers. The tangential faces have additional cellulose deposited upon their surfaces, making these walls strongly pigmented and much thicker than the walls of the remainder of the cell. The thinner portions of the walls of the participating cells disintegrate as the teeth mature, isolating the teeth from each other, since only a portion of the tangential face is thickened, especially in cells close to the apex of the columella. Each tooth is thus made up of a linear series of articulated segments with their surfaces variously ornamented. In these Bryidae the outer face of each segment of each tooth is formed of a single plate of thickening across the tangential face of one cell. The inner surface of the same segment of that tooth is formed from parts of two adjacent cells of the inner concentric circle, and thus is formed of two inner plates. Since the outer face is formed of a single plate (or scale), this type of peristome is described as haplolepidous (Fig. 8-19C,D).

The other means of peristome tooth production essentially reverses the nature of formation of the outer face of the tooth. Each longitudinal segment of a tooth is formed from parts of two adjacent cells of the outer concentric circle, while the inner face is formed from the single plate. Such a peristome is termed diplolepidous. The inner faces of the teeth tend to have jagged projections (Fig. 8-19A,B). In many mosses with diplolepidous peristome teeth, there are two concentric rows of teeth. The outer row is usually diplolepidous, while the inner row is either diplolepidous or haplolepidous. Such double peristomes are thus derived from three concentric cell layers.

The peristome teeth usually number 16 per circle; sometimes they are deeply forked so that 32 teeth seem present; also, pairs of teeth may be fused or may be entirely absent (Figs. 8-20–8-23).

Peristome teeth of the Bryidae are always articulated (Figs. 8-19 and 8-20). The degree of thickening on the outer or inner surface of the tooth determines its behavior when it is moistened. If the thickening is greatest on the outer surface, the tooth tends to curve outward when it dries, but if the thickening is greatest on the inner surface, the tooth usually curves inward when moistened. The inner surface tends to bear jagged projections, and spores from the interior of the sporangium attach to this surface as the tooth curves inward and are thrown off as the tooth curves outward. These pulsating movements of teeth assist in spore dispersal. The entire process of flicking takes a few seconds. After the tooth curves inward and outward many times it tends to become brittle and

FIGURE 8-19
Haplolepidous (right) and diplolepidous (left) peristome teeth of Bryidae. *(A, B) Splachnum ampullaceum. (A)* Longitudinal section through apex of sporangium (×300). *(B)* Transverse section, showing three peristome teeth; note that the outer faces of the teeth are formed from plates derived from parts of the walls of two cells (×225). *(C, D) Dicranum scoparium* (×300). *(C)* Longitudinal section through upper portion of sporangium. *(B)* Transverse section showing two teeth; note that the outer face is formed from the face of a single cell. (After Janzen, 1918.)

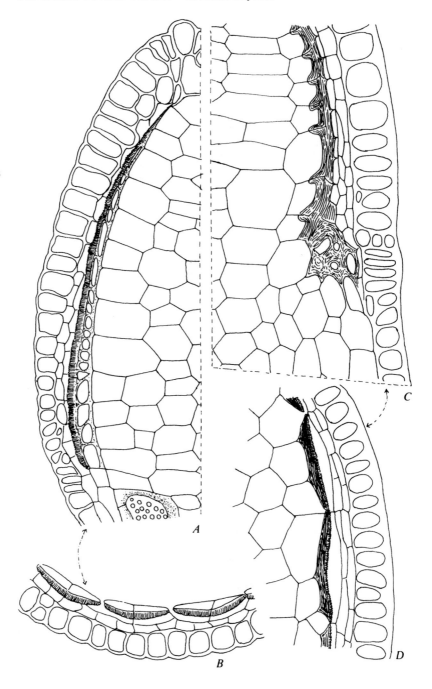

break off. Very small changes in humidity initiate hygroscopic responses in peristome teeth.

Bryidae with peristomes made up of two circles of teeth have outer teeth that

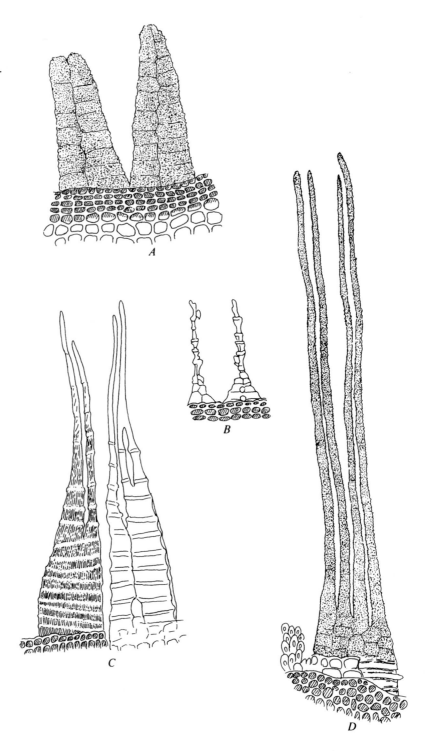

FIGURE 8-20
Peristome teeth of Bryidae.
(A) Macromitrium zimmer-
mannii (×240). *(B) Mielichho-*
feria mielichhoferi (×200).
(C) Dicranum sp., showing
external view (left), and view
of internal surface (right)
(×110). *(D) Rhacomitrium*
lanuginosum (×200). *(A, C,*
D, after Fleischer,
1900–1902; *B,* after Janzen,
1918.)

A

B

C

D

110

tend to be hygroscopic while the inner teeth infrequently respond to any change in moisture (Figs. 8-21, 8-22*A*, and 8-23*A,C*). The outer teeth (exostome) often curve inward upon moisture increase and thrust themselves between the inner teeth (endostome). A loss in moisture content causes the teeth to curve outward, to uncoil rapidly, and to dislodge any spores caught on their surfaces. In some double peristomes the endostome forms a remarkable dome (as in *Cinclidium*) or a latticed truncate cone (as in *Fontinalis*, Fig. 8-23*A*), but in most cases slender hyaline teeth bend inward and prevent water droplets from entering the sporangium and rapidly wetting the spores.

Since the mouth of the sporangium is often oriented downward, the spores fall by gravity upon the inner faces of the teeth and are gradually scattered out of the sporangium as the peristome teeth respond to moisture changes (Figs. 8-22 and 8-23). In some Bryidae, the spores must sift through twisted peristome teeth as the sporangium is shaken by any disturbance.

Haplolepidous teeth are diverse in both gross morphology and pattern of ornamentation on the surface. Often there are 16 teeth, but these teeth may be deeply or shallowly forked. Occasionally the teeth are attached to a long fused sleeve (as in some species of *Tortula*) with the teeth spirally twisted when dry. Rarely (as in *Octoblepharum* and some species of *Orthotrichum*), there are eight teeth, owing to fusion of pairs.

Sometimes an operculum is not differentiated. In such cases, peristome teeth are absent except in unusual circumstances where they appear as rudimentary fragments. Sporangia that lack an operculum shed spores when the sporangium jacket is broken; in most of these mosses the seta is short or nearly absent, and the entire gametophore surrounding the sporangium is often composed of a few vegetative leaves. Diminutive mosses of this kind usually grow on soil in open sites and, on drying, break loose from the soil and can be blown or washed to a new site. The substratum, when mud, can be inadvertently transported by animals to new sites, thus dispersing the embedded spores. Occasionally the nonoperculate sporangium has a long seta; here the spores can escape only when the sporangium wall disintegrates or is broken. The great diversity in peristome structure is useful in identification since it shows considerable uniformity among some families.

GENERALIZATIONS

There appear to be certain evolutionary trends within the Bryidae. A number of these assumptions are controversial and should not be treated as proven:

1. Unicellular spores are more generalized than multicellular spores.
2. Heterotrichous protonemata are more generalized than protonemata with specialized lenticular cells; these are more generalized than protonemata that lack a caulonema stage; the most specialized is a multicellular spore in which the protonemal stage is completely bypassed.

FIGURE 8-21
Peristome teeth of Bryidae, showing both exostomial and endostomial teeth. *(A) Bryum caespiticium,* showing external surface of exostomial tooth (far left), internal surface of exostomial tooth (far right), and endostomial teeth between the exostomial teeth (×160). *(B) Orthodontium infractum* (×200). *(A,* after Janzen, 1918; *B,* after Fleischer, 1900–1902.)

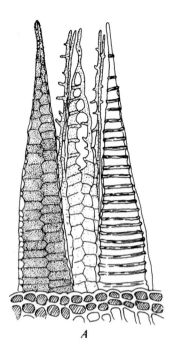
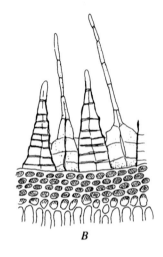

A

B

FIGURE 8-22
Peristomes of Bryidae as revealed by scanning electron microscopy. *(A) Encalypta procera* (×20). *(B) Encalypta brevicolla* (×35). *(C) Coscinodon cribrosus* (×60). *(D) Hypnum recurvatum* (×40). *(A, B,* provided by Diana G. Horton; *C, D,* provided by Dale H. Vitt.)

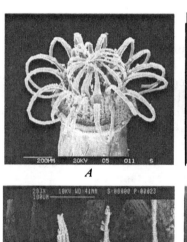
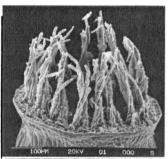

A

B

C

D

FIGURE 8-23
Peristomes of Bryidae as revealed by scanning electron microscopy. (*A*) *Fontinalis antipyretica* (×55). (*B*) *Macrocoma orthotrichoides* (×100). (*C*) *Cinclidium stygium* (×50). (Provided by Dale H. Vitt.)

A

B

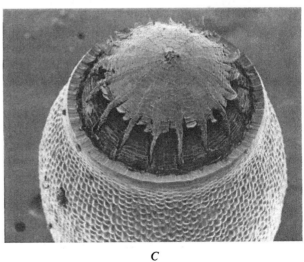

C

3. An apical cell with three cutting faces is more generalized than one with two cutting faces.
4. Leaves in three ranks are more generalized than leaves in two ranks.
5. Radial symmetry in the leafy shoot is more generalized than bilateral symmetry.
6. Complanate leafy shoots with reduced underleaves, in spite of their arrangement in three ranks, are more specialized than complanate shoots in which the leaves are radially arranged.
7. The presence of rhizoids on stems is more generalized than their absence.
8. The presence of sex organs in a terminal cluster is more generalized than sex organs in a lateral cluster.
9. Branching by innovations is more generalized than regular production of lateral branches on a stem of indeterminate growth.
10. A smooth, cucullate calyptra is the most generalized type.
11. A sporophyte with a long seta is more generalized than one with a short seta.
12. An erect cylindric sporangium is more generalized than any other shape or orientation.
13. A smooth sporangium wall is more generalized than one that is regularly grooved or ornamented,
14. The presence of an operculum is more generalized than its absence.
15. The absence of an extensive intercellular cylindric layer in the sporangium is more generalized than its presence.
16. The absence of an apophysis is more generalized than its presence.
17. A smooth seta is more generalized than one that is papillose or bears appendages.
18. Numerous spores in a sporangium indicates a more generalized state than few spores.
19. A foveolate spore is the most generalized while those that are elaborately ornamented or smooth are specialized.

A number of features are difficult to interpret. Whether diplolepidous or haplolepidous teeth are the more generalized is difficult to determine. It appears that, if the exostome is diplolepidous, then the absence of an endostome is more specialized than its presence. Stems with a central strand can be interpreted as either generalized or specialized; in some stems it appears that the lack of differentiation of distinctive layers has been selected for existence in a wet environment, and is thus specialized; in other cases the lack of such differentiation could be interpreted as generalized.

Considerable detailed research on the life histories is still necessary. Even a detailed understanding of confident identification of many Bryidae remains in a very unsatisfactory state. Rapid destruction of moss habitats, especially near and in the tropics, may eliminate many mosses before they are discovered and studied.

FURTHER READING

Allsop, A., and G. C. Mitra. 1958. The morphology of protonema and bud formation in the Bryales. *Ann. Bot.* **22**:95–115.

Anderson, L. E. 1963. Modern species concepts: mosses. *Bryologist* **61**:107–119.

——— 1964. Biosystematic evaluations in the Musci. *Phytomorphology* **14**:27–51.

Banerji, J., and S. Subir. 1956. The life history of *Barbula indica*. *Proc. Natl. Inst. Sci. India* **22B**:304–316.

Bequaert, J. 1921. On the dispersal by flies of the spores of certain mosses of the family Splachnaceae. *Bryologist* **24**:1–4.

Berthier, J. 1972. Recherches sur la structure et le developpement de l'apex du gametophyte feuillé des mousses. *Rev. Bryol. Lichénol.* **38**:421–551.

——— 1978. Analyse des capacités morphogenes du filament des Bryales. *Bryophyt. Bibl.* **13**:223–241.

Blomquist, H. L., and L. L. Robertson. 1941. The development of the peristome in *Aulacomnium heterostichum*. *Bull. Torrey Bot. Club* **68**:569–584.

Bonnot, E. G. 1967. Sur la structure de l'apex du gametophyte feuillé de la mousse *Anomodon viticulosus* (L.) Hook. & Tayl. *Bull Soc. Bot. France* **114**:4–11.

——— 1967. Sur la valeur et la signification des paraphylles chez les Bryales. *Bull Soc. Bot. France Mem.* **1967**:236–248.

Bopp, M. 1956. Die Bedeutung der Kalyptra für die Entwicklung der Laubmoossporogone. *Ber. Deutsch. Bot. Gesell.* **69**:455–468.

——— 1961. Morphogenese der Laubmoose. *Biol. Rev.* **36**:237–280.

——— 1963. Development of the protonema and bud formation in mosses. *J. Linn. Soc. (Bot.)* **58**:305–309.

Bristol. B. M. 1916. On the remarkable retention of vitality of moss protenema. *New Phytol.* **15**:137–143.

Bunger, E. 1890. Beitrage zür Anatomie der Laubmooskapsel. *Bot. Centralbl.* **42**:193–199, 225–230, 257–262, 289–299, 321–326, 353–356.

Burr, J. L. 1939. The development of the antheridium, archegonium and sporangium of *Cyathophorum bulbosum* (Hedw.) C.M. *Trans. Roy. Soc.* **68**:437–456.

Castaldo, R., R. Ligrone, and R. Gambardella. 1979. A light and electron microscope study on the phylloids of *Leucobryum candidum* (P. Beauv.) Wils. *Rev. Bryol. Lichénol.* **45**:345–360.

Coesfeld, R. 1892. Beiträge zur Anatomie und Physiologie der Laubmoose. *Bot. Zeitung* **50**:153–164, 169–176, 185–193.

Correns, C. 1899a. Über Scheitelwachsthum, Blattstellung und Anlagen des Laubmoose-stämmchens. Schwendener Festschrift, 28 pp. Berlin.

——— 1899b. *Untersuchungen über die Vermehrung der Laubmoose durch Brutorgane und Stecklinge*, 472 pp. Jena.

Crosby, M. R. 1980. The diversity and relationships of mosses, in Taylor, R. J. and R. E. Leviton (eds.), *The Mosses of North America,* pp. 115–129. San Francisco: AAAS.

Crundwell, A. C. 1979. Rhizoids and moss taxonomy, in Clarke, G. C. S., and G. Duckett (eds.), *Bryophyte Systematics,* pp. 347–363. London: Academic Press.

Davy de Virville, A. 1927–1928. L'action du milieu sur les mousses. *Rev. Gen. Bot.* **39**:711–726, 767–783. Ibid. **40**:156–173.

Denning, K. 1935–1936. Untersuchung über sexuellen Dimorphismus der Gametophyten bei heterothallischen Laubmoosen. *Flora* **30**:57–86.

Derschau, M. van. 1900. Die Entwickelung der Peristomzähne des Laubmoossporangiums. *Bot. Centralbl.* **82**:161–168, 193–200.

Edwards, S. R. 1979. Taxonomic implications of cell patterns in haplolepidous moss peristomes, in G. C. S. Clarke and G. Duckett (eds.), *Bryophyte Systematics,* pp. 317–346, London: Academic Press.

Erlanson, C. O. 1930. The attraction of carrion flies to *Tetraplodon* by an odoriferous secretion of the hypophysis. *Bryologist* **33**:13–14.

Ernst-Schwarzenbach, M. 1939. Zur Kenntnis des sexuellen Dimorphismus der Laubmoose. *Arch. Julius Klaus-Stiftung Zurich* **14**:361–474.

——— 1944. La sexualité et le dimorphisme des spores des mousses. *Rev. Bryol. Lichénol.* **14**:105–113.

Evans, A. W., and H. D. Hooker. 1913. Development of the peristome in *Ceratodon purpureus. Bull. Torrey Bot. Club* **40**:97–109.

Forman, R. T. T. 1965. A system for studying moss phenology. *Bryologist* **68**:289–300.

Frey, W. 1971. Blattentwicklung bei Laubmoosen. *Nova Hedwigia* **21**:119–136.

——— 1974. Vergleichende entwicklungsgeschichtliche Untersuchungen an Laubmoosblattern als Beitrag zur Systematik der Laubmoose. *Soc. Bot. France. Coll. Bryol.* **1974**:29–34.

Gorton, B. S., and R. E. Eakin. 1957. Development of the gametophyte in the moss *Tortella caespitosa. Bot. Gaz.* **199**:31–38.

Greene, S. W. 1960. The maturation cycle, or the stages of development of gametangia and capsules in mosses. *Trans. Br. Bryol. Soc.* **3**:736–745.

Haberlandt, G. 1886. Beiträge zur Anatomie und Physiologie der Laubmoosen. *Pringsheims Jahrb. Wiss. Bot.* **17**:457–359 and Plates 22–27.

Heald, F. D. 1898. A study of regeneration as exhibited by mosses. *Bot. Gaz.* **26**:169–210.

Herzog, T. 1911. Parallelismus und Konvergenz in den Stammreihen der Laubmoose. *Hedwigia* **50**:86–99.

Hirohama, T., and Z. Iwatsuki. 1980. Surface ornamentation on rhizoids of the species of Bartramiaceae (Musci). *J. Hattori Bot. Lab.* **48**:259–275.

Hörmann, H. 1959. Zur Morphologie and Anatomie von *Climacium dendroides* Web. & Mohr und *Thamnium alopecurum.* B.S.G. *Nova Hedwigia* **2**:201–208.

Ingold, C. T. 1959. Peristome teeth and spore discharge in mosses. *Trans. Bot. Soc. Edinburgh* **38**:76–88.

Ireland, R. R. 1971. Moss pseudoparaphyllia. *Bryologist* **74**:312–330.

Iwatsuki, Z., and T. Kodama. 1961. Mosses in Japanese gardens. *Econ. Bot.* **15**:264–269.

Janzen, P. 1913. Die Jugendformen der Laubmoose und ihre Kultur. *Ber. Westpreuss. Bot. Zool. Vereins.* (Danzig) **35**:1–62.

―――― 1916. Die Haube der Laubmoose. *Hedwigia* **58**:156–280.

―――― 1921. Die Blüten der Laubmoose. *Hedwigia* **62**:163–281.

Kavina, K. 1915. Die Verzweigung der Laubmoose. *Hedwigia* **56**:308–332.

Koponen, A. 1977. The peristome and spores in Splachnaceae and their evolutionary and systematic significance. *Bryophytorum Biblioth.* **13**:535–568.

Kreulen, D. J. W. 1972. Spore output of moss capsules in relation to ontogeny of the archesporial tissue. *J. Bryol.* **7**:61–74.

Kuhlbrodt, H. 1922. Über die phylogenetische Entwicklung des Spaltöffnungsapparatus und Sporophyten der Moose. *Beitr. allg. Bot.* **2**:363–402.

Lal, M. 1961. In vitro production of apogamous sporogonia in *Physcomitrium coorgense* Broth. *Phytomorphology* **11**:263–269.

Lanzius-Beninga, S. 1850. Beiträge zur Kenntnis des innern Baues der ausgewachsenden Mooskapsel, insbesondere des Peristoms. *Nova Acta Leop. Car. Acad.* **22**:560–604.

Lazarenko, A. S. 1955. Beiträge zur Artbildung der Laubmoose. *Mitt. Thuring Bot. Gesells.* **1**:31–46.

―――― 1957. On some cases of singular behavior of the moss peristome. *Bryologist* **60**:14–17.

―――― 1963. Dynamics of quantitative variation in the sporophyte of *Desmatodon randii* (Kenn.) Lazar. in a natural population and monosporic cultures. *Bull. Moscow obs. isp. Priody* **68**:133–148.

Lodge, E. 1959. Effects of certain cultivation treatments on the morphology of some British species of *Drepanocladus*. *J. Linn. Soc. London (Bot.)* **56**:218–224.

―――― 1960. Studies of variation in British material of *Drepanocladus fluitans* and *Drepanocladus exannulatus*. *Svensk Bot. Tidskr.* **54**:368–393.

Loeske, L. 1910. *Studien zur vergleichenden Morphologie und phylogenetischen Systematik der Laubmoose*. Berlin: Max Lande.

Longton, R. E. 1976. Reproductive biology and evolutionary potential of bryophytes. *J. Hattori Bot. Lab.* **41**:205–223.

Lorch, W. 1931. Anatomie der Laubmoose, in Linsbauer, K. (ed.), *Handbuch der Pflanzenanatomie* **7**(Lief. 28):i–viii, 1–358.

Lorenz, P. G. 1867. Grundlinien zu einer vergleichenden Anatomie der Laubmoose. *Pringsh. Jahrb. f. wiss. Bot.* **6**:363–466.

McClymont, J. W. 1955. Spore studies in the Musci, with special reference to the genus *Bruchia*. *Bryologist* **58**:287–306.

―――― and D. A. Larsen. 1964. An electron microscope study of spore wall structure in the Musci. *Am. J. Bot.* **51**:195–200.

Merl, E. M. 1917. Scheitelsegmentierung und Blattstellung der Laubmoose. *Flora* **109**:189–212.

Mogensen, G. 1978. Spore development and germination in *Cinclidium* (Mniaceae, Bryophyta), with special reference to spore mortality and false anisospory. *Can. J. Bot.* **56**:1032–1060.

———— 1981. The biological significance of morphological character in bryophytes: the spore. *Bryologist* **84**:187–207.

Nehira, K. 1976. Protonema development in mosses. *J. Hattori Bot. Lab.* **41**:157–165.

Nishida, Y. 1978. Studies on the sporeling types in mosses. *J. Hattori Bot. Lab.* **44**:371–454.

Paton, J. A. 1957. The occurrence, structure and functions of the stomata in British bryophytes I. Occurrence and structure. *Trans. Br. Bryol. Soc.* **3**:228–242.

———— and J. V. Pearce. 1957. The occurrence, structure and functions of the stomata in British bryophytes II. Functions and physiology. *Trans. Br. Bryol. Soc.* **3**:242–259.

Patterson, P. M. 1953. The aberrant behavior of the peristome teeth of certain mosses. *Bryologist* **56**:157–159.

Pfaehler, A. 1904. Etude biologique et morphologique sur la dissemination des spores chez les mousses. *Bull. Soc. Vaudoise Sci. Nat.* **40**:41–132.

Pottier, J. 1920. Recherches sur le developpement de la feuille des mousses. Thèse, Paris.

———— 1925. Nouvelles recherches sur le developpement de la feuille des muscinées. *Bull Soc. Bot. France* **1925**:1–60.

Proctor, M. C. F. 1979. Surface wax on the leaves of some mosses. *J. Bryol.* **10**:531–538.

———— 1979. Structure and eco-physiological adaptations in bryophytes, in G. C. S. Clarke and G. Duckett (eds.), *Bryophyte Systematics,* pp. 479–509. London: Academic Press.

Proskauer, J. 1958. On the peristome of *Funaria hygrometrica. Am. J. Bot.* **45**:460–463.

Roth, D. 1969. Embryo und Embryotheca bei den Laubmoosen. *Bibl. Bot.* **129**:1–49 and 9 plates.

Saito, K., and T. Hirohama. 1974. A comparative study of spores or taxa in the Pottiaceae by use of the scanning electron microscope. *J. Hattori Bot. Lab.* **38**:475–488.

Salmon, E. S. 1899. On the genus *Fissidens. Ann. Bot.* **13**:103–130.

Schoenau, K. 1911. Zur Verzweigung der Laubmoose. *Hedwigia* **51**:1–56.

Schofield, W. B. 1981. Ecological significances of morphological characters in the moss gametophyte. *Bryologist* **84**:145–161.

Smirnova, Z. N. 1948. The significance of transplant experiments in the study of polymorphic species of mosses. *Bot. Zh. S.S.R.* **33**:446–457.

Stalfelt, M. G. 1937. Der Gasaustausch der Moose. *Planta* **27**:30–60.

Steere, W. C. 1947. A consideration of the concept of genus in Musci. *Bryologist* **50**:247–258.

———— 1958. Evolution and speciation in mosses. *Am. Nat.* **92**:5–20.

Steinbrinck, C. 1897. Die hygroscopische Mechanismus des Laubmoosperistoms. *Flora* **84**:131–158.

Stone, I. G. 1961. The gametophore and sporophyte of *Mittenia plumula* (Mitt.) *Lindb. Austral. J. Bot.* **9**:124–151.

———— 1961. The highly refractive protonema of *Mittenia plumula* (Mitt.) Lindb. (Mitteniaceae). *Proc. Roy. Soc. Vict.* **74**:119–124.

Taylor, E. C. 1962. The Philibert peristome articles. An abridged translation. *Bryologist* **65**:175–212.

Vaizey, J. R. 1888. On the anatomy and development of the sporangium of the mosses. *J. Linn. Soc.* **24**:262–285.

———— 1890. On the morphology of the sporophyte of *Splachnum luteum. Ann. Bot.* **5**:1–10.

Vitt, D. H. 1981. Adaptive modes of the moss sporophyte. *Bryologist* **84**:166–186.

Watson, W. 1913. Xerophytic adaptations of bryophytes in relation to habitat. *New Phytol.* **13**:149–169, 181–189.

Whitehouse, H. L. K. 1966. The occurrence of tubers in European mosses. *Trans. Br. Bryol. Soc.* **5**:103–116.

Woesler, A. 1933. Entwicklungsgeschichtliche und cytologische Untersuchungen an den Vorkeimen einiger Laubmoose sowie einige Beobachtungen über die Entwicklung ihrer Stämmchen. *Beitr. Biol. Pflanz.* **21**:59–116.

Zielinski, F. 1909. Beiträge zur biologie des Archegoniums und Haube der Laubmoose. *Flora* **100**:1–36.

9

The Large-Spored Mosses — Subclass Archidiidae

The subclass contains a single genus, *Archidium,* in the family Archidiaceae and order Archidiales (Fig. 9-1). *Archidium* contains 26 species, many of which have been infrequently collected, undoubtedly as a result of their exceedingly small size (usually less than 2 mm tall), and their brief seasonal appearance each year. The genus is widespread in temperate climates at middle latitudes, with one species, *Archidium ohioense,* widely scattered from southeastern North America through Eurasia and in New Caledonia. All species are terrestrial and commonly occur on bare clayey, sandy, or gravelly moist soil of disturbed habitats. Based on gametophytic features, the subclass is most closely related to the Bryidae, but on sporophytic features it is entirely isolated from all other mosses. As in other mosses, the gametophore morphology appears to have been strongly determined by the restricted ecology of the species; these same features also characterize genera in the Bryidae that occupy the same habitat type and share the same ecological features.

THE GAMETOPHYTE

Spores are usually large (50–130 μm in greatest diameter), and can be seen with the naked eye. The number varies from 4 to 176 per sporangium but is usually less than 40. Spore germination has been studied by J. A. Snider. He noted that germination success was low (less than 30%). The first cell division in a germinating spore occurs as the spore coat is ruptured, cutting off a small cell that emerges through the germ pore. Further division of this emergent cell results in a quadrant of cells of unequal size. One of these four cells becomes the apical cell of the new gametophore, while the remaining cells produce rhizoidal or protonematal filaments. This type of germination is highly specialized and seems to be unique to the Archidiidae, and yet it is similar to that of some Bryidae.

The gregarious gametophores are annual or perennial, with the erect stems

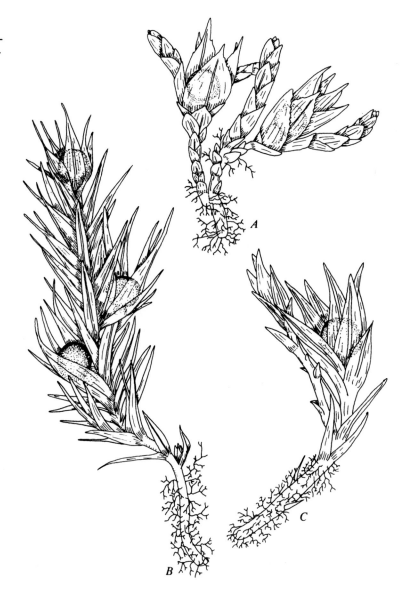

2–20 mm high. Leaves are spirally arranged on vegetative shoots. The clasping to channeled leaves vary from ovate to narrowly lanceolate (Fig. 9-2). Leaf cells are smooth, thin- or thick-walled, and chlorophyllose. The unistratose leaves have a multistratose costa in which the cells are uniformly thick-walled. The stem is structurally simple, with the inner cells thin-walled and larger than the outer cells. Rhizoids are smooth and tend to emerge from the bases of branches. Branching is irregular and by innovations. Vegetative shoots formed during one year often recline and produce branches by innovations in the fol-

FIGURE 9-2
Details of the gametophore of *Archidium*. (*A*) Upper two-thirds of leaf of *A. minus* (×66). (*B*) Leaf of *A. ohioense* (×100). (*C–E*) *A. stellatum*. (*C*) Transverse section of leaf (×100). (*D*) Transverse section of stem (×100). (*E*) Antheridium-producing gametophore (×35). (*A*, after Crum and Anderson, 1981; *B*, after Sullivant 1864; *C–E*, after Stone, 1973.)

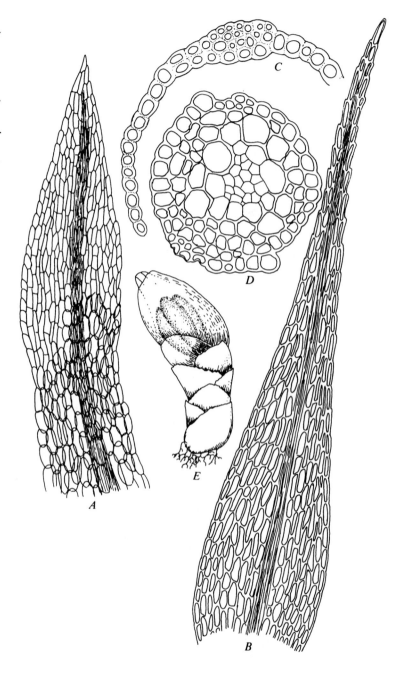

lowing year. Rhizoids arise at the bases of these new branches. Since the branches arise from a single epidermal cell of the stem, they are readily detached. This appears to be its only means of vegetative reproduction and seems to be highly effective in expanding a population. Gametophores are monoicous; antheridia are few, axillary with filiform paraphyses, and associated with the perichaetia or borne on an apical or axillary perigonium; the perichaetia are apical or lateral and possess 2–3 archegonia and abbreviated paraphyses.

THE SPOROPHYTE

Sporophyte development is unique to the subclass, but as in most mosses, the endothecium produces the spores. In the Archidiidae, however, no columella is ever present. In differentiation of amphithecium in the embryo, *Archidium* is unique among the mosses (Figs. 9-3 and 9-4). The first division of the zygote is transverse, as in other mosses (Fig. 9-5A). The foot cell then undergoes numerous cell divisions and produces an enormous foot (in comparison to the size of the entire sporophyte, (Figs. 9-4E–G and 9-6). The upper cell of the zygote undergoes numerous divisions to form the remainder of the sporophyte. In the differentiation of endothecium and amphithecium, however, the quadrant of cells (present in other mosses), is not observed in cross-sectional view. The four cells that result from the second anticlinal division are unequal in size; a succeeding periclinal division results in the differentiation of endothecium and amphithecium. Between the endothecium and amphithecium, a dome-shaped intercellular space appears, beneath which the sporogenous sac forms from the endothecium (Fig. 9-4F).

All of the sporogenous sac cells have the potential to become spore mother cells. In most species the number of spore mother cells is less than 15 per

FIGURE 9-3
Schematic diagram (transverse section) showing early developmental stages occurring in the *Archidium* sporangium. (*A*) Two-celled embryo. (*B*) Unequal division in each segment of two-celled embryo. (*C*) Second division at right angles to initial wall, forming two-celled endothecium and four-celled amphithecium. (*D–F*) Further divisions of amphithecium to produce multistratose jacket. (G) Division within the endothecium resulting in an outer spore sac layer (*ss*) and an inner tissue layer (*st*), from which the sporogenous tissue is formed. (Endothecium is stippled, amphithecium is not stippled; after Snider, 1975.)

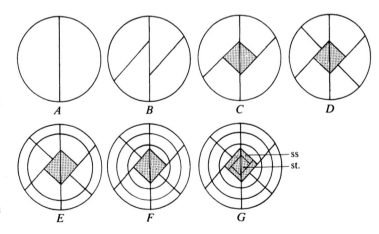

FIGURE 9-4
Development of the sporo-
phyte in *Archidium*. (*A*)
Longitudinal section showing
the embryo with the apical
cell and foot cell differentiated; a transverse section is
shown below. (*B*) Longi-
tudinal section of a later stage,
in which endothecium and
amphithecium have formed in
the embryo; the transverse
section below it shows that
the endothecium is the central
layer of two cells and the
amphithecium is unistratose.
(*C*) Longitudinal section,
showing the continued growth
of the amphithecium, which is
shown to be bistratose in the
transverse section shown
below, while the endothecium
remains two cells in diameter.
(*D*) Longitudinal section
showing further growth of the
amphithecium. (*E*) Longitudinal section, showing the
three-layered amphithecial
tissue, the intercellular
chamber differentiating
around the endothecium, the
further growth of the amphithecium, and the growth of the
foot. (*F*) Longitudinal section,
showing growth of the embryo
to rupture the calyptra. (*G*)
Longitudinal section of a later
stage, showing the appearance
of two spore mother cells
(shown enlarged in the endothecium). (After Snider,
1975.)

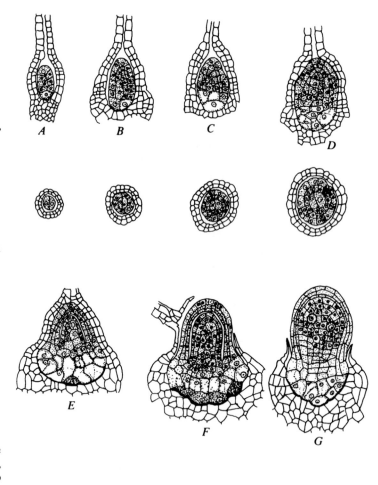

sporangium, and sometimes there is only one (in which case a single tetrad represents all of the spores formed in that sporangium). The cells that do not form
spore mother cells disintegrate and apparently serve as nutritive tissue for the
developing spores. In one species, *A. dinteri,* all cells inside the spore sac serve
as spore mother cells, and the final result of meiosis is numerous small spores
(up to 176). The inner jacket cells of the sporangium also disintegrate as the
spores mature until the jacket is unistratose (Fig. 9-6*B*). Spores are shed with

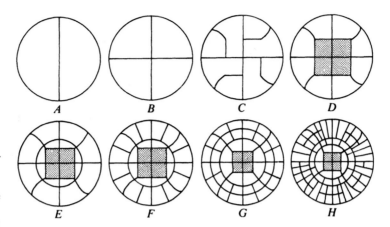

FIGURE 9-5
Schematic diagram (transverse section) showing early developmental stages in the typical bryidean sporangium. (*A*) Two-celled embryo. (*B, C*) Further divisions of undifferentiated embryo. (*D*) Formation of endothecium (within) and amphithecium (outside). (*E–H*) Further divisions of the amphithecium to produce a multistratose jacket; further divisions of the endothecium produce the internal columella and the outer sporogenous layer. (After Snider, 1975.)

irregular rupture of this jacket. The spores are often bright yellow, partly caused by the spore coat color and partly the large oil body within.

Calyptra expansion sheathes the sporangium as it enlarges, but when it nears maturity, the calyptra is torn irregularly and falls away or persists as an evanescent tattered sheath near the base of the sporangium.

EVOLUTION AND RELATIONSHIPS

It is possible that *Archidium* is a relatively ancient genus, perhaps originating before the massive continent of Pangaea began its breakup. The large spores suggest that long-distance dispersal, at least by wind, would be unlikely. Since the plants grow on open, often muddy soil, however, the possibility of bird dispersal (as spores in mud) cannot be discounted.

The weak distinction between acrocarpy and pleurocarpy suggests a general-

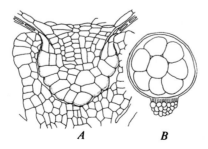

FIGURE 9-6
Detail of the sporophyte of *Archidium stellatum*. (*A*) Longitudinal section through the foot penetrating the apex of the gametophore (×225). (*B*) Longitudinal section through sporophyte, showing the small number of enlarged spores, the absence of a columella, and the essential absence of a seta (×75). (After Stone, 1973.)

ized condition. The genus shows several features of considerable specialization, including its very reduced number of spores, the reduced size of the entire plant, the very simple means of spore dispersal, and the means of simple vegetative propagation. Most of these features appear to be adaptations to a specialized ecology, and thus are difficult to assess in an interpretation of its primitiveness.

In structure of the gametophore, the subclass is most like the subclass Bryidae, with which it probably shared a common progenitor. In sporophyte structure and development, it is unique among the mosses. It appears that those Bryidae that share a similar ecology have achieved a similar structure through reduction, while *Archidium* may have possessed ancestors that already were well-adapted to this open habitat.

GENERALIZATIONS

Features that make the Archidiidae a highly distinctive evolutionary line among the mosses include:

1. The unusually large spores
2. The few spores produced in each sporangium
3. The unique mode of spore germination
4. The mode by which the amphithecium arises
5. The enormous foot of the sporophyte
6. The presence of a domelike intercellular space in the sporangium into which the sporogenous cells expand

FURTHER READING

Bruch, P., W. P. Schimper, and T. Gümbel. 1850. *Bryologia Europaea, seu Genera Muscorum Europaeorum Monographice Illustrata. Archidium I* (fasc. 43):33–35.

Crum, H. A., and L. E. Anderson. 1981. *Mosses of Eastern North America. Vol. I. Subclass Archidiidae*, pp. 70–79. New York: Columbia University Press.

Lietgeb, H. 1879. Das Sporogon von *Archidium. Sitzungsber. Kaiserl. Akad. Wiss., Math,-Naturwiss. Cl. Abt.* 1:447–460.

Snider, J. A. 1975. Sporophyte development in the genus *Archidium* (Musci). *J. Hattori Bot. Lab.* 39:85–104.

——— 1975. A revision of the genus *Archidium* (Musci). *J. Hattori Bot. Lab.* 39:105–201.

Stone, I. G. 1973. Two new species of *Archidium* from Victoria, Australia. *Muelleria* 2:191–213.

10 Evolutionary Trends and Interrelationships among the Mosses

The different evolutionary lines among the mosses are remarkably clear. In sporophytic features there is only slight overlap in the features that separate the lines, while gametophytic features are basically similar among the subclasses. Comparisons are made in Tables 10-1 and 10-2.

THE GAMETOPHYTE

In the gametophyte there appears to be a tendency toward structural reduction. It is possible that the ancestral moss produced a thallose "protonematal" phase similar to that still present in the genus *Sphagnum* and, to a lesser degree, in Tetraphidae. The production of uniseriate and multiseriate branched protonematal systems, as in the Andreaeidae, would be derivative. Heterotrichous protonemata would be considered even more specialized, and the production of very specialized caulonematous systems, as is represented in the Diphysciaceae of the Buxbaumiidae and in *Schistostega* and *Mittenia* in the Bryidae, would be an elaboration of this type. The greatest degree of specialization is in the production of multicellular spores and the complete absence of a protonematal phase as in the Dicnemonaceae in the subclass Bryidae. Each of these Bryidae that have specialized protonemata also show ecological specialization.

In the gametophore itself it is difficult to make any confident assumptions, especially concerning stem anatomy. It is possible that a hydrome-leptome system can be considered generalized, and its absence is the result of reduction. Certainly within the gametophores of the Polytrichidae this appears to be the case. Since a leptome system is presently unknown in gametophores of other subclasses of mosses, it is not possible to draw any secure conclusions concerning the evolutionary significance of this feature. The possession of a conducting system in the gametophore appears to have enabled the Polytrichidae to produce rather tall free-standing plants. It should be noted that such large

TABLE 10-1.
Gametophytic Features of
the Subclasses of Mosses

Gametophyte	*Andreaeidae*	*Sphagnidae*	*Tetraphidae*	*Polytrichidae*
Spores	Precocious endosporic germination	Exosporic	Exosporic	Exosporic
Protonema	Multiseriate branched	Thallose or filamentous	Usually uniseriate filament	Uniseriate filament
Protonemal flaps	Present	Absent	Present	Absent
Branching of gametophore	Irregular	Fascicles	Essentially unbranched	Essentially unbranched in most
Stem anatomy	Little differentiated	Much differentiated	Little differentiated	Much differentiated (hydroids and leptoids)
Hydroids	Absent	Absent	Present	Present
Leptoids	Absent	Absent	Absent	Usually present
Leaf cell variety in one leaf	Little	Considerable	Little	Often between base and lamina
Leaf lamellae	Absent	Absent	Absent	Usually present
Costa of leaf	Present or absent	Absent	Usually present	Usually present
Paraphyllia	Absent	Absent	Absent	Absent
Pseudopodium	Present	Present	Absent	Absent
Rhizoids	biseriate or uniseriate	Uniseriate, in protonometal phase only	Uniseriate	Uniseriate
Rhizoidal gemmae	Absent (flaps as gemmae?)	Absent?	Absent (flaps as gemmae?)	Absent or rarely present
Gametophore gemmae	Absent	Absent	Present (*Tetraphis* only)	Rare (*Alophosia* only)

free-standing gametophores generally possess "splash-cup" perigonia that enhance wider dispersal of sperms to reach the tall perichaetium-bearing gametophores. Other moss gametophores, often of greater length, are generally supported by water or reclining or pendent. In some epiphytic mosses the stems of rigid gametophores have considerable supportive tissue but lack any elaborated conducting system.

Buxbaumiidae	Bryidae	Archidiidae
Exosporic	Generally exosporic	Exosporic
Uniseriate filament	Generally uniseriate filaments	Absent
Often present	Absent	Absent
Essentially unbranched or abbreviated branches	Often much branched	By innovations (irregular)
Little differentiated	Little to slightly differentiated	Little differentiation
Present?	Often present	Absent?
Absent	Absent	Absent
Little in most	Little in most, can be considerable	Little
Absent	Rarely present	Absent
Present or absent	Present or absent, often double	Present
Absent	Present or absent	Absent
Absent	Usually absent	Absent
Uniseriate	Uniseriate	Uniseriate
Present(?) or absent	Present or absent	Absent or present
Absent	Present or absent	Absent

The earliest basic moss gametophore possibly possessed the following features:

1. A stem with thickened cortical cells and lacking an elaborated leptome or hydrome system
2. An erect gametophore with the leaves in three ranks

TABLE 10-2.
Sporophytic Features of the
Subclasses of Mosses

Sporophyte	*Andreaeidae*	*Sphagnidae*	*Tetraphidae*	*Polytrichidae*
Calyptra when capsule mature	Small and evanescent or large	Usually not apparent, and evanescent	Large, usually pleated	Large, often hairy
Sporangium shape	Elliptic, cylindric, or turbinate	Subspherical	Cylindric	Diverse
Seta	Usually absent	Absent	Present	Present
Seta anatomy	Not described or seta absent	Absent	Central strand (hydroids)	Sometimes hydroids and/or leptoids
Stomata	Absent	Rudimentary	Present or absent	Present or absent
Annulus	Absent	Absent	Absent	Small
Sporangium dehiscence	Longitudinal lines	Operculum: explosive	Operculum + teeth	Operculum + teeth (usually)
Peristome teeth	None	None	Four, multicellular (of whole cells)	16, 32, many, or none, multicellular (of whole cells)
Epiphragm	Absent	Absent	Absent	Present or absent
Columella	Dome-shaped	Dome-shaped	Cylindric	Cylindric
Air chambers in sporangium	Absent	Absent	Absent	Present or absent

3. Leaves costate but lacking lamellae
4. Leaf cells without surface ornamentation
5. Leaf cells all chlorophyllose
6. Rhizoids abundant, serving to attach the base of the gametophore to the substratum
7. Gametophores monoicous, with the archegonia and antheridia on separate branches
8. Gametophores branching by innovations, or essentially unbranched
9. Calyptra smooth, cucullate, or mitrate
10. Gametophores acrocarpous

It is probable that the immediate progenitor of the Sphagnidae was not the same as that of any of the other subclasses. The fact that all subclasses share

Buxbaumiidae	Bryidae	Archidiidae
Small and smooth	Large or small, usually smooth	Evanescent
Cylindric, often with oblique face	Diverse	Subspherical
Present or absent	Present or absent	Absent
Sometimes hydroids and/ or leptoids	Sometimes hydroids and/ or leptoids	Absent
Present or absent	Usually present	Absent
Small	Present or absent	Absent
Operculum + teeth (usually)	Operculum + teeth (usually)	Decomposition
16 or 32 endostomial pleats or none, exostome variable; mature teeth of parts of many cells	8, 16, or none; exo- and/ or endostome; mature teeth of parts of many cells	Absent
Absent	Absent	Absent
Cylindric	Cylindric	Absent
Present	Present or absent	Early in development

many gametophytic features may reflect the limited morphogenetic potential that restricts their diversity. The common possession of an apical cell that has three cutting faces, the way in which leaves originate from the stem segments derived from this apical cell, and the way in which leaves differentiate their cells are all such constraints.

Each of the subclasses shows trends in specialization or simplification of the gametophore. In the subclass Andreaeidae there is great simplicity in leaf arrangement, details of the leaf structure, and position of the sex organs. Yet, in some species a distinctly differentiated and specialized perichaetium is developed. Such a feature is probably more derived than its absence.

It is possible that the gametophore of the subclass Tetraphidae most resembles that of the hypothetical ancestral type. In *Tetraphis*, however, the stem produces of a central conducting strand of hydroids, a specialized feature. The unusual sex-change condition in the gametophore of *Tetraphis* and some species of *Atrichum* would appear to be a remarkable specialization as well. In *Tetrodontium* the extremely reduced gametophore is interpreted as highly specialized.

Considerable specialization exists in the gametophore in the subclass Polytrichidae. The highly elaborated conducting system, with both hydrome and leptome, accompanied by such features as sheathing leaf bases, leaf lamellae, and "rhizome," all represent highly specialized features. It is from ancestors with these specialized characteristics, however, that such reduced gametophores of genera like *Racelopus* were derived; these genera lack lamellae and conducting systems but retain a perennial protonema. It is impossible to determine whether a nearly smooth calyptra, such as that of *Atrichum* and *Oligotrichum*, is derived or generalized.

The gametophore shows considerable reduction in all genera in the Buxbaumiidae, raching its extreme reduction in the perigonium of *Buxbaumia*, which is composed of a single antheridium surrounded by a unistratose sheath. Even in the Diphysciaceae, within this subclass, there appears to be a reduction series with *Diphyscium* and *Muscoflorschuetzia* having a gametophore with multistratose chlorophyllose leaves, while *Theriotia* leaves exhibit considerable specialization, being composed of chlorophyllose and hyaline cells. In all genera the stems show little tissue differentiation.

There is great variety in the gametophore in the Bryidae. Even within a single large family, as in the Dicranaceae, there are varying degrees of diversification in leaf structure. In the Bryidae, however, stem structure shows less tissue differentiation than in the Polytrichidae. Within the Bryidae lamellae reappear, this time in genera of drier habitats, suggesting that this feature may have adaptive advantage in such habitats. It appears that physiological tolerance rather than morphological modifications may be of greater significance to survival of mosses.

In the Bryidae a number of gametophytic features occur in greater frequency than in other subclasses, the significance of which is not apparent:

1. Monoicous gametophores
2. Dorsiventrally flattened gametophores
3. Production of multiple costae
4. Elaborate ornamentation on leaf cell surfaces
5. Diversification of the cells in the alar region of the leaf
6. Production of a diversity of gemmae
7. Elaboration of various branching patterns, resulting in diverse growth forms, some of which appear to be of adaptive value
8. A strong tendency toward specialized perichaetia and perigonia.

Within the Bryidae there are also examples of extreme reduction in the gametophore, as, for example, the genus *Ephemeropsis* where the protonema is persistent and the gametophore is reduced to few-leafed perichaetia and perigonia.

A similar type of reduction exists in *Discelium, Ephemerum,* and *Micromitrium.* In the Archidiidae the gametophore shows similar structural simplicity.

Finally, in the Sphagnidae the specialization in the gametophore contrasts markedly with the presumed generalized condition of the thalloid protonematal phase. The elaboration of hyaline porose cells and chlorophyllose cells in the unistratose leaves suggests a very specialized condition. The same is true for the elaboration of cortical stem cells, reaching their extreme of specialization as retort cells. The unusual branching pattern with fascicles of lateral branches is also specialized, as is the nature of the perigonial branches and a pseudopodium associated with sporophyte growth. The latter characteristic appears to have evolved independently in the Andreaeidae and Sphagnidae but also appears in a modified form in rare individuals in the Bryidae (e.g., *Neckeropsis*).

Within the Sphagnidae, species with fibrillose cortical stem cells may be the most generalized, while those with retort cells are more specialized. There are, within the genus, several other trends: some species of seepage sites as well as in submerged sites tend to show a considerable simplification in branching pattern.

THE SPOROPHYTE

In the mosses the sporophyte also shows several clear evolutionary trends. The marked reduction in the number of cell layers involved in production of the peristome teeth is especially significant.

The Andreaeidae and Sphagnidae are special cases. In both of these, probably from entirely independent evolutionary lines, a columella overarched by the sporogenous layer was produced. The distinctive sporangium dehiscence in the Sphagnidae is not repeated in the rest of the mosses and thus must have arisen only once and in this single genus. The variation in sporangia within *Sphagnum* seems to be mainly in size and color, and even these features are relatively constant. All species in the Andreaeidae lack an operculum, and dehiscence of the sporangium is always by longitudinal lines, with most of the sporangium contents being exposed by this dehiscence. In *Andreaeobryum* the dehiscence lines somewhat resemble the behavior of peristome teeth.

Peristome tooth production in the subclass Tetraphidae is the most generalized, with all cells immediately enclosed by the operculum being utilized to produce the four multicellular teeth.

There is a decided reduction in the number of concentric cell rows involved in production of the teeth in the Polytrichidae. In most genera, three to four cell layers are utilized, and the resulting teeth are small, multicellular, and in a single circle. In *Dawsonia,* however, the teeth are in several irregular circles, are multicellular and elongate, and are derived from more than six concentric cell rows, in which the apex of the columella participates, as in the Tetraphidae.

The peristome teeth in the Buxbaumiidae are sometimes derived from many concentric cell rows, as in *Dawsonia* of the Polytrichidae. In the Buxbaumiidae, however, most of the circles of teeth are of remnants of cell walls and not of whole cells. There is often an outer row of jagged irregular multicellular toothlike processes that project from the upper edge of a ring of cells extremely similar to the ring of cells from which the teeth in the Polytrichidae project. This suggests that these subclasses may have had a common ancestry, especially since basic structure of the gametophore would be consistent with this assumption.

Even greater reduction of the peristome has occurred in the subclass Bryidae, with the teeth derived from never more than three concentric cell rows and the teeth themselves consisting of only cell wall fragments. Furthermore, within the Bryidae the teeth are articulated and hygroscopic and therefore important in spore dispersal.

In the Archidiidae the absence of a peristome or operculum, the absence of a columella, the reduction in number of spores, the absence of a seta, and the unique genesis of the cell layers in the embryonic sporophyte all point to an independent evolutionary line. Structure of the gametophore appears to have been strongly controlled by the habitat, and its gross morphology is very similar that of some Bryidae, which show a similar ecology. In both the Polytrichidae and the Bryidae some taxa show no peristome teeth whatsoever, and in the Archidiidae and in some taxa of the Bryidae no operculum is differentiated.

In spore number the Polytrichidae shows taxa in which spores can exceed 65 million within a single sporangium, while in the Bryidae, the number is in the many thousands. In some Bryidae there may be a difference in spore size in the same sporangium. In some species of *Macromitrium,* for example, two spores of a tetrad tend to be smaller than the other two. These smaller spores produce perigonial plants, while the larger ones produce perichaetial plants. This phenomenon is termed anisospory. Multicellular spores appear to be unique to the Bryidae, although in *Andreaea* and *Andreaeobryum* the spores germinate endosporously after they have been shed. The extreme reduction in spore number is shown by the Archidiidae.

Among the various subclasses of the mosses there are conspicuous trends in elaboration of peristome teeth, while in gametophytic features there are trends within each of the subclasses. At the present time (1984) there is a renewed interest in research in interrelationships among the mosses. Electron microscopy continues to reveal details concerning peristome structure, and studies of the genesis of the peristome contribute further data upon which to assess interrelationships. Such details are available for a very limited number of taxa, and consequently, such assessments are likely to remain tentative for a considerable period of time. Phytochemical, cytological, biogeographical, and evolutionary patterns provide further evidence for analysis and interpretation of interrelationships.

11 The Liverworts or Hepatics — Class Hepaticae

In earlier times, when man's understanding of nature was even less than it is now, it was believed that the Creator marked various plants to identify the ailments they could cure. Since the thallus of some liverworts resembles a liver, such plants were considered useful in making a concoction that would aid in curing liver ailments. Hence the name "liver-plant," or liverwort. Unfortunately, there is no evidence that liverworts possess curative properties. Furthermore, most plants now recognized as liverworts (or hepatics) are leafy and show not the slightest resemblance to a liver.

The class Hepaticae includes the thallose liverworts and leafy liverworts or "scale mosses." The class consists of approximately 8,000 species in at least 330 genera. It is sometimes divided into two subclasses, Jungermanniidae and Marchantiidae, which show the two main divergent evolutionary lines, but the features of these subclasses overlap to a certain degree. The more detailed evolutionary trends are clearly represented by the six orders:

Order Calobryales
Order Jungermanniales
Order Metzgeriales
Order Sphaerocarpales
Order Monocleales
Order Marchantiales

These orders are based essentially on structure of the gametophyte, particularly the gametophore. The nature of the sporophyte is very similar among orders, but a few critical sporophytic features characterize some of them.

The hepatics possess a number of features that distinguigh them from the mosses:

1. The "protonema" is usually reduced to two or three cells of the uniseriate germ tube, and the apical cell of the gametophore is differentiated early.
2. The rhizoids are generally unicellular and are not branched, although the tip is sometimes knobbed.

3. The protonematal phase produces no gemmae.
4. The gametophore is either leafy or thallose and leaves are arranged in two or three rows.
5. Sex organs lack paraphyses among them, but mucilage filaments are usually present.
6. Leaves are generally unistratose and lack a costa.
7. Leaf cells are commonly isodiametric and frequently possess trigones.
8. Gametophore cells often have complex oil bodies.
9. Leaves are often lobed.
10. The jacket of the sporangium never has stomata.
11. Cells of the sporangium jacket often have transverse or nodular thickenings.
12. The sporangium jacket is sometimes unistratose.
13. The sporangium usually opens by four longitudinal lines.
14. Even in the rare cases where there is an operculum, peristome teeth are absent.
15. Within the sporangium there are often sterile threadlike hygroscopic cells with helical wall thickenings; these are elaters; a columella is absent.
16. The sporophyte usually produces a colorless seta of thin-walled cells and is held rigid by turgor pressure in the component cells.
17. The sporophyte generally persists for a very brief period after the spores are shed.
18. The calyptra is ruptured and remains at the base of the seta when the seta elongates rapidly, and protects the maturing sporangium before seta elongation.
19. The seta elongates after the sporangium is completely differentiated with the included spores and elaters, and the seta is rarely a photosynthetic organ.
20. Generally all spores in a sporangium are shed at the same time.

THE GAMETOPHYTE

The gametophyte shows considerable variety in both thallose and leafy members, and spore germination patterns are more diverse than in the mosses. Generalizations are difficult to make concerning gametophores of all hepatics, since the morphology differs considerably among the orders. In leafy hepatics, both stem and leaves are green when young, and the stem usually shows little internal tissue differentiation. In a few thallose members a conducting system exists, but it is never so complex as in the mosses. Nutrient absorption is usually a very simple process, but in a few thallose hepatics there is a complex gas exchange system, with pores opening to intercellular chambers and a distinct layer of green tissue making up these chambers.

In leafy forms, the apical cell usually has three cutting faces. In thallose forms, the number of cutting faces of the apical cell is two, three, or four.

The gametophore, on drying, often changes form. Such changes usually slow rapid water loss and involve imbrication of leaves or incurving of thallus margins and occasionally incurving of branches. Many hepatics show no apparent morphological changes on drying, and some cannot survive long periods of dehydration. A few hepatics, when subjected to long periods of drying, form a protective coat of dried-up cells that surround a cluster of starch and oil-rich cells, including an apical cell, and this "tuber" can tolerate desiccation and produce a new thallus when favorable conditions return.

The leaves in leafy hepatics are always sessile and generally arranged oblique to the long axis of the stem. Lateral leaves are frequently in two rows, and the third row of leaves, when present, is on the underside of the stem; these underleaves (or amphigastria) often differ in size and shape from the lateral leaves. Marginal cells are toothed, ciliate, or entire, whether in a thallose or leafy gametophore; sometimes the surface of the thallus is covered by hairs, but this is not common. Simple papillae are sometimes present on the cell surfaces. Lamellae are occasionally present.

The rhizoid position varies in leafy forms; they are restricted to the underside of the stem, where they are sometimes abundant, or to the underleaves, or they even replace the underleaves. They are sometimes confined to the smaller lobe of lateral leaves, or are entirely absent. Paraphyllia occur in a few genera of leafy hepatics (e.g., *Trichocolea*).

The rhizoids in thallose gametophores are often most numerous on the thickest part of the thallus and confined to the under surface. In most simple thalli all rhizoids emerge perpendicular to the surface and extend downward to the substratum; such rhizoids are smooth-walled. In some complex thalli, however, rhizoids may be smooth-walled, with pronounced peglike thickenings on the inner face of the wall, or with regularly spaced, blunt thickenings that give the rhizoid a sinuous appearance (Fig. 17-13). Such rhizoids are often oriented in two patterns, those perpendicular to the thallus and scattered over the entire undersurface, and those diverging backward from the apex of the thallus and arranged longitudinally parallel to it; these are mainly on the thickened central part of the thallus.

Sometimes scales are in two or more rows on the undersurface of the thallus, among the rhizoids (Fig. 17-14); these are especially conspicuous near the thallus apex. Occasionally, scales are on the upper surface of the thallus. Such scales and rhizoids aid in capillary conduction of water. Rhizoids are usually colorless, but they, and the scales, are sometimes deeply pigmented, especially with red-purple.

The gametophores often produce gemmae. In leafy hepatics the gemmae are often on leaf margins or at the apex of a stem, completely replace the leaves (Fig. 13-14). Often they bud off as chains and may be unicellular or multicellular. In thallose gametophores the gemmae are sometimes produced at the thallus margin, within cells of the thallus, or in a specialized gemma-cup or flask (Fig. 17-15).

More than 90% of the hepatics are dioicous. The sex organs are generally enclosed in a protective envelope. In leafy gametophores the archegonia are surrounded by a perichaetium and/or a chlorophyllose unistratose tube, the perianth. In thallose members the archegonia are often surrounded by a chlorophyllous multistratose or unistratose tube of thallus tissue; this is the perigynium. Occasionally, in thallose liverworts, the archegonia are embedded in a chamber within the thallus or sometimes exposed on the surface of the stem or thallus. Mucilage threads usually accompany sex organs.

The antheridia in leafy hepatics are spherical and often in axils of reduced perigonial leaves of specialized branches. In thallose gametophores the antheridia are sometimes spherical and exposed on the stem or thallus surface, or they are in a sleevelike chamber (perigonium). Sometimes, in thallose forms, antheridia are elongate and somewhat tapered to the apex; these are embedded within the thallus.

THE SPOROPHYTE

The sporophyte is made up of an expanded foot that penetrates the gametophore (Fig. 13-16*B*), usually a hyaline seta, and a sporangium that is sheathed by the calyptra before the seta elongates. The seta consists of thin-walled cells, is usually fragile, and disintegrates soon after the sporangium sheds the spores. The seta varies considerably in the number of cells that constitute its circumference (Fig. 13-17); the number of cells, however, is relatively constant within a species, and is a useful character for separation of some species. The sporangium jacket lacks stomata and is usually deeply pigmented with brown or black that is deposited in nodular or transverse thickenings of the cell walls, especially on the outer faces (Fig. 13-18). When sheathed by the calyptra, the sporangium has considerable chlorophyll and is a photosynthetic organ. The jacket is multistratose in most hepatics, and dehiscence is usually by four longitudinal lines that split the sporangium jacket into four equal parts.

Within the mature sporangium there are tetrads of spores and usually hygroscopic elaters. Elaters are elongate sterile cells scattered among the spores; they have brownish to black pigmented helical wall thickenings that vary from one to three coils in a single elater. The elaters tend to have relatively tightly coiled helices; when the elaters dehydrate to a given level, the coils unwind abruptly, and the elaters spring into the air, dislodging spores from the sporangium. In most hepatics, all spores are shed as soon as each sporangium opens.

EVOLUTIONARY TRENDS AND FOSSIL RECORD

There is some controversy concerning which features in hepatics are generalized and which are specialized. However, there is an agreement among many hepaticologists that the following features appear to be the most generalized:

1. A unicellular spore is more generalized than a multicellular spore.

2. An elongate protonema is usually more generalized than the absence of a protonematal stage.
3. A radially symmetrical gametophore is more generalized than a bilaterally symmetrical gametophore.
4. Leaves in three clear ranks are more generalized than a two ranked arrangement.
5. A thallus not differentiated into diverse tissue layers is more generalized than one in which the thallus possesses pores, chlorenchyma layer, and a parenchyma layer.
6. Cells that have numerous oil bodies and chloroplasts in the same cells are more generalized than those in which oil bodies are restricted to scattered cells that have no chloroplasts.
7. Smooth rhizoids are more generalized than those with peglike inner thickenings.
8. Exposed sex organs are more generalized than those that are protected by a sheath or are embedded in the thallus.
9. A massive seta is more generalized than a slender seta, and this is more generalized than the absence of an apparent seta.
10. A multistratose sporangial jacket is more generalized than a unistratose jacket.
11. A sporangial jacket with ornamented thickenings on transverse or radial walls is more generalized than one that lacks thickenings.
12. Elongate tapered elaters are more generalized than stubby elaters, and these are more generalized than the absence of elaters.
13. A sporangium with numerous spores is more generalized than one with few spores.
14. A sporangium with longitudinal lines of dehiscence is more generalized than one that opens by irregular rupturing.

The fossil record for the hepatics is very unsatisfactory both because of its scarcity and because presumed gametophores rarely have sporophytes attached, and thus attribution of the fossil material to hepatics remains tentative. Fossils presumed to be thalli, but of unknown affinities, are named *Thallites*. If the material is interpreted as an hepatic of uncertain affinities, it is named *Hepaticites*. Other more explicit generic names are created to signify material in which the relationships are seemingly apparent or in which the morphology is so distinctive that a unique generic name is desirable.

The most ancient fossil material determined as an hepatic is *Hepaticites devonicus* of Devonian age, at least 350 million years ago. It consists of fragments of what appears to be a thallus with toothed margins, a distinct midrib, and rhizoids. This material impressed R. M. Schuster sufficiently that he proposed the genus *Pallaviciniites* to indicate its strong resemblance to the Metzgerialean hepatic, *Pallavicinia*. This organism was contemporaneous with some of the earliest of the lignified land plants. Other fossil material of apparent hepatics appears in the Carboniferous and succeeding time periods.

SUBCLASSES OF HEPATICAE

Two major evolutionary lines have been recognized in the hepatics: Jungermanniidae and Marchantiidae. These two subclasses show a degree of overlap; in consequence, the orders, rather than subclasses, more clearly illustrate evolutionary lines.

The Jungermanniidae are characterized as follows:

1. They are predominantly leafy, but a few are thallose.
2. Oil bodies usually are numerous in each cell, and each cell has chloroplasts.
3. Rhizoids are smooth-walled.
4. Thallose members are always essentially of thin-walled cells and lack air chambers and ventral scales.
5. The apical cell usually has three cutting faces, but rarely has two or four.
6. The antheridium develops with a primary cell that does not undergo two centric vertical divisions, and in this early stage does not form four daughter cells from this primary cell.
8. The sporangium usually has four longitudinal lines of dehiscence.
9. The sporangium jacket is 4–10 cells thick in most.
10. The spore mother cells are usually four-lobed before meiosis.

The Marchantiidae are characterized as follows:

1. They are predominantly thallose; few are leafy.
2. Oil bodies, when present, are usually one per cell, and chloroplasts are absent in that cell.
3. Rhizoids frequently possess peglike thickenings on the inner surface of the wall.
4. Thallose members frequently have air chambers and pores, often have distinct chlorenchyma and parenchyma layers, and often have ventral scales.
5. The apical cell usually has two or four cutting faces.
6. The antheridium develops with the primary cell undergoing two centric vertical divisions at right angles, forming four daughter cells from this primary cell.
7. The seta is usually very short.
8. The sporangium often opens irregularly, or sometimes by four longitudinal lines.
9. The sporangium jacket is unistratose.
10. The spore mother cells are usually unlobed before meiosis.

Hepatics show distributional patterns similar to those of mosses, and occupy a parallel variety of habitats. They appear to be most richly represented in the humid tropics and become uncommon in arctic environments where they are generally intermixed among mosses.

FURTHER READING

Barkman, J. 1950. Over de Phylogenie der Levermossen. *Buxbaumia* **4**:32–55.

Bopp, M., and F. Feger. 1961. Das Grundschema der Blattentwicklung bei Lebermoosen. *Rev. Bryol. Lichénol.* **29**:256–273.

Buch, H. 1911. *Über die Brutorgane der Lebermoose,* 69 pp. Helsinki.

———— 1932. Morphologie und Anatomie der Hepaticae, in Verdoorn, F. (ed.), *Manual of Bryology,* pp. 41–72. The Hague: Martinus Nijhoff.

———— 1932. Experimentelle Morphologie, in Verdoorn, F. (ed.), *Manual of Bryology,* pp. 73–88. The Hague: Martinus Nijhoff.

Campbell, D. H. 1904. Resistance to drought by liverworts. *Torreya* **4**:81–86.

———— 1936. The relationships of the Hepaticae. *Bot. Rev.* **2**:53–66.

Cavers, F. 1903. On asexual reproduction and regeneration in Hepaticae. *New Phytol.* **2**:121–133, 155–166.

———— 1903. Explosive discharge of antherozoids in Hepaticae. *Torreya* **3**:179–182.

———— 1903. On saprophytism and mycorrhiza in Hepaticae. *New Phytol.* **2**:30–35.

———— 1910–1911. The interrelationships of the bryophytes. I–XI. *New Phytol.* **9**:81–112, 157–186, 196–234, 269–304, 341–353. Ibid. **10**:1–46, 84–86.

Engel., J. J. 1982. Hepaticopsida in Parker, S. P. (ed.) *Synopsis and Classification of Living Organisms.* pp. 271–305. New York: McGraw-Hill.

Evans, A. W. 1939. The classification of the Hepaticae. *Bot. Rev.* **5**:49–96.

Fulford, M. 1948. Recent interpretations of the relationships of the Hepaticae. *Bot. Rev.* **14**:127–173.

Garjeanne, A. J. M. 1911. Der Verpilzung der Lebermoose-rhizoiden. *Flora* **102**:147–185.

Goebel, K. von. 1930. *Organographie der Pflanzen. 2nd part: Bryophyten-Pteridophyten* (3rd ed.). Jena: Gustav Fischer.

Herzog, Th. 1925. Anatomie der Lebermoose, in Linsbauer, K. (ed.), *Handbuch der Pflanzenanatomie,* Vol. 7, pp. 1–112. Berlin: Borntraeger.

Kreh, W. 1909. Über die Regeneration der Lebermoose. *Nova Acta Acad. Caes. Leop.-Carol.* **90**:217–301.

Mielinski, K. 1926. Über die Phylogenie der Bryophyten mit besonderer Berücksichtigung der Hepaticae. *Bot. Arch.* **16**:23–118.

Mueller, K. 1951. Die Lebermoose Europas, in *Rabenhorst's Kryptogamenflora* (3rd ed.), Vol. 6, pp. 1–290. Leipzig: Geest and Portig K.-G.

Parihar, N. S. 1965. *An Introduction to Embryophyta I. Bryophyta* (5th ed.). Allahabad: Central Book Depot.

Savicz-Ljubitzkaja, L. I., and I. I. Abramov. 1959. The geological annals of Bryophyta. *Rev. Bryol. Lichénol.* **28**:330–342.

Schiffner, V. 1917. Die systematisch-phylogenetische Forschung in der Hepaticologie. *Progr. Rei Bot.* **5**:387–520.

Schuster, R. M. 1966. *The Hepaticae and Anthocerotae of North America, East of the Hundredth Meridian,* Vol. I. New York: Columbia University Press.

———— 1979. The phylogeny of the Hepaticae, in Clarke, G. C. S., and J. G. Duckett (eds.), *Bryophyte Systematcs. Systematics Assoc. Sp. Vol.* **14**:41–82.

———— 1984. Comparative anatomy and morphology of the Hepaticae, in Schuster, R. M. (ed.), *New Manual of Bryology.* pp. 760–891. Nichinan, Japan: Hattori Bot. Lab.

———— 1984. Evolution, phylogeny and classification of the Hepaticae, in Schuster, R. M. (ed.), *New Manual of Bryology.* pp. 892–1070. Nichinan, Japan: Hattori Bot. Lab.

Underwood, L. M. 1894. The evolution of the Hepaticae. *Bot. Gaz.* **19**:347–361.

Verdoorn, F. 1934. Bryologie und Hepaticologie, ihre Methodik und Zukunft. *Ann. Bryol.* **4**:1–39.

Watson, E. V. 1971. *The Structure and Life of Bryophytes.* London: Hutchinson University Library.

12 The Mosslike Hepatics — Order Calobryales

The radial symmetry of the erect axes of the Calobryales makes them resemble mosses in a superficial way. Some researchers retain only the genus *Haplomitrium* (approximately 12 species) in this order and place *Takakia* (2 species) in its own order, Takakiales. In the present text the order is treated as containing two families, Haplomitriaceae and Takakiaceae, each with a single genus (Fig. 12-1). *Takakia* shows an interrupted distribution from the Himalayas to North Borneo and Japan, thence along the Aleutian chain and southward to south coastal British Columbia, Canada. *Haplomitrium* is most richly represented in the Indian subcontinent and adjacent island groups in the Indian Ocean and southeast Asia. Species occur also in the Southern Hemisphere. Only one species, *H. hookeri* (Fig. 12-1E), is widely scattered in the Northern Hemisphere both near sea level and in alpine sites. *Takakia* (Fig. 12-2) generally grows on moist humus from near sea level to subalpine elevations, while *Haplomitrium* is on both humic substrata and raw mineral soil, also generally in moist sites. Both genera are most frequent on acidic substrata in the mist of waterfalls and at the margins of rapid streams.

The Order shows the following characteristic features:

1. Gametophores possess erect stems with spirally arranged leaves.
2. The stem sometimes possesses a central strand.
3. Branching is intercalary (initiated from lateral cells well below the stem apex) and exogenous (from the outer cells of the stem).
4. Rhizoids are absent, but a "rhizomatous" system is present that produces both erect and subterranean stems.
5. Leaf cells have simple or complex oil bodies and many chloroplasts in each cell.
6. Sex organs are lateral and lack any specialized protective bracts.
7. The archegonium usually has a thick stalk and long neck.

FIGURE 12-1

Diversity in gametophores of Calobryales: *Haplomitrium*. (*A−C*) *H. intermedium* (×5). (*A, C*) Archegonial gametophores; note the continued growth of the shoots after archegonium production. (*B*) Antheridium-producing shoot, also showing growth after antheridium production. (*D*) *H. mnioides,* the left gametophore showing stem-calyptra with emergent sporophyte (×3). (*E*) *H. hookeri* (×5). (*E*) Antheridium-producing gametophore with terminal cluster of antheridia on left; sporophyte-bearing gametophore on right. (*F*) *H. gibbsiae* (×5); note stem-calyptra with leaves and archegonia attached. (*A−D, F,* after Schuster, 1966.)

144

FIGURE 12-2

Takakia lepidozioides gametophore. (*A*) Habit sketch, showing erect leafy axis and naked rhizomatous axes with clusters of beaked slime papillae (×12). (*B*) Detail of several leaves (×70). (*C*) Shoot apex, showing a single archegonium among the leaves (×70). (*D*) Detail of a single archegonium (×350). (*C, D,* after Hattori et al., 1958.)

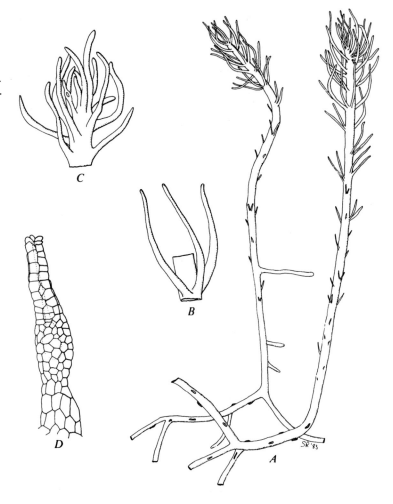

Since *Takakia* sporophytes are unknown, sporophytic features for the order are based on *Haplomitrium:*

8. An extensive stem-calyptra (Fig. 12-1*D–F*) is produced that protects the elongating embryo. This is composed of both archegonial cells and numerous stem cells from the stem apex, and is a chlorophyllose structure that sometimes has diminutive, widely scattered leaves on its surface.

9. The seta is massive and chlorophyllose when young, even when the sporangium is exserted well above the stem-calyptra.

10. The sporangium is elongate-cylindric, and the jacket is unistratose.

11. The cells of the jacket possess a longitudinal, thickened, annular band on their transverse faces. (This feature is unique to *Haplomitrium.*)

12. The sporangium opens by one to four longitudinal lines.

13. Elaters are uni- or bispirally thickened and are scattered longitudinally among the spores. They are extremely long and narrowly tapered at both ends.

As is apparent in the preceding list, and as will become apparent in succeeding pages, this order possesses features that characterize most hepatics. *Takakia* and *Haplomitrium* differ from each other in some fundamental structures, and are discussed separately.

TAKAKIA: THE GAMETOPHYTE

Only archegonium-bearing gametophores have been discovered in *Takakia* (Fig. 12-2). Discovery of antheridia and sporophytes might considerably alter the interpretation of the relationships of the genus. Some authors place it in an independent Order Takakiales because it possesses the following features that are unique (among the hepatics) to it:

1. The erect stems appear to be cutinized.
2. The "rhizomes" act as absorptive "roots."
3. The "rhizomes" bear clusters of multicellular, often branched, mucilage-producing flask-shaped cells.
4. The leaves are often structurally reminiscent of stems with determinate growth.
5. The leaves are usually in three rows, but all leaf segments are identical in morphology, and none can be interpreted as ventral.
6. Leaves are sometimes in groups of four at one level on the stem, and are sometimes fused at the base.
7. The leaf cells possess simple oil bodies, as in many mosses.
8. The archegonia arise singly without relation to a leaf, and supplant the position of a leaf.
9. The archegonium is sometimes on a distinct pedestal (as in some mosses, e.g., *Andreaea*).
10. Neck cells of the archegonia are in six vertical rows, similar to the mosses and different from hepatics other than the Marchantiidae.
11. The chromosome number is four or five (dependent on species or populations).

This impressive list of features emphasizes the distinctiveness of this genus. Because of basic similarities to the Calobryales, however, it is treated with this group.

Erect shoots of *Takakia* are rarely taller than 1 cm. These form turflike mats of green leafy stems that arise from a horizontal system of much branched pallid shoots. This system is often embedded in humus built up by the partly decomposed accumulation of past years' stems. The turfs may grow on bare rock faces in humid shaded sites, on humus in cliff crevices, and on somewhat sheltered humus of banks in tundralike habitats. The vivid green plants at first glance resemble a moss or green alga.

The leafy erect shoots of *Takakia* are generally unbranched. If branches appear, they emerge at right angles to the erect axes and form pallid rhizomatous shoots. The stem anatomy (Fig. 12-3G) shows epidermal cells with somewhat thickened outer walls, a region of parenchyma cells, and a central strand of cells of reduced size. The central strand is important in endohydric conduction. Chloroplasts and oil bodies are in outer cells of the stem. The pallid rhizomatous shoots structurally resemble the erect shoots, except that they lack leaves. They possess at irregular intervals, especially near the bases of branches, clusters of mucilage cells in which the terminal cell is roughly flask-shaped; the mucilage is exuded through the tip of the tapered neck of the cell (Fig. 12-3A−C). This mucilage enhances growth of fungal hyphae. It is possible that the *Takakia* requires the fungus as a symbiotic partner and that its presence enhances mineral and water absorption through the rhizomatous system.

Leaves are usually less than 1 mm long and are arranged spirally in irregular rows as one, two, or four long-tapered, conic projections that emerge from the stem cortex. In the axils of the young leaves there are usually short, uniseriate mucilage hairs. Sometimes paired leaves are fused at the base, producing a bifid structure, sometimes the four projections emerge from a broad fused base and from this dichotomize as two pairs, diverging equally (bisbifid). These patterns are also found in some members of the Order Jungermanniales. The leaves and stems are extremely brittle in *Takakia;* this appears to be the only reproductive method available since these fragments can form new gametophores.

Archegonia are lateral and resemble the leaves (Fig. 12-2C). When they become older they are often brownish pigmented, as in the mosses. Each stem may bear three or four, and these are usually near the stem apex. The archegonia have a thickened pedestal, a somewhat expanded venter, and a neck that consists of six vertical rows of cells.

When *Takakia* was first discovered in Japan, its relationships even to bryophytes were uncertain, since archegonia were absent in the first collection. The whimsical Japanese name for *Takakia* (nanjamonja-goke = "impossible moss") reflects this.

HAPLOMITRIUM: THE GAMETOPHYTE

In the older literature two genera are recognized for the Haplomitriaceae: *Haplomitrium* and *Calobryum*. The minor features that have been used to distinguish the genera appear to be unreliable, and so the two genera have been combined into a single genus (Fig. 12-1).

The spores of *Haplomitrium* are unicellular and richly chlorophyllose (Fig. 12-4). The germinating spore usually undergoes at least one transverse cell division within the spore coat. When the spore coat is ruptured, further cell divisions result in a globose cell mass. Reclining branches emerge from this cell

FIGURE 12-3
Details of *Takakia lepidozi-oides* gametophore. (*A–C*) Stages in the development of beaked slime papillae from the rhizomatous branches (×640). (*C*) Showing the mature papilla exuding slime. (*D*) Slime papilla from near leaf base, exuding slime. (*E*) Cells from axis and leaf primordium, showing chloroplasts and nuclei (×1400). (*F*) Transverse section of a leaf (×140). (*G*) Transverse section of stem, showing central conducting strand (×240). (*H*) Longitudinal section through archegonium (×220). (*A–E*, after Proskauer, 1962; *F–H*, after Hattori and Mizutani, 1958.)

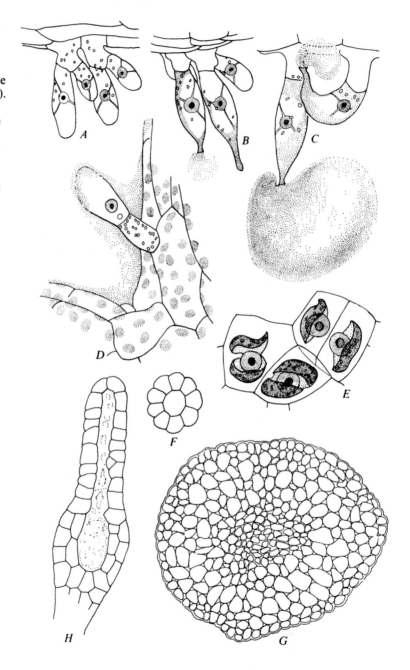

mass and produce a reduced "rhizomatous" system. From this system, erect branches emerge and grow into leafy gametophores. Further growth of the creeping portion produces a much branched reclining system that bears further perpendicular branches. All branches emerge at right angles to the axis. The

FIGURE 12-4
Spores and spore development in *Haplomitrium*. (*A*) *H. mnioides* spore (×1200). (*B*) *H. gibbsiae* spore (×800). (*C*) *H. intermedium* spore (×1530). (*D–K*) Germination of spore and formation of sporeling (×300) in *H. rotundifolium* (diagrammatic). (*A–C*, after Schuster, 1966, *D–J*, after Nehira, 1961; *K*, after Yang, 1967.)

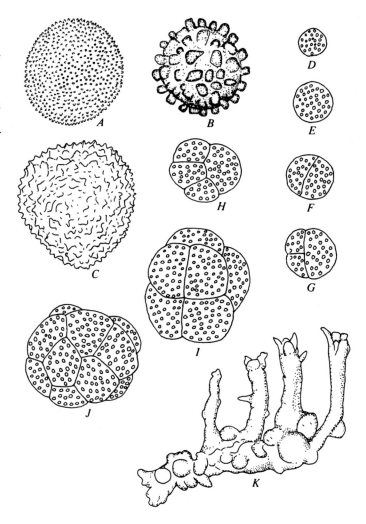

gametophore is somewhat succulent, and the erect leafy stems are bright green; all cells are thin-walled and most contain many chloroplasts and spherical or long-elliptic complex oil bodies (Fig. 12-5*A–C*).

The pallid creeping system has no leaves and few chloroplasts. Besides producing numerous horizontal branches at right angles, it gives rise to branches that extend downward, like roots, into the substratum. Many oil droplets are in enlarged outer cells of these naked stems and in a zone beneath. Sometimes a central strand made up of narrow cells also contains a few oil droplets. The cells of the strand strongly resemble the hydroids of mosses and act as an internal conducting system for the aerial shoots, in which the leaf surfaces are relatively impermeable to external uptake of water.

In the leafy axes the stem shows an internal structure similar to that of the

FIGURE 12-5
Gametophytic details in *Haplomitrium.* (*A–C*) Oil bodies and plastids (stippled) in leaf cells; plastids not shown in most cells. (*A*) *H. intermedium* (×600). (*B*) *H. hookeri* (×840). (*C*) *H. gibbsiae* (×600). (*D, E*) Transverse sections of stems. (*D*) *H. hookeri* (chlorophyllose tissue stippled) (×160). (*E*) *H. intermedium* (stippled area densely filled with starch grains) (×300). (*F*) Transverse section of a leaf of *H. mniodes,* one-eighth of the length from the leaf base (×25). (After Schuster, 1966.)

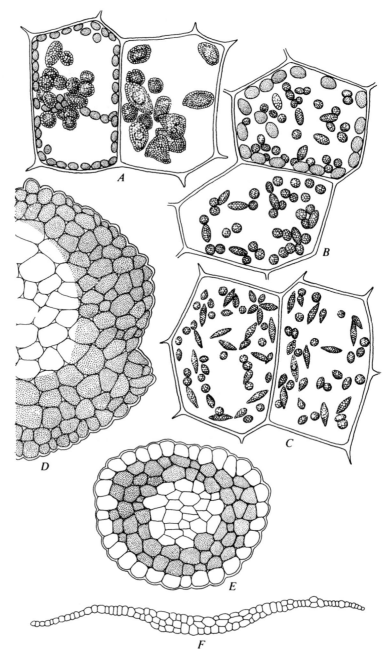

rhizomatous system. A central conducting strand is usually present and is continuous with that of the rhizomatous system but does not extend into the leaves. The leafy stems are usually less than 2 cm tall but reach 6 cm in *H. giganteum.* The leaves are arranged in three ranks with those in one rank sometimes somewhat smaller than the other two. It is of particular interest that this row of smaller leaves is on the dorsal, rather than the ventral side of the stem. Leaves are usually less than 5 mm wide, but in *H. giganteum* they are up to 9 mm wide. Leaves are elliptic to broadly orbicular and predominantly unistratose, except near the base, where they are sometimes multistratose. Margins are entire, although in *H. hookeri* the outline is often so irregular that the leaf appears to have shallow lobes. The leaves taper to their transverse attachment on the stems.

Haplomitrium sometimes produces gemmalike cell masses on the margins and surfaces of leaves in culture, but these have not been observed in nature (Fig. 12-7*E*). The plant appears to be intolerant of desiccation.

The plants are dioicous. The early stages of the antheridia and archegonia are similar (Fig. 12-6). The male sex organs are lateral and axillary on antheridium-bearing shoots. Each antheridium is subspherical and has a stout stalk (Fig. 12-7*C*). Since many antheridia are often condensed near the shoot apex and the apical leaves form a cuplike rosette (Fig. 12-1*B,E*), it is possible that raindrops can rupture the antheridial stalks and throw the antheridia away from the gametophore and nearer to the archegonium-bearing gametophores.

The archegonia are also lateral in most species and in the axils of the leaves near the shoot apex. In some species, if fertilization takes place, this erect shoot tends not to produce any further leaves. If fertilization does not take place, the stem continues to grow and, in time, produces more vegetative leaves and archegonia. The archegonia are partially protected by the leaves that surround them.

When fertilization takes place, usually only a single archegonium is fertilized on each erect shoot. As the embryo grows there is growth of both the surrounding archegonium and the leaf-bearing stem beneath the archegonium. This results in an extensive stem-calyptra, which elongates to accommodate the enlarging embryo (Fig. 12-1*D−F*). This stem-calyptra can reach more than 5 mm in length before its apex is ruptured and the sporophyte emerges. Such a stem-calyptra sometimes bears small, widely spaced leaves (Fig. 12-1*F*).

FIGURE 12-6
Ontogeny of sex organs in the Calobryales, showing the unique development of the antheridium, which greatly resembles the development of the archegonium in early stages. (After Schuster, 1966.)

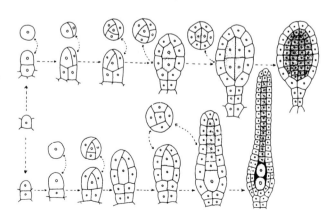

FIGURE 12-7
Haplomitrium hookeri. (*A*)
Elater and spores (×230). (*B*)
Surface ornamentation of
cells of sporangium jacket,
showing unique longitudinal
bands (×350). (*C*) Antheridium (×80). (*D*) Archegonium
(×80). (*E*) A single leaf with
gemmalike marginal clusters
of cells, as occurring in artificial culture (×20). (*E*, after
Fulford and Diller, 1956.)

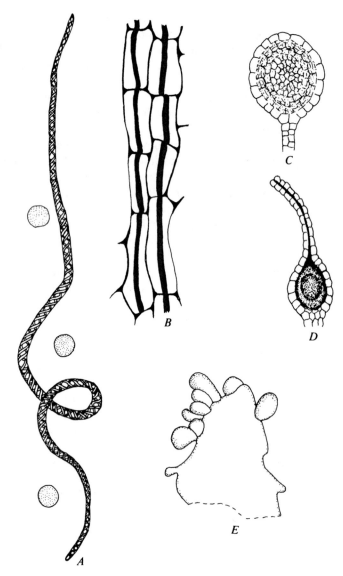

HAPLOMITRIUM: THE SPOROPHYTE

The massive seta is green when it emerges; it elongates rapidly and sometimes reaches lengths of 8–30 mm. The sporangium is always elliptical (Fig.
12-1*D–F*). The jacket wall is unistratose; each of its cells possesses a longitudinal straplike dark brown band on the transverse walls (Fig. 12-7*B*). Within
the sporangium are tetrads of spores and long, tapered elaters, each with two
to three loosely wound, pale-brown helices (Fig. 12-7*A*). The jacket of the
sporangium produces one to four longitudinal lines of weakness. As the

sporangium dries, it shrinks somewhat in length and gapes open along the lines of weakness, exposing the elaters that, on drying, are expelled from the sporangium, throwing out the spores as well. The seta then collapses and disintegrates rapidly.

EVOLUTIONARY TRENDS AND RELATIONSHIPS

Within the genus *Haplomitrium,* the generalized condition appears to be symmetrical radial arrangement of the leaves and a relatively poorly developed "rhizomatous" system. In more specialized species there is a distinctly differentiated row of smaller leaves, and the "root" system is often very pronounced. This system is similar in *Takakia,* and the internal conducting system is also present in both genera.

The Order shows greatest resemblances to some of the more generalized genera of Jungermanniales and shares some gametophytic features with the fossil genus *Naiadita* of the Sphaerocarpales and some sporophytic features with the monotypic order Monocleales. These appear to be the results of convergent evolution and do not imply close relationships. A number of apparent resemblances are also shared by the Order Metzgeriales, with which the Calobryales may share a common ancestry.

The most distinctive features of the Order Calobryales are as follows:

1. The spore germinates to produce a multicellular mass that produces the rhizomatous system.
2. The rhizomatous system sometimes has a central conducting strand.
3. Rhizoids are absent.
4. The archegonium-bearing shoot has indeterminate growth.
5. Sex organs are lateral on an essentially leafy axis.
6. A rootlike system develops in many species.
7. The stem sometimes has a central strand.
8. The unistratose jacket of the sporangium has longitudinal thickened bands.

In a number of respects the order can be considered specialized; for example, the unistratose sporangium wall is generally considered to be specialized. The development of a rootlike system and the presence of a central conducting strand can also be interpreted as specialized. On the other hand, the stem-calyptra, radial symmetry, and shoots of indeterminate growth are usually interpreted as generalized features. The Calobryales are unknown in the fossil record.

FURTHER READING

Asakawa, Y., S. Hattori, M. Mizutani, N. Tokunaga, and T. Takemoto. 1979. Chemosystematics of bryophytes III. Terpenoids of the primitive Hepaticae *Takakia* and *Haplomitrium. J. Hattori Bot. Lab.* **46**:77–90.

Campbell, D. H. 1920. Studies on some East Indian Hepaticae. *Calobryum blumii* N. ab. E. *Ann. Bot.* **34**:1–12, figs. 1-6.

Campbell, E. O. 1959. The structure and development of *Calobryum gibbsiae* Steph. *Trans. Roy. Soc. New Zealand* **87**:243–254.

Fulford, M., and V. Diller. 1956. Studies on the growth of *Haplomitrium* I. Organic media. *Rev. Bryol. Lichénol.* **25**:239–246.

———, J. Taylor and R. Hatcher. 1958. The "calyptra" of *Calobryum blumii* Nees. *Phytomorphology* **8**:298–302.

Goebel, K. 1891. Morphologische und biologische Studien IV. Ueber Javanische Lebermoose 2. *Calobryum blumii* Nees. *Ann. Jard. Bot. Buitenz.* **9**:11–25.

Grubb, P. J. 1970. Observations on the structure and biology of *Haplomitrium* and *Takakia,* hepatics with roots. *New Phytol.* **69**:303–326.

Hattori, S., Z. Iwatsuki, M. Mizutani, and S. Inoue. 1974. Speciation in *Takakia. J. Hattori Bot. Lab.* **38**:115–121.

——— and M. Mizutani. 1958. What is *Takakia lepidozioides? J. Hattori Bot. Lab.* **20**:295–303.

———, A. J. Sharp, M. Mizutani, and Ż. Iwatsuki. 1968. *Takakia ceratophylla* and *T. lepidozioides* of Pacific North America and a short history of the genus. *Miscell. Bryol. Lichénol.* **4**:137–149.

Inoue, H. 1961. Supplements to the knowledge on *Takakia lepidozioides.* Hatt. et Inoue. *Bot. Mag. Tokyo* **74**:509–513.

Inoue, S. 1973. Karyological studies in *Takakia ceratophylla* and *T. lepidozioides. J. Hattori Bot. Lab.* **37**:275–286.

Lilienfeld, F. 1911. Beiträge zur Kenntnis der Art *Haplomitrium hookeri. Bull. Acad. Sci. Cracovie* **1911**:315–399.

Nehira, K. 1961. The germination of spores in Heapticae I. *Calobryum rotundifolium* (Mitt.) Shiffn., *Bazzania albicans* Steph. and *Heteroscyphus planus* (Mitt.) Shiffn. *Hikobia* **2**:185–189.

Persson, H. 1958. The genus *Takakia* found in North America. *Bryologist* **61**:359–361.

Proskauer, J. 1962. On *Takakia,* especially its mucilage hairs. *J. Hattori Bot. Lab.* **25**:217–223.

Schuster, R. M. 1966. Studies on Hepaticae XV: Calobryales. *Nova Hedwigia* **13**:1–63, 12 figs.

——— 1971. Two new antipodal species of *Haplomitrium* (Calobryales). *Bryologist* **74**:131–143.

Yang, B. Y. 1966. The geographical distribution and growth habits of *Haplomitrium. Taiwania* **12**:9–20.

——— 1967. Spore germination and leafy gametophyte of *Haplomitrium rotundifolium* developed in culture. *Taiwania* **13**:57–71.

———, F. M. Hsu, and S. M. Lee. 1968. Spore germination and leafy gametophyte of *Haplomitrium blumii* developed in antiseptic culture. *Taiwania* **14**:73–80.

13 The Scale-Mosses— Order Jungermanniales

A quick glance might suggest a strong resemblance of a scale-moss (Order Jungermanniales) to a true moss of the subclass Bryidae. Jungermanniales show many growth forms that resemble the Bryidae, and they occupy similar habitats. Indeed mosses and Jungermanniales often grow intermixed. In spite of these superficial similarities, the Jungermanniales have features that set them apart from mosses.

The order contains more than two-thirds of all known hepatics. There are 43 families, including at least 280 genera and approximately 7,000 species (Figs. 13-1 and 13-2). The gametophore shows considerable diversity, and it is possible to note tendencies from leafy shoots through leafless shoots to an essentially thallose condition (Figs. 13-1, 13-2, and 13-20). The order is found throughout the world from frigid climates into humid tropical climates. The greatest floristic diversity is attained in humid climates where some genera dominate the bryophyte vegetation. Most species grow in moist environments, especially in somewhat shaded sites, but others thrive in full sunlight and can tolerate extreme desiccation. They grow on most substrata, from rock and mineral soil to humus, either as pure colonies or intermixed with other plants; many are epiphytes on woody plants, and some are epiphytes on other bryophytes. Several species grow in water.

Features that characterize the order are as follows:

1. Some members produce a short filamentous protonema.
2. Most genera are leafy.
3. Leaves are spirally arranged and are frequently in three rows with two rows of lateral leaves and one row of underleaves (amphigastria).
4. Leaves are commonly lobed.
5. Leaf cells often possess trigones.
6. Complex oil bodies are frequent in cells that include many chloroplasts.
7. Smooth rhizoids are common on the underside of the stem.

FIGURE 13-1
Diversity in Jungermanniales.
(*A*) *Blepharostoma trichophyllum,* showing deeply divided leaves, and a terminal perianth (×20). (*B, C*) *Frullania tamarisci* ssp. *nisquallensis.* (*B*) Habit, dorsal view, showing incubous leaf arrangement and three perianths in different stages of development (×6). (*C*) Detail, ventral view, showing helmet-shaped lobules of lateral leaves, the symmetric bilobed amphigastria, and the perichaetia surrounding a perianth (×18). (*D*) *Douinia ovata,* viewed from above showing unequally bilobed leaves with dorsal lobules, and a sporangium emerging from a perianth (×15).

8. Archegonia are generally enclosed within a cylindric unistratose chlorophyllose tube, the perianth.
9. Archegonium formation involves the apical cell of the stem, and succeeding growth of the stem is by innovations.

FIGURE 13-2
Diversity in Jungermanniales.
(*A*) *Plagiochila porelloides*
with single sporophyte emerging from flattened perianth;
leaves are succubously arranged (×6). (*B*) *Calypogeia
muelleriana*, showing mar-
supia in soil, with emergent
sporophytes; the left
sporangium shows that the
dehiscence lines are helical
(×12). (*C, D*) *Mastigophora
woodsii*. (*C*) Showing pinnate
branching; viewed from dorsal surface (×4). (*D*) Detail of
undersurface showing un-
derleaves and paraphyllia
(×15). (*E*) *Lophozia incisa*,
with inflated perianth and dehisced sporangium (×10). (*F*)
Gymnocolea inflata, with
swollen perianth (×13).

10. The sporophyte commonly has an elongate nonchlorophyllose seta.
11. The sporangium jacket is multistratose, and the transverse walls are commonly ornamented with transverse thickenings.
12. The sporangium usually opens by four longitudinal lines.

13. Elaters are among the spores.
14. Spore mother cells are lobed.

SPORES AND THEIR SURFACE ORNAMENTATION

Most spores are spherical, and ornamentation of the surface is diverse (Figs. 13-3 and 13-4). Ornamentation appears to show relative uniformity within a genus and sometimes within a family, but many apparently unrelated genera share a similar ornamentation, and thus spore ornamentation has not been used as a taxonomic character. Most spores are unicellular when shed, but sometimes, as in *Porella,* are multicellular (Fig. 13-4*F*). In this genus, as well as in some other genera, spores are elongate. Spores are usually small (6–24 μm as a general rule), and thus are the smallest among the hepatics. Spores tend to be one to two times the diameter of the associated elaters. Spore number within a sporangium varies from as low as 250 to more than a million.

SPORE GERMINATION PATTERNS

K. Nehira (1974) summarized and illustrated 17 spore germination patterns in the Jungermanniales. In the following discussion, this classification has been simplified, but general trends should be apparent (Figs. 13-5 and 13-6). The spore is generally unicellular when shed; sometimes it germinates to produce a germ tube that, by growth and successive transverse cell divisions, produces a uniseriate, sometimes branched, filament of elongate cells. Ulti-

FIGURE 13-3
Spores of Jungermanniales as revealed by scanning electron microscopy; all are distal views. (*A*) *Cuspidatula monodon* (×2000). (*B*) *Chandonanthus squarrosus* (×1250). (*C*) *Schistochila appendiculata* (×2500). (*D*) *Lepidolaena clavigera* (×500). (*E*) *Porella elegantula* (×500). (*F*) *Lepicolea attenuata* (×850). (Provided by Jane Taylor.)

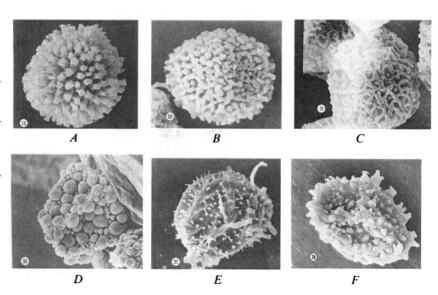

A   *B*   *C*

D   *E*   *F*

FIGURE 13-4
Spores of Jungermanniales.
(A) *Brachiolejeunea sand-
vicensis* (×450). (B) *Trichoco-
leopsis saccatula* (×450). (C)
Mylia verrucosa (×800). (D)
Leptolejeunea subacuta
(×500). (E) *Frullania kago-
shimensis* (×350). (F) *Porella
ulophylla* (×450). (After
Miyoshi, 1966.)

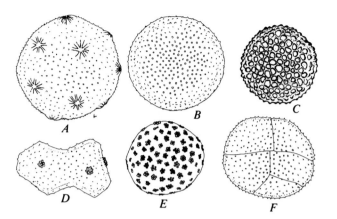

mately the apical cell produces a budlike clump of cells sometimes termed
a sporeling. This produces rhizoids and differentiates an apical cell and
subsequently a leafy gametophore. This is the most generalized protonema
type in the Jungermanniales (Fig. 13-5*A–E*).

Often when the unicellular spore germinates, a short unbranched protonema
of stout cells emerges, and the apical cell soon produces a sporeling that
subsequently forms a leafy gametophore (Fig. 13-5*F–I*). In a third type, the
filament of short, stout cells undergoes longitudinal divisions, and an apical cell

FIGURE 13-5
Spore germination and sporeling variation in Jungermanniales. (*A–E*) *Cephalozia
otaruensis*, showing formation
of filamentous protenema
(*A–D*, ×150) and production
of leafy gametophore (*E*,
×105). (*F–I*) *Plagiochila fru-
ticosa*, showing formation of
very abbreviated protonema.
(*J–M*) *Radula kojana*, showing formation of disc-pro-
tonema (*J–L*, ×200) and production of leafy sporeling (*M*,
×180). (*N–O*) *Fullania moni-
liata* (×120), globose protonema formed within the
spore. (*P–Q*) *Drepanole-
jeunea japonica* (×130), massive protonema formed within
the spore. (After Nehira,
1966.)

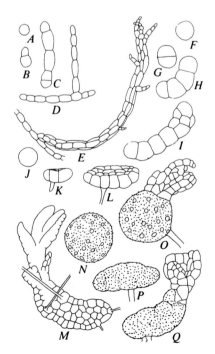

FIGURE 13-6
Spore germination and sporeling variation in Jungermanniales. (*A−F*) *Jubula javanica*
(*A−E*, ×200; *F*, ×85), production of a thallose protonema. (After Nehira, 1966.)

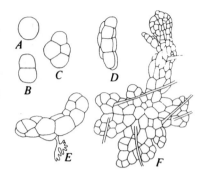

produces the first cell of the leafy gametophore. The sporeling thus arises from
all of the early cells, and no discrete protonematal phase exists. Sometimes instead of producing a protonema, the spore produces a small unistratose plate of
cells that attaches itself to the substratum by rhizoids which emerge from the
undersurface. One of the cells of the plate differentiates to produce an apical
cell, which then initiates the leafy gametophore (Figs. 13-5*J−L* and 13-6).

Further reduction of the protonematal phase produces a spore that is multicellular before it is shed (Fig. 13-5*N−Q*). This multicellular mass differentiates an apical cell, which then proceeds to form the leafy gametophore.

As apparent in this very simplified summary, patterns of spore germination
and reduction of the protonematal phase in the Jungermanniales are similar to
those present in the mosses. A filamentous protonema tends to characterize
those hepatics in which the leafy gametophore is also interpreted as generalized, while those interpreted as specialized usually grow from very reduced
protonemata. Ecologically, too, the more generalized hepatics tend to occupy
terrestrial sites while the more specialized are in more extreme environments,
such as epiphytic sites or exposed rock faces.

STEM STRUCTURE AND BRANCHING PATTERNS

Stem anatomy is remarkably simple (Figs. 13-7 and 13-8). The cortical cells
are enlarged and thinner-walled than those of the central axis, or they are
thicker-walled and form a multistratose region. Often some cells of the underside of reclining stems have endogenous fungi of possible importance in nutrient and moisture uptake in the stem. In gross pattern, lateral branching patterns fall into three categores:

1. Dichotomous, where the apex of a main stem produces two equal
   branches, and the branching stem is Y-shaped
2. Irregular, in which lateral branches arise at irregular intervals on the stem
3. Pinnate, in which the lateral branches arise regularly on either side and in
   one plane on the main stem

FIGURE 13-7
Anatomy of the stem in Jungermanniales: transverse sections. (*A*) *Lophozia ventricosa* (×160). (*B*) *Lophozia muelleri* (×110). (*C*) *Cephalozia leucantha* (×105). (*D*) *Herbertus aduncus* (×210). (*E*) *Cololejeunea diaphana* (×115). (After Schuster, 1966.)

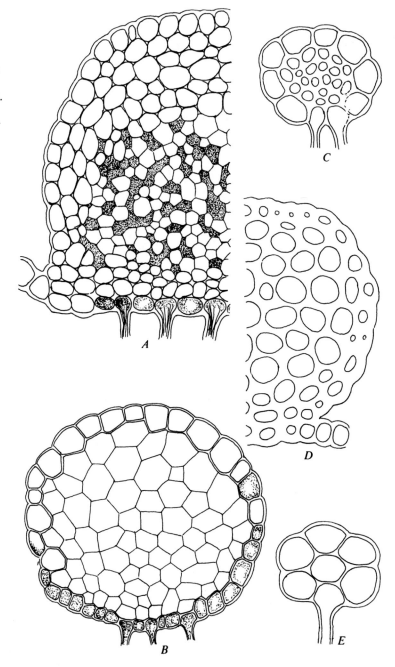

In the Jungermanniales the branches originate in diverse ways. Sometimes the branches arise from the outer cortical cells, but they occasionally supplant a lateral leaf entirely, or replace part of a lateral leaf; such branches are termed

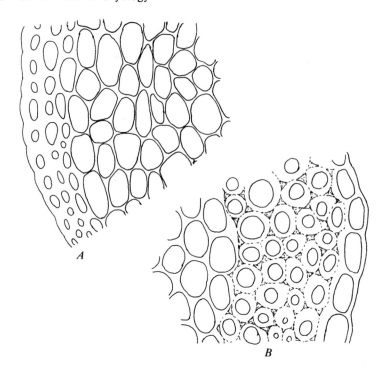

axillary. Sometimes branches originate in the axil of an underleaf or replace the underleaf; these are termed postical branches. Finally, some branches arise deep within the cortex of the stem on a mature shoot and ultimately emerge through the cortex; these are endogenous branches.

RHIZOID STRUCTURE AND POSITION

The rhizoids in most hepatics are unicellular and smooth-walled, but in a few of the Jungermanniales, including *Schistochila,* the rhizoids are sometimes multicellular, with their branched or swollen tips possessing many transverse walls; occasionally these terminal branches have biseriate cells (Fig. 13-9). The rhizoids in most Jungermanniales are composed of a single elongate cell, and the apex is blunt, but occasionally the tip is knoblike or branched. Sometimes fungi are especially abundant in these knobbed tips. Rhizoids are usually colorless, but are sometimes purplish.

Rhizoids vary somewhat in position among the Jungermanniales, but are generally associated with the underside of the stem. The rhizoids in the most generalized of the Jungermanniales are widely distributed on the underside of the stem, while in more specialized genera the underleaves are completely replaced by clusters of rhizoids. Occasionally rhizoids emerge from the lobules, and sometimes there are no rhizoids, at least in the mature gametophore.

FIGURE 13-9
Rhizoids of Jungermanniales.
(*A, B*) *Calypogeia muellerana*
(×100). (*C*) *Schistochila cun-
ninghamii* (×100). (*C*, after
Schuster and Engel, 1977.)

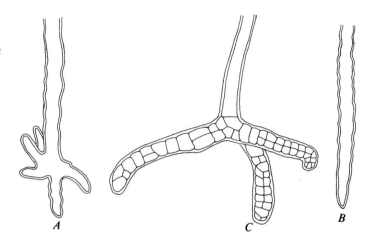

LEAF ARRANGEMENT ON THE GAMETOPHORE

The apical cell is similar to that in the mosses, and thus in most genera the
fundamental arrangement of leaves is in three ranks (Fig. 13-10); however, the
three-ranked arrangement is not as distorted by differential growth of the stems
as is apparent in most mosses. Rarely the leaves are derived from an apical cell

FIGURE 13-10
Longitudinal sections through
stem apices of Jungerman-
niales, showing apical cells
and derivative segments (outlined in darker lines). (*A*)
Cephaloziella rhizantha
(×50). (*B, C*) *Plagiochila
porelloides* (*B*, ×35; *C*, ×25;
after Schuster, 1966.)

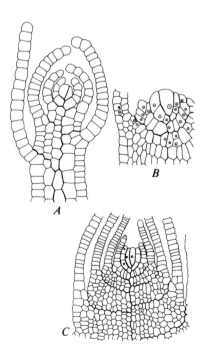

with two cutting faces (as in *Pleurozia*), and the leaves are in two rows. Sometimes the leaves are in two rows since one of the segments cut from the apical cell consistently produces a series of stem segments on the ventral side that initiate no leaves. Differential growth of the stem in leafy hepatics, instead of affecting the three-ranked arrangement of leaves around the stem, alters the orientation of the lateral leaf attachment on the stem so that the sessile base is either transversely or obliquely oriented (Fig. 13-11). The underleaves are always transversely attached.

Three basic patterns of lateral leaf orientation are represented (Fig. 13-6):

1. Transverse, where the leaf base attachment is perpendicular to the long axis of the stem (Fig. 13-11*A,B*). This is usually considered the most generalized type of leaf orientation and also characterizes the Calobryales and, incidentally, the fossil genus *Naiadita* of the Sphaerocarpales.
2. Incubous (Fig. 13-11*I–L*), where the upper border of the lateral leaf (i.e., the border facing the apex), when viewed from the dorsal surface, overlaps the lower border of the leaf immediately above it on the stem. This type of leaf arrangement is also considered generalized, and tends to be in genera in which other features are also generalized. Genera showing this type of orientation include *Bazzania*, *Lepidozia*, and *Calypogeia*.
3. Succulous (Fig. 13-11*C–H*). This type of leaf arrangement is the predominant type in the Jungermanniales. Viewed from the dorsal surface of the stem, the lower border of a leaf overlaps the upper border of the leaf imme-

FIGURE 13-11
Leaf insertion and orientation in the Jungermanniales (diagrammatic). (*A, B*) Transverse. (*C, D*) Simple succubous. (*E, F*) Succubous-transverse, with adaxial end of insertion transverse. (*G, H*) Succubous complicate bilobed, with insertions becoming decurrent. (*I, J*) Simple incubous. (*K, L*) Incubous complicate bilobed. (After Schuster, 1966.)

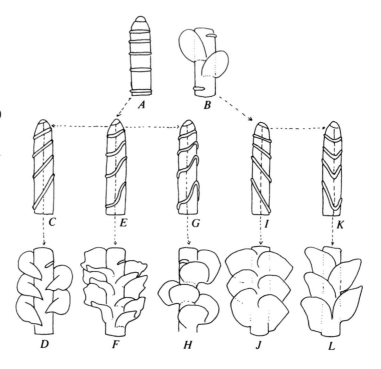

FIGURE 13-12
Diversity of leaf shapes in Jungermanniales. (*A*) *Colura inuii,* showing remarkable development of the pouchlike lobule (×9). (*B*) *Porella platyphylla,* viewed from undersurface showing incubous insertion, complicately bilobed lateral leaf, and unlobed underleaf (×15). (*C*) *Goebeliella cornigera,* showing remarkable lobule (×20). (*D*) *Cololejeunea macounii,* showing pouchlike lobule (×15). (*E*) *Cephalozia lunulifolia,* showing bilobed leaf (×15). (*F*) *Metacalypogeia schusterana,* showing unlobed incubously inserted leaf, viewed from upper surface (×15). (*G*) *Schistochila berteroana,* showing flaplike lobule (×12). (*H*) *Ptilidium ciliare,* showing several lobed leaf with ciliate margins (×15). (*B, H,* after Schuster, 1966.)

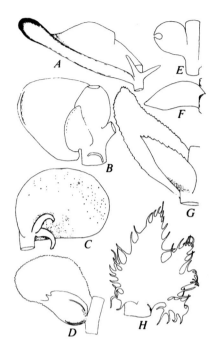

diately below it on the same side. Thus, in a succubous leaf, it is necessary to view both sides of the stem in order to see the upper border of the leaf, while in an incubous leaf, most of the upper margin of a leaf is exposed when viewed from the upper side of the stem. As in all apparently neat categorizations, it is sometimes difficult to determine to which orientation pattern a particular plant belongs.

LEAF STRUCTURE AND SHAPE

Leaf outline varies considerably: leaves are simple or lobed. Some leaves that are otherwise simple have teeth at the leaf tip. Lobed leaves show great variety in the number of lobes; two or four lobes are common, but occasionally three or five or more lobes are produced (Fig. 13-12).

In bilobed leaves, when the lobes are essentially alike, the leaves are termed simply bilobed (Fig. 13-12*E*). If lobes differ conspicuously in size or structure, the leaves are termed complicately bilobed (Fig. 13-12*B*). The smaller lobe of a complicately bilobed leaf is called a lobule. Within a species, and often within a genus, the lobules are consistently of the same shape on the entire shoot, and serve as a useful feature to distinguish some genera and species. The lobule is sometimes highly specialized in form, as in the genus *Frullania,* where it is a little helmet-shaped structure (Fig. 13-1*C*). Some species of *Schistochila* have one or several lamellae on the leaf surface. Rarely the leaf lobes and lobules are remarkably complex, and produce convoluted internal chambers, as in the genus *Pleurozia.*

Leaves are predominantly unistratose, and most leaf cells contain numerous chloroplasts and complex oil bodies. Complex oil bodies are made up of several oil globules held together as a single unit (Fig. 13-13*A*−*C*). Simple oil bodies, in contrast, are made up of a single oil globule (Fig. 13-13*D,H*). Some genera have the cell walls deeply pigmented with brown, red-brown, or black, but the walls are usually colorless. The exposed surfaces of the cells are usually smooth, but sometimes papillae or regular longitudinal ridges cross the exposed face of the cell. Walls between adjacent cells sometimes show thick and thin areas and are perforated by plasmodesmata. Walls at corners of three adjoining cells are often markedly thickened; these thickenings are called trigones (see Fig. 13-13*B,C,E*−*G*).

Within a leaf there is often variation in cell shape. The marginal cells sometimes differ conspicuously from the rest of the leaf cells and sometimes bear teeth or cilia. Sometimes a row of conspicuously elongate or colored cells forms a line down the center of the leaf; this vitta resembles a midrib, except that the vitta is unistratose.

VEGETATIVE REPRODUCTIVE STRUCTURES

Specialized vegetative diaspores are frequent in the order (Fig. 13-14). Sometimes diminutive gametophores form, usually on the surfaces of leaves, which can produce independent gametophores.

Gemmae are the most common vegetative diaspores. These are often produced on leaf tips and margins, especially near the tips of shoots. Sometimes the upper leaves and the entire shoot apex are devoted to production of gemmae (Fig. 13-14*F*). Gemmae are usually one- or two-celled and, when they are shed, germinate to produce protonematal structures much like those formed by spores of the same species.

Most hepatics have rather brittle stems, especially when dry, and the gametophore is readily fragmented. Sometimes the leaves are also brittle. Each living fragment has the potential to produce a new gametophore.

SPECIALIZED STRUCTURES SURROUNDING SEX ORGANS

A remarkable array of specialized structures protect the archegonia and young sporophyte; sometimes a stem-calyptra protects the developing sporophyte (Figs. 13-1 and 13-2). This is usually considered a generalized feature found in genera considered "primitive" based on additional structural features;

FIGURE 13-13
Oil bodies of Jungermanniales. (*A*) *Diplasiolejeunea rudolphiana* (×665). (*B*) *Cryptocolea imbricata* (×870). (*C*) *Radula tenax* (×820). (*D*) *Chonocolea doellingeri* (×960). (*E*) *Bazzania tricrenata* (×1250). (*F*) *Brachiolejeunea bahamensis* (×850). (*G*) *Lophozia silvicola* (×775). (*H*) *Lejeunea minutiloba* (×768). (After Schuster, 1966, 1969, 1980.)

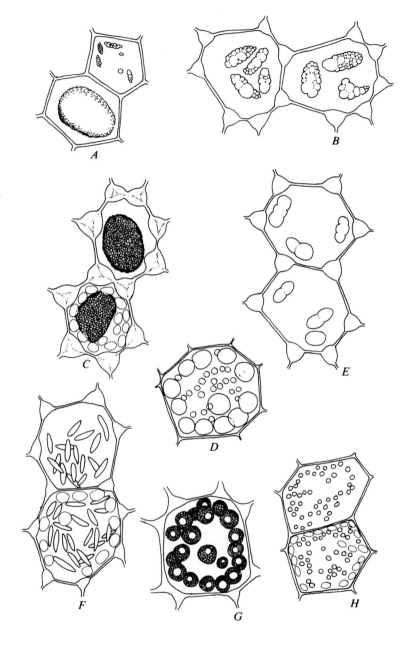

in such genera the archegonia are initially protected by somewhat modified perichaetial leaves; they are chlorophyllose and usually of the same shape as vegetative leaves, as in *Gymnomitrion*.

Somewhat modified, perigonial leaves surround the spherical antheridia, each of which is attached by a slender stalk at the axils of these leaves (Fig.

FIGURE 13-14
Diversity of vegetative diaspores in Jungermanniales.
(*A*) *Plagiochila tridenticulata*
shoot with portions barren of
leaves that have fallen off
(×3). (*B*) A leaf of *P. tridenticulata* producing a new gametophore (×6). (*C–E*)
Lophozia ascendens. (*C*)
Apex of shoot with gemma-
bearing leaf tips (×10). (*D, E*)
Details of gemmae (×150),
(*F, G*) *Lophozia capitata.* (*F*)
Shoot with terminal clusters
of gemmae (×8). (*G*) Detail of
a lobe apex with gemmae
(×50). (*H*) *Plagiochila cadu-
ciloba*, fragmentation of
leaves (×30). (*I*) *Blepharostoma trichophyllum*, showing
leaf apices "budding" off gemmae (×60). (After Schuster,
1966, 1969.)

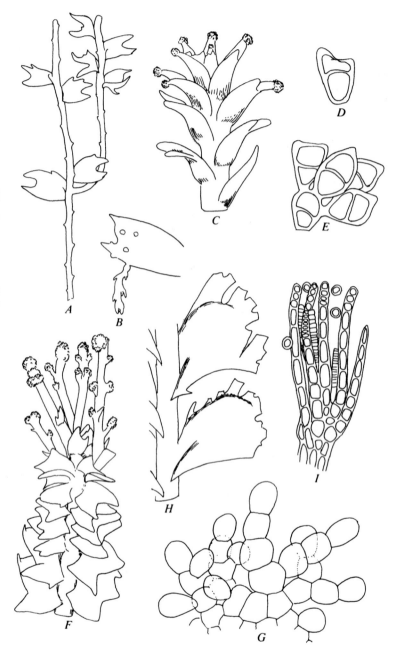

13-15). Perigonial leaves often differ in shape from the vegetative leaves and
terminate the main axis or branches. Several antheridia often occur in each
axil, and short mucilage filaments are among them. To accommodate the
spherical antheridia, the lower part of the perigonial leaf is often swollen as a
small pouch, and the whole antheridial branch is sometimes catkinlike.

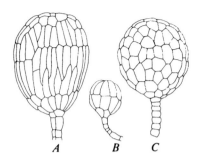

FIGURE 13-15
Antheridia of Jungermanniales (×170). (*A*) *Calypogeia* sp. (*B*) *Aphanolejeunea cornutissima.* (*C*) *Lophozia incisa.* (After Schuster, 1966.)

Of more general occurrence is a tubular chlorophyllose structure termed the perianth. This originates in much the same manner as the leaves except that the perianth is formed from a ring of cells that encircles the apex of the archegonium-bearing branch, immediately surrounding the archegonia, and grows upward as a sleeve to sheathe the archegonia. The detailed morphology of perianths serves as a useful taxonomic character since its shape is consistent within a species and often within a genus. Perianths are of four general shapes in cross-sectional view: circular, triangular, flattened, or pleated. The mouth of the perianth may be tapered to a beak, fringed with ciliar or teeth, puckered with many pleats, or simply compressed. All of these structural modifications appear to assist in keeping the archegonia moist and protecting the young sporophyte within.

Another structure that protects the archegonia in some genera is a subterranean multistratose pouch constructed of stem tissue, so that the archegonia are on the floor within this pouch or tube, as in *Calypogeia* (Fig. 13-2*B*). It reaches extraordinary proportions as an elongate tube in the Southern Hemisphere genus *Goebelobryum.* This pouch is often termed a marsupium or marsupidium. These pouches often possess numerous rhizoids, and bore into the substratum like a root.

Generally the gametophores are dioicous, and the archegonia terminate either the main stem or a reduced lateral branch; in both cases when the apical cell has produced the archegonia, it produces no further cells. This type of archegonial production is termed acrogyny. In all other orders of hepatics the archegonia arise laterally, without using up the apical cell of either the main shoot or a reduced lateral branch; these are termed anacryogynous.

SPOROPHYTES AND SPORE DISPERSAL

Generally the sporophyte consists of a foot that penetrates the gametophore, a seta which, when mature, has little or no chlorophyll, and a sporangium (Fig. 13-16). The seta is often 2–3 cm long, and its circumference is usually made up of numerous cells, although the number of cells can sometimes be reduced to as few as four (Fig. 13-17). Occasionally the seta is very short, and the sporangium is barely exserted beyond the protective leaves or perianth of the

FIGURE 13-16
Longitudinal section through
sporophyte of Jungerman-
niales: *Ptilidium pulcher-
rimum,* with most of seta
omitted (×50). (*A*) Note multistratose sporangial jacket
with elaters and spores within
the sporangium. (*B*) Lower
portion of seta with foot penetrating apex of gametophore.
(After Schuster, 1966.)

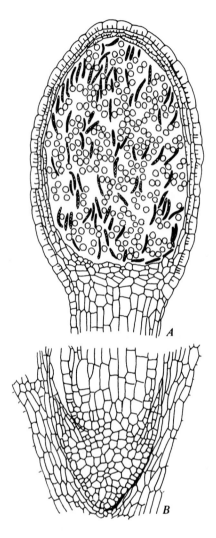

gametophore. Under favorable temperature and moisture, elongation of a seta
is rapid in a mature sporophyte and is held rigid by turgor pressure for a brief
period.

Sporangium shape varies from cylindric to spherical and is constant within a
species. The sporangium jacket is two to eight cell layers in thickness, and the
outer cells are usually ornamented with nodular thickenings on the radial walls
or bandlike thickenings across the transverse face of the cells (Fig. 13-18).
These thickenings are usually heavily pigmented with brown or black, and the
mature sporangium is also pigmented. Usually the sporangium is ruptured by
four longitudinal lines, so that it opens by four flaps (or "valves") of jacket tissue. Sometimes these lines run in helices around the sporangium, as in
Calypogeia and *Gyrothyra.*

FIGURE 13-17
Seta anatomy of Jungermanniales. (*A*) *Mylia anomala,* transverse section (×60). (*B, C*) *Jubula pennsylvanica* (×60). (*B*) Surface view. (*C*) Transverse section. (*D*) *Diplasiolejeunea rudolphiana,* transverse section (×230). (*E*) *Telaranea nematodes,* transverse section (×80). (*F*) *Cephaloziella divaricata,* transverse section (×230). (*G*) *Zoopsis argentea,* transverse section (×80). (*H*) *Cephaloziella rubella,* transverse section (×185). (After Schuster, 1966.)

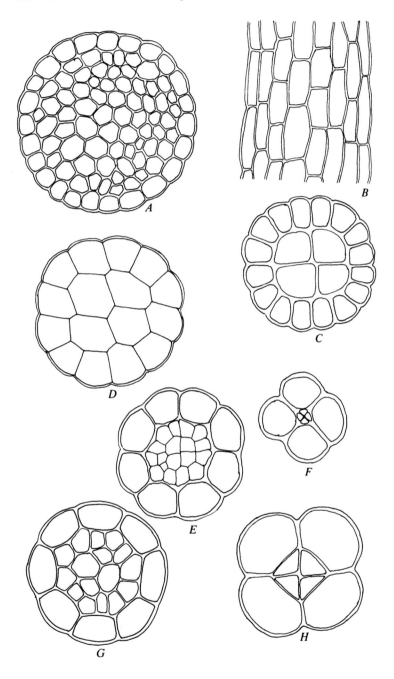

Inside the sporangium are numerous unicellular elaters. These are mixed with the spores and possess at least one or sometimes two or three helical wall thickenings; (Fig. 13-19*C*). The coils tighten as the elaters dry out. When tension becomes too great, the coils abruptly loosen, presumably in response to

FIGURE 13-18
Ornamentation of cells in jacket of sporangia in Jungermanniales. (*A*) *Lepidozia reptans*, transverse section through jacket (×950). (*B*) *Nowellia curvifolia*, epidermal cells. (*C*) *Temnoma pulchellum*, transverse section through jacket (×400). (*D*) *Arnellia fennica*, epidermal cells (×365). (*E*) *Frullania oakesiana*, epidermal cells (×350). (*F*) *Arnellia fennica*, inner cells (×365). (*G*) *Radula complanata*, epidermal cells (×300). (After Schuster, 1966.)

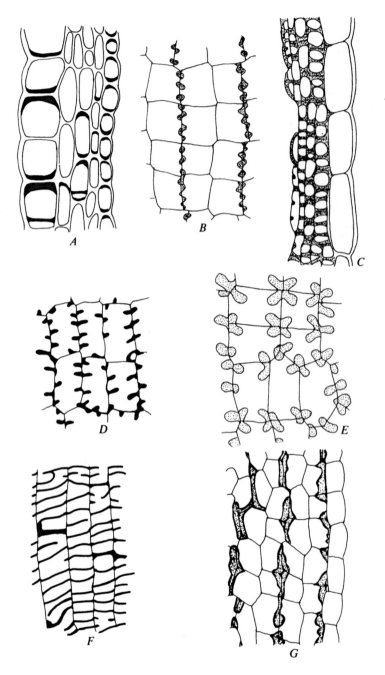

the rupturing of a capillary water column within the elater; the elater springs loose, jumping into the air, loosening the spores near it, and throwing them into the air. Thus, as a sporangium opens, the spores are rapidly released from the sporangium (Fig. 13-19*A–D*).

FIGURE 13-19
Spore dispersal in Jungermanniales. (*A–E*) *Cephalozia bicuspidata.* (*A*) Diagrammatic longitudinal section through sporangium showing elaters coated with spores. (*B*) Sporangium, showing longitudinal dehiscence lines. (*C*) Elater and spores. (*D*) One of the four divisions of sporangium jacket with elater attached showing pattern of movement of the hygroscopic elater, shedding spores as it moves. (*E*) Dehiscing sporangium throwing off spores and elaters. (*F–L*) *Frullania dilatata.* (*F*) Unopened sporangium. (*G*) Diagrammatic longitudinal section of sporangium, showing attachment of elaters. (*H*) Opened sporangium, diagrammatic longitudinal section with stretched elaters. (*I*) Elaters detached from "floor" of sporangium. (*K*) View of opening sporangium throwing off spores. (*L*) Opened sporangium with elaters still attached. (After Ingold, 1939.)

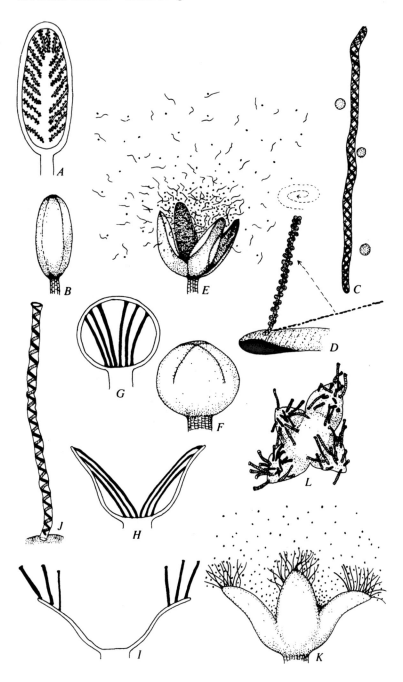

In some hepatics, as in *Frullania,* elaters are attached to both the "floor" and "ceiling" of the sporangium. When the four divisions of the sporangium jacket curve outward as the sporangium opens, these elaters are hinged to the four divisions and are stretched until one end of each elater is abruptly broken from the floor of the sporangium. This causes the elaters to flick outward and throw out spores (Fig. 13-19*E–K*). All spores are released within a few seconds after the sporangium opens. Since all sporangia on a gametophore do not mature simultaneously, the season of spore dispersal may perist for several days or even weeks within a clone; in most species it of relatively short duration and decidedly seasonal.

EVOLUTIONARY TRENDS

Based on assumptions concerning generalized versus specialized features within the Jungermanniales, several evolutionary trends have been suggested:

1. Unicellular spores are more generalized than multicellular spores.
2. An extensive protonema is more generalized than a short protonema or the absence of a protonematal phase.
3. The erect habit is more generalized than a prostate or creeping habit.
4. Radial symmetry is more generalized than bilateral symmetry.
5. Leaves changeable in their morphology are more generalized than leaves that are constant in morphology.
6. Deeply divided leaves are more generalized than undivided leaves.
7. Symmetric leaf shape is more generalized than asymmetry.
8. Asymmetric leaves with simple lobules are more generalized than those with complex lobules.
9. The presence of complex oil bodies is more generalized than their absence.
10. Distribution of rhizoids on most of the ventral surface of the stem is a more generalized pattern than their restriction to specific patches.
11. Irregular branching is more generalized than regular branching in a single plane.
12. The lack of specialized vegetative diaspores is more generalized than their presence.
13. The presence of sex organs on an unspecialized axis is more generalized than those restricted to reduced branches.
14. The presence of numerous archegonia or antheridia on a sexual shoot is more generalized than the presence of few.
15. The absence of specialized protective sheaths around the archegonia is more generalized than the presence of such sheaths.
16. A massive foot on the sporophyte is more generalized than a reduced foot.
17. A massive seta numerous cells in circumference is more generalized than a reduced seta four cells in circumference.
18. An elongate seta is more generalized than a very short one.
19. A sporangium jacket that is 6–8-stratose is more generalized than one that is 2–3-stratose.
20. Simple sporangial dehiscence is more generalized than complex sporangial dehiscence.

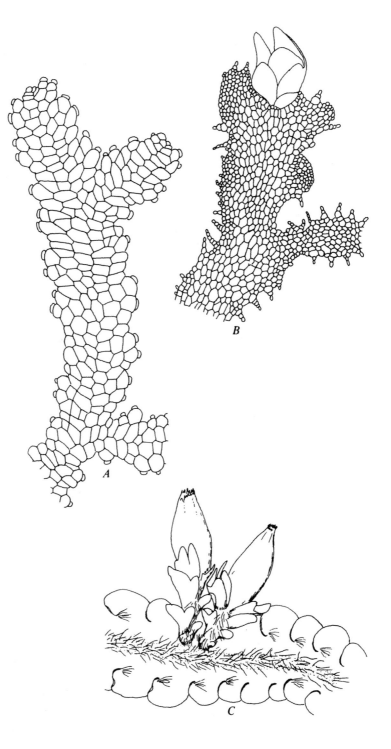

FIGURE 13-20

Thalloid gametophores in Jungermanniales. (A) *Zoopsis flagelliforme* (×50). (B) *Metzgeriopsis pusilla* showing leafy perigonium (×30). (C) *Schiffneria hyalina* showing two perianths and rhizoids on undersurface of thallus (×20). (A, B, after Schuster, 1966; C, after Inoue, 1974.)

21. Straight lines of dehiscence are more generalized than helical lines of dehiscence.
22. Elaters that are tapered and elongate are more generalized than those that are blunt and stout.
23. Elaters that are free in the sporangium are more generalized than those that are attached to the jacket walls.

TENDENCIES TOWARD THE THALLOSE CONDITION

In several genera and some species of the Jungermanniales, these are remarkably reduced leaves on the main axis, as in some species of *Cephalozia*. In *Zoopsis,* the axis is flattened and leaves are reduced to a few cells or even to a small bulge (Fig. 13-20*A*). Some genera, including *Metzgeriopsis,* produce only sexual branches that are leafy, and the remainder of the gametophore is a unistratose thallus (Fig. 13-20*B*).

In *Schiffneria,* a lobate flattened thallus produces leafy sexual branches, the archegonial branches of which possess perianths (Fig. 13-20*C*). In *Pteropsiella,* the gametophore is decidedly thallose with a central axis and lateral wings. It possesses underleaves and thus clearly belongs to the Jungermanniales. Finally, the genus *Phycolepidozia* has a much-branched leafless stem that produces rhizoids and leafy sexual branches (Fig. 13-21).

Each of these trends towards a thallose condition appears to represent a tendency toward extreme reduction, and these trends are sometimes used to support the hypothesis that the thallose hepatics were derived from leafy ancestors.

FURTHER READING

Andreas, J. 1899. Über den Bau der Wand und die Öffnungsweise des Lebermoos-sporogons. *Flora* **86**:161–213.

Basile, D. V. 1969. Toward an experimental approach to the phylogeny of leafy liverworts, in Gunkel, J. E. (ed.), *Current Topics in Plant Science,* pp. 121–133. New York: Academic Press.

Berrie, G. K. 1963. Cytology and phylogeny of liverworts. *Evolution* **17**:347–357.

Bischler, H. 1961. Recherches sur l'anatomie de la tige chez le Lejeuneaceae Paradoxae. *Rev. Bryol. Lichénol.* **30**:232–252.

Blomquist, H. L. 1929. The relation of capillary cavities in the Jungermanniaceae to water absorption and storage. *Ecology* **10**:556–557.

Bopp, M., and F. Feger. 1961. Das Grundschema der Blattentwicklung bei Lebermoosen. *Rev. Bryol. Lichénol.* **29**:256–273.

Buch, H. 1911. *Über die Brutorgane der Lebermoose,* 69 pp., 3 plates. Helsingfors.

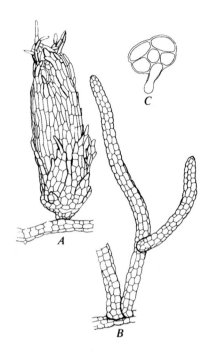

FIGURE 13-21
Unusual Jungermanniales: *Phycolepidozia exigua.* (*A*) Shoot sector with perianth (×70). (*B*) Shoot sector showing branching and leafless axes (×70). (*C*) Transverse section of shoot with a single rhizoid (×230). (After Schuster, 1966.)

———— 1919. Über den Einfluss von Licht und Feuchtigkeit auf die Wachstumsrichtung der Lebermoos Gametophyten. *Översikt. Finsk. Vetensk. Soc. Forhandl.* **61**(10):1–8.

———— 1920. Physiologische und experimentell-morphologische Studien an beblatteren Lebermoosen I-II. *Översikt. Finsk. Vetensk. Soc. Forhandl.* **62**(Afd A. Nr 6):1–46.

———— 1932. Morphologie und Anatomie der Hepaticae, in Verdoorn, F. (ed.), *Manual of Bryology,* chap. II, pp. 41–72. The Hague: Chronica Botanica.

———— 1932. Experimentelle Morphologie, in Verdoorn, F. (ed.), *Manual of Bryology,* chap. III, pp. 73–88. The Hague: Chronica Botanica.

Buchloch. G. 1951. Symmetrie und Verzweigung der Lebermoose. Ein Beiträg zur Kenntnis ihrer Wuchsformen. *Sitzb. Heidelberger Akad. Wiss. Math-Naturw. Klasse.* **1951**(4):1–71.

Cavers, F. 1903. On asexual reproduction and regeneration in Hepaticae. *New Phytol.* **2**:121–133, 155–166.

Clee, D. A. 1937. Leaf arrangement in relation to water conduction in the foliose Hepaticae. *Ann. Bot.* **1**:325–328.

Crandall, B. J. 1969. Morphology and development of branches in the leafy Hepaticae. *Nova Hedwigia, Beihefte.* **30**:1–261.

Crandall-Stotler, B. J. 1972. Morphogenetic patterns of branch formation in leafy Hepaticae—a résumé. *Bryologist* **75**:381–403.

Degenkolbe, W. 1938. Brutogane bei beblatteren Lebermoosen. *Ann. Bryol.* **10**:43–96.

Douin, C. 1908. Le pedicelle de la capsule des hépatiques. *Bull. Soc. Bot. France* **55**:195–202, 270–276, 360–366, 368–376.

–––––– 1916. Le pedicelle de la capsule des hépatiques. *Rev. Gen. Bot.* **28**:129–132.

–––––– 1929. La theorie des initiales chez les hépatiques a feuilles. *Bull. Soc. Bot. France* **72**:565–591.

Evans, A. W. 1905. Diagnostic characters in the Jungermanniaceae. *Bryologist* **8**:57–62.

–––––– 1912. Branching in the leafy Hepaticae. *Ann. Bot.* **26**:1–37.

–––––– 1935. The anatomy of the stem in the Lejeuneaceae. *Bull. Torrey Bot. Club* **62**:187–214, 259–280.

–––––– 1939. The classification of the Hepaticae. *Bot. Rev.* **5**:49–96.

Fulford, M. 1957. The young stages of the leafy Hepaticae: A resumé. *Phytomorphology* **6**:199–235.

–––––– 1963. Continental drift and distribution patterns in the leafy Hepaticae. Soc. Econ. Palaentol. Mineral., Spec. Paper 1D: 140–145.

Gavaudan, P. 1930. Recherches sur la cellule des hépatiques. *Botaniste,* Ser 22: 105–294.

Goebel, K. 1893. Archegoniatenstudien 3. Rudimentare Lebermoose. *Flora* **77**:82–103.

–––––– 1893. Archegoniatenstudien 5. Die Blattbildung der Lebermoose und ihre biologische Bedeutung. *Flora* **77**:423–459.

Heckman, C. A. 1970. Spore wall structure in the Jungermanniales. *Grana* **10**:109–119.

Ingold, C. T. 1939. *Spore Discharge in Land Plants.* Oxford: Clarendon Press.

–––––– 1956. Cinematographic observations on spore and elater discharge in *Lophocolea. Trans. Br. Bryol. Soc.* **3**:121–123.

Knapp, E. 1930. Untersuchungen über die Hullorgane um Archegonien und Sporogonen der akrogynen Jungermanniaceen. *Bot. Abh.* **16**:1–168.

–––––– 1930. Hepaticologische Studien I. Ist die Entwicklung des Lebermoosperianths von der Befruchtung abhängig? *Planta* **12**:354–361.

Müller, K. 1939. Untersuchungen über die Ölkorper der Lebermoose. *Ber. Deutsch. Bot. Gesell.* **57**:325–370.

–––––– 1948. Die systematische Wert von Sporophytenmerkmalen bei den beblatterten Lebermoosen. *Sv. Bot. Tisdkr.* **42**:1–16.

–––––– 1951–58. Die Lebermoose Europas, in *Tabenhorst's Kryptogamenflora* (3rd ed.), Vol. 6, 1365 pp. Leipzig: Geest and Portig K. G.

Nehira, K. 1966. Sporelings in the Jungermanniales. *J. Sci. Hiroshima Univ. ser. b.* **1**:1–49.

–––––– 1971. Evolution of the sporeling type in Hepaticae. *Hikobia* **6**:76–84.

–––––– 1974. Phylogenetic significance of the sporeling pattern in Jungermanniales. *J. Hattori Bot. Lab.* **38**:151–160.

Schuster, R. M. 1953. Boreal Hepaticae, a manual of the liverworts of Minnesota and adjacent regions. *Amer. Midl. Nat.* **49**:257–684.

–––––– 1965. Studies on Hepaticae XXVI. The *Bonneria-Paracromastigum-*

Hyalolepidozia-Zoopis-Pteropsiella complex and its allies—a phylogenetic study (Part I). *Nova Hedwigia* **10**:19–61.

———— 1966. *The Hepaticae and Anthocerotae of North America,* Vol. I. New York: Columbia Univ. Press.

———— 1966. Studies on Hepaticae XXVIII. On *Phycolepidozia,* a new, highly reduced genus of Jungermanniales of questionable affinity. *Bull. Torrey Bot. Club* **93**:437–449.

———— 1969. Problems of antipodal distribution in lower land plants. *Taxon* **18**:a46–91.

———— 1970. Studies on Hepaticae XVII. The Family Jungermanniaceae, *s. lat.:* a reclassification. *Trans. Br. Bryol. Soc.* **6**:86–107.

———— 1972. Evolving taxonomic concepts in Hepaticae, with special references to circumpacific taxa. *J. Hattori Bot. Lab.* **35**:169–200.

———— 1972. Phylogenetic and taxonomic studies on Jungermanniidae. *J. Hattori Bot. Lab.* **36**:321–405.

14 The Multiform Thallose Hepatics — Order Metzgeriales

It is possible that the Order Metzgeriales is the most ancient of the hepatics. Within the fossil record, *Hepaticites devonicus (= Pallaviciniites devonicus)* is the earliest presumed hepatic, and in the successive fossil record other presumed hepatics also show a morphology decidedly similar to the Metzgeriales (Fig. 14-1*B,C*). *Hepaticites kidstonii (= Treubiites kidstonii)* of upper Carboniferous age is extremely similar to the extant genus *Treubia* (Fig. 14-1*A*), and *Hepaticites lobatus (= Blasiites lobatus)* of the same age is similar to the extant genus *Blasia*. Finally, from the same age, *Hepaticites metzgerioides* is morphologically similar to the extant genus *Metzgeria*. Thus, among the earliest presumptive hepatics, the major morphological types of the Metzgeriales appear to be represented. Since none of these fossils possess sporophytes, their true relationships remain uncertain, but the circumstantial information is impressive.

In many publications, the Order Metzgeriales is designated "Jungermanniales Anacrogynae," in contrast to the acrogynous Jungermanniales. This designation, besides being clumsy, is nomenclaturally aberrant. Unfortunately not all taxa are anacrogynous. Although the order shows clear relationships with the Jungermanniales, it represents an independent evolutionary line.

The order contains 12 families with 28 genera and approximately 550 species. The order is widely distributed from the arctic to the tropics, and from sea level to alpine elevations; diversity and abundance are greatest in the humid subtropics and tropics. Most species are terrestrial, some are epiphytic, and one is a subterranean saprophyte. Nearly half of the genera possess a single species and tend to show a relatively restricted geographic distribution.

SPORES

Spores are usually unicellular, but in *Pellia* and several other genera they are multicellular (Fig. 14-2*A*). Surface ornamentation is diverse; indeed, in *Fossombronia,* where gametophores show a remarkable uniformity within the

FIGURE 14-1
Fossil plant material thought
to belong to the Metzgeriales.
(*A*) *Hepaticites* (*Treubiites*)
kidstonii (×40). (*B, C*) *Hepaticites* (*Pallaviciniites*) *devonicus* (×15). (*B*) central portion
of thallus. (*C*) margin of
thallus. (*A*, drawn from a photograph from Walton, 1925;
B, C, drawn from a pho-
tograph from Hueber, 1961.

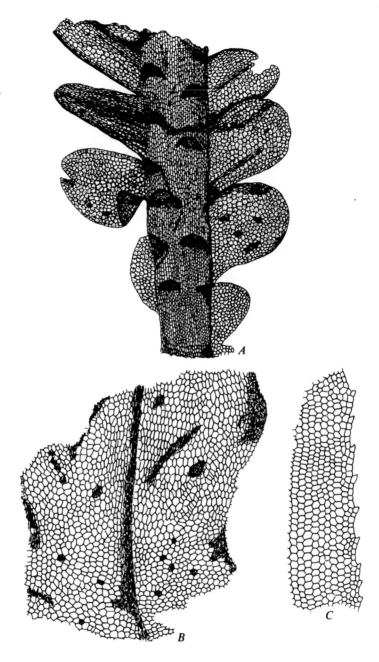

genus, the difference in spore ornamentation is important in distinguishing
species. Unicellular spores vary from 12 to 50 μm in diameter, but some multicellular spores reach lengths to 100 μm. Spores are generally fewer than in
sporangia in the Jungermanniales; this is essentially a reflection of the larger
size of Metzgerialean spores.

FIGURE 14-2
Spores of Metzgeriales. (*A*)
Pellia neesiana (×300). (*B*)
Cavicularia densa (×300). (*C,
D*) *Fossombronia japonica.*
(*C*) Distal view. (*D*) Proximal
view (×300). (After Miyoshi,
1966.)

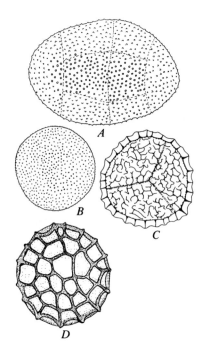

SPORE GERMINATION PATTERNS

A detailed analysis of spore germination patterns by M. Fulford (1975) recognizes four categories:

1. A filamentous protonema, in which the germinating spore produces a germ tube that grows into a short uniseriate filament, often a single cell in length (Fig. 14-3*A,B,E*). The apical cell of this filament (or of a branch) ultimately divides to produce a multistratose cell mass, and from this mass an apical cell is differentiated that gives rise to the gametophore.
2. A strap-shaped unistratose protonemal phase, in which the short uniseriate germ tube produces a biseriate or multiseriate protonema, and its apical cell produces a biseriate or multiseriate protonema, and its apical cell produces a multiseriate mass that initiates the gametophore.
3. A cylindric or globose exosporic protonemal phase, in which the germinating spore ruptures the spore coat and produces a multistratose cell mass from which the gametophore is initiated (Fig. 14-3*F–J*).
4. A globose multicellular endosporic protonemal phase, in which the spore undergoes several cell divisions before the spore coat is ruptured (Fig. 14-3*C,D*). Indeed, in some cases the spores are multicellular within the capsule. This multicellular spore initiates an apical cell that produces the gametophore.

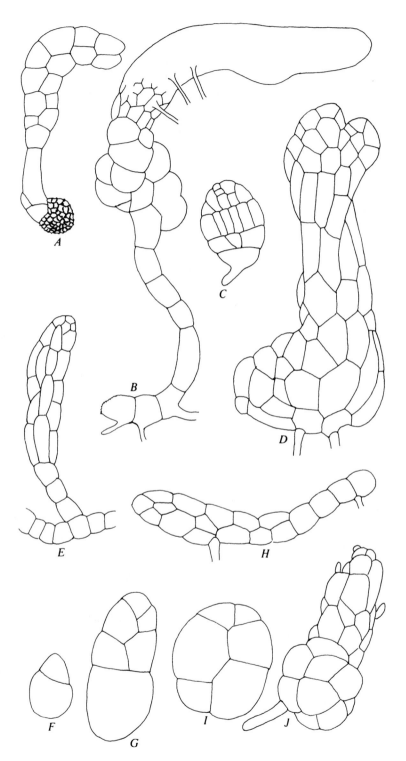

FIGURE 14-3
Spore germination patterns and sporeling production in Metzgeriales. (*A, B*) *Fossombronia pusilla* (×210). (*C, D*) *Pellia epiphylla* (×250). (*E*) *Aneura nagasakiensis* (×210). (*F–H*) *Metzgeria* spp. (×210). (*I, J*) *Cavicularia densa* (×210). (After Fulford, 1975.)

A

C

B

D

E

H

F

G

I

J

All patterns are subject to modification in culture conditions. Whether this is true in nature would be difficult to determine. The available information is insufficient to make any broad generalizations concerning either the ecologic or taxonomic significance of these germination patterns.

VARIATION IN GAMETOPHORE STRUCTURE

There is a great diversity in morphology of the gametophore (Figs. 14-4, 14-5, and 14-6), from a clearly thallose condition to one that is essentially leafy. In most individuals the archegonia are lateral; the apex of the stem continues indeterminate growth, and is not utilized directly in the production of the archegonia. This condition where a lateral cell, instead of an apical cell, initiates archegonium production, is termed anacrogynous. In some genera (e.g., *Metzgeria, Podomitrium*) the sex organs are confined to a reduced branch. The gametophore is usually constructed mainly of relatively thin-walled cells and never possesses complex pores or air chambers (Fig. 14-7). Occasionally, however, an internal conducting strand is present (Fig. 14-7*B*). Creeping gametophores vary from 1 cm to more than 10 cm in length and are up to 2 cm wide. Erect gametophores reach heights of up to 3–4 cm.

Although lenticular apical cells are most frequent, tetrahedral, wedge-shaped, and hemidiscoid apical cells are represented in some genera. The apical cell (with other cells of the growing point) is generally protected by mucilage papillae or filaments.

The structurally simplest gametophore is probably that of *Pellia* or *Riccardia,* in which the thickened central band of the thallus thins gradually toward the margins. All cells are parenchymatous. In *Pellia* (Fig. 14-4*D,E*), where the thallus reclines, ventral rhizoids are abundant along the thickened central band and the cells of the upper surface are more richly chlorophyllose than those deeper within the thallus. In *Riccardia* the thallus is often chlorophyllose throughout, and the succulent thallus often has few rhizoids when mature.

Some thalli (e.g., in *Metzgeria*), have a terete axis or midrib with unistratose wings on either side (Figs. 14-4*B* and 14-5*B,C*). Rhizoids or hairs are confined to the margin and to the ventral surface of the midrib, and all cells are parenchymatous. *Apometzgeria* has unicellular hairs on both surfaces of the wings (Fig. 14-4*B*), while *Petalophyllum* has lamellae on the upper surface of the thallus. Some genera, including *Symphyogyna* and *Pallavicinia* (Figs. 14-6*E* and 14-7*B*), have a central conducting strand in the thallus. This is made up of elongate pitted cells and serves to conduct water internally in the thallus.

The thalli of *Blasia* (Figs. 14-5*E,F* and 14-8*A*) and *Cavicularia* possess endogenous colonies of the blue-green alga, *Nostoc.* Most thalli contain endogenous fungi; in *Cryptothallus* they are especially abundant.

In *Hattorianthus, Podomitrium,* and *Hymenophyton,* the thallus is structurally similar to sporophytes of the fern family Hymenophyllaceae (Fig. 14-6*B,C*). Since these plants occupy similar habitats, it is apparent that this growth form is adapted to such sites. Such parallel evolution of a bryophytic

FIGURE 14-4
Diversity in Metzgeriales. (*A*) *Apotreubia nana* (×4). (*B*) *Apometzgeria pubescens* (×5). (*C*) *Aneura pinguis* with young sporophyte (×4). (*D, E*) *Pellia neesiana*. (*D*) Antheridium-producing thallus, showing antheridial chambers on thickened middle portion of thallus (×3). (*E*) Sporophyte-bearing thallus, with involucre and calyptra shown at seta base (×3).

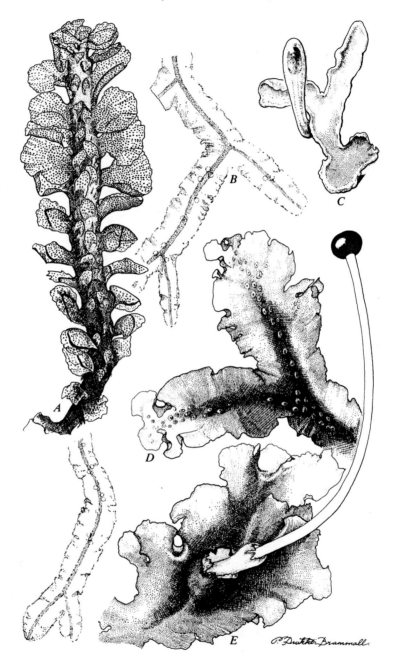

life form in lignified plants is unusual. In these genera, as well as others, the thallus often possesses a creeping stem with rhizoids, and arising from this stem are the erect thalli that may have several pseudodichotomizing "blades," each wing of which has its central axis.

FIGURE 14-5
Diversity in Metzgeriales.
(*A–C*) *Metzgeria conjugata.*
(*A*) General habit sketch (×2).
(*B*) Sporophyte-bearing
thallus, undersurface view
(×10). (*C*) Antheridium-
producing thallus, undersurface view, showing reduced
antheridial branches enclosing
the antheridia (×10). (*D*)
Moerckia blyttii (×15). (*E, F*)
Blasia pusilla (×8). Vegetative thallus, showing lobate
margins. (*F*) sporangium-
bearing thallus.

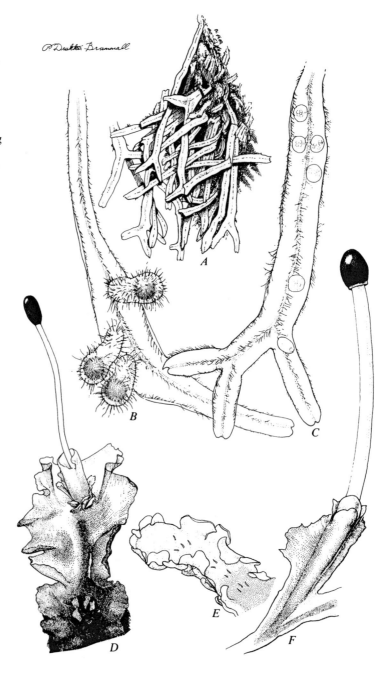

Several genera have leafy thalli. Lateral leaflike lobes are bilaterally arranged
on the reclining axis of *Fossombronia* (Fig. 14-8*F*). Distinctive paired lobes are
on either side of the stem of *Phyllothallia* (Fig. 14-6*A*). *Treubia* bears fleshy, al-
ternately arranged lobes; a scale on the dorsal surface of the stem is associated

with each lobe (Fig. 14-4*A*). In some thalli there are numerous rows of scales on the thickened central line, as in *Hattorianthus* and (to a lesser degree), in *Moerckia* (Fig. 14-5*D*). Occasionally the thallus margin is regularly lobate, as in some species of *Symphyogyna* (Fig. 14-6*B*); sometimes there are marginal teeth, as in *Pallavicinia;* in still others there are ruffles, as in *Moerckia* and *Blasia* (Fig. 14-5*D,E*).

Branching is irregular in many thalli, but in several, including *Metzgeria* and *Hymenophyton*, branching is pseudodichotomous (Figs. 14-4*B,* 14-5*A−C,* and 14-6*C*). In several species of *Riccardia* branching is clearly pinnate; sometimes it is bipinnate. Branches are usually terminal but may be intercalary, and they

FIGURE 14-6
Metzgeriales: diversity in gametophore morphology. (*A*) *Phyllothallia nivicola* (×3). (*B*) *Symphyogyna brogniartii* (×3). (*C*) *Hymenophyton flabellatum* (×3). (*D*) *Cryptothallus mirabilis* (×3). (*E*) *Pallavicinia lyellii* (×4). (*A,* after Schuster, 1968; *B,* after Evans, 1925; *C,* after Scagel et al., 1982; *E,* after Steere, 1940.)

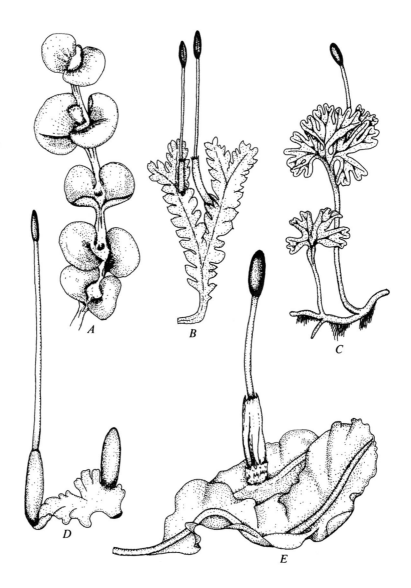

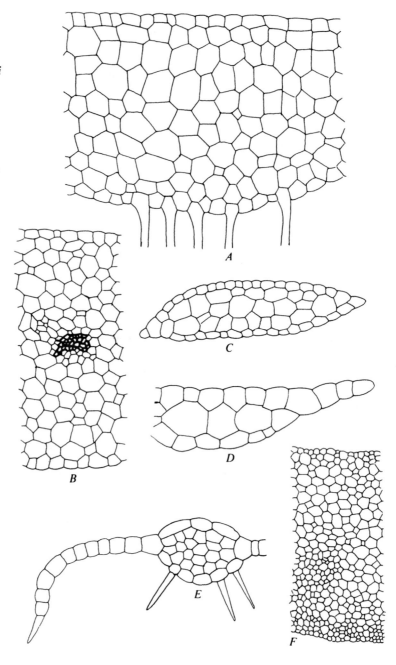

originate ventrally some distance behind the thallus apex. Considerable variety
in oil body number and structure is shown also in the cells of the thallus.
Complex oil bodies are absent in some genera, while in others they are
numerous. In *Treubia* single large, complex oil bodies occur in scattered cells,

as in the Marchantiales. In this same genus, occasional internal constrictions appear in the rhizoids, also reminiscent of the Marchantiales.

All members are bilaterally symmetrical, and scales, when present, are in two or more dorsal rows (Figs. 14-4*A* and 14-5*D*,*E*). Most genera have unicellular smooth rhizoids that are colorless, but occasionally rhizoids are brightly pigmented *(Fossombronia, Allisonia, Makinoa)*. Some genera lack rhizoids (e.g., *Metzgeria*).

FIGURE 14-8
Vegetative diaspores in Metzgeriales. (*A*, *C*) *Blasia pusilla*. (*A*) Thallus with gemma flasks (×20). (*B*) Single gemma from flask (×250). (*C*) Stellate gemma from thallus surface (×50). (*D*, *E*) *Riccardia palmata*. (*D*) Thallus tip, showing endogenously produced gemmae at lobe tips (×15). (*E*) Single gemma (×35). (*F*) *Fossombronia himalayensis*, thallus with tuber-producing branch (×4). (*G*, *H*) *Metzgeria furcata*. (*G*) Thallus segment, showing marginal gemmae (×30). (*H*) Detail of a single gemma (×60). (*D–H*, after Scagel et al., 1982.)

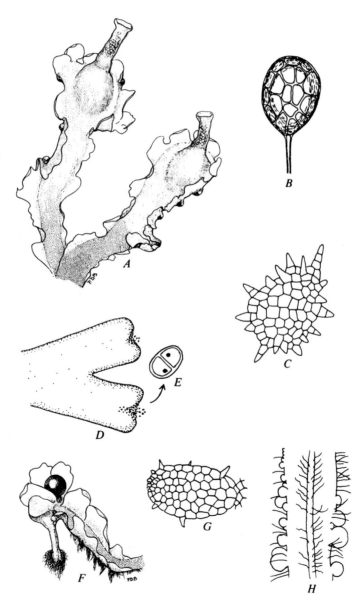

VEGETATIVE REPRODUCTION

In the Metzgeriales, as in most bryophytes, when the older portion of a branched gametophore decays, the branches are isolated as separate gametophores. Thus, progressively a thallus extends outward from an initial point where the diaspore originally established itself. Under adverse conditions, almost any epidermal cell of the thallus can produce a vegetative diaspore.

There are several means of specialized vegetative reproduction. Gemmae sometimes originate within cells of the dorsal surface of the thallus in some species of *Riccardia*. The cell contents round up and divide to form an ovoid two-celled gemma. The endogenous gemma is released when the wall of the cell containing the gemma is ruptured.

Gemma production is exogenous in *Blasia* but occurs within a long-necked flask on the dorsal surface of the thallus (Fig. 14-8*A*). The gemmae are budded off from cells within the flask. When mature, mucilage within the flask imbibes moisture, and the gemmae are extruded through the neck. *Blasia* also produces stellate gemmae on the thallus surface (Fig. 14-8*C*).

Some genera, including *Metzgeria,* bud off gemmae on the thallus margins (Fig. 14-8*G*). As in some mosses, gemma production appears to be more frequent in juvenile stages than in thalli with sporophytes. *Fossombronia himalayensis* produces "tubers" at the stem tips that burrow into the ground (Fig. 14-8*F*). The tuber is essentially an apical cell surrounded by mainly dead cells, and this structure is able to survive the unfavorable season and, upon return of favorable conditions, produces a new gametophore.

THE SEX ORGANS AND THEIR PROTECTIVE SHEATHS

The ontogeny of sex organs is shown in Fig. 14-9; this is the same as in the Jungermanniales. Both monoicous and dioicous species are found in some genera, but the dioicous condition is more common. In some genera, as in *Fossombronia* and *Treubia,* the sex organs are exposed on the dorsal surface of the stem, but preceding fertilization, the archegonia are usually protected by immature leaflike flaps of the thallus. In most genera, however, they are protected by a sheath of tissue or are embedded in the thallus.

Each antheridium of *Cavicularia densa* is sheathed in a small flask, and these flasks are scattered on the dorsal surface of the midrib, usually near the apex of the thallus lobe. The antheridia in *Makinoa* are in small lunate pockets near the tips of thallus lobes. Each antheridium in *Hattorianthus* is protected by a unistratose scale; these are arranged in irregular rows along the midrib of the erect thallus lobes. The antheridia are embedded in the thickened central portion of the thallus in *Pellia* (Fig. 14-4*E*). *Aneura* has them embedded in the thallus but confined to reduced lateral branches. In *Metzgeria* they are sheathed in a unistratose, hemispherical, reduced postical "branch" on the thallus (Fig. 14-5*C*).

The archegonia show a similar diversity of protective structures. Leaflike

FIGURE 14-9
Ontogeny of sex organs in Jungermanniales and Metzgeriales (longitudinal sections); antheridia in upper portion of figure and archegonia below with transverse sections shown for some stages. (After Schuster, 1966.)

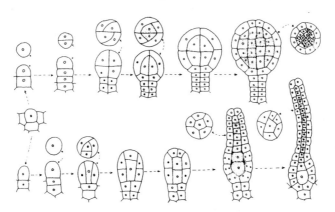

lobes protect several archegonia in *Fossombronia,* while in *Pellia* and other genera a sleeve of protective tissue, usually termed a perigynium or involucre, forms a flap of tissue around the upper surface pocket of the thallus (Figs. 14-4*E,* 14-5*F,* and 14-6*E*). The archegonia are on a reduced branch in *Aneura* and *Riccardia,* but after fertilization a distinct sleeve forms an extensive sheath constituted of both calyptra and thallus tissue, which encloses the elongating embryo; a similar situation exists in *Metzgeria* (Fig. 14-4*C,* 14-5*B,* and 14-6*D*).

SPOROPHYTES

Sporophytes usually possess a massive seta many cells in circumference and thickness. As in most other hepatics, the seta elongates rapidly and pushes the sporangium well above the thallus when it is mature. The seta occasionally contains some chlorophyll in earlier stages of elongation, but is generally colorless when mature. The sporangium varies from spherical to cylindric (Fig. 14-10). The sporangial jacket is two to five cells thick, and the epidermal cells often show semiannular darkly pigmented thickenings; sometimes such thickenings are lacking (Fig. 14-11). Elaters are usually elongate and tapered, and spirals vary from one to three. In some species of *Fossombronia* the elaters are short and blunt and possess annular thickenings.

The sporangium usually opens by longitudinal slits. In *Moerckia* there is a single slit, and when the sporangium dries out and shrinks in diameter, the sporangium gapes open to expose the elaters and spores, which are rapidly dispersed (Fig. 14-10*F,G*). Generally there are four longitudinal slits, as in the Jungermanniales (Fig. 14-10*A,B,D,E*). The Metzgeriales, however, sometimes have a small cylinder of sterile tissue within the sporangium to which some elaters are attached; this is the elaterophore. In several genera, including *Riccardia,* the elaterophore hangs down inside the apex of the sporangium, and thus, when the sporangium ruptures by four slits, each of the divisions is terminated by a tuft of elaters attached to the quartered ela-

FIGURE 14-10
Dehiscence of sporangia in
Metzgeriales. (*A, B*) *Pellia
neesiana* with longitudinal
lines and internal elaterophore
(×20). (*C*) *Fossombronia
longiseta* with irregular rupture of the sporangium wall
(×20). (*D, E*) *Aneura pinguis*
showing the elaters at the tips
of the divisions, representing
the splitting of an apical elaterophore (×15). (*F, G*)
Moerckia blyttii, dehiscence
by two longitudinal lines
(×15).

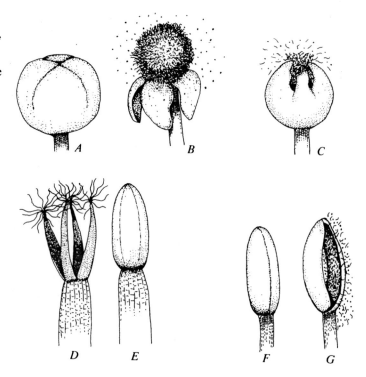

terophore (Fig. 14-10*D*). The elaterophore is at the junction of the seta and
sporangium in *Pellia;* thus when the sporangium opens it exposes free elaters
plus a woolly mass in the center (Fig. 14-10*B*). The sporangium ruptures irregularly in *Fossombronia* (Fig. 14-8*C*). Elaters show considerable variety in
the order (Fig. 14-12).

HABITAT DIVERSITY

Although most species grow on moist earth banks, several (including many
species of *Metzgeria*), are commonly epiphytic. Some species of *Metzgeria* are
tolerant of desiccation and occur also on rock; a few grow on evergreen leaves
of lignified plants. Some species of *Riccardia* grow in extremely wet sites, but
none are strict aquatics.

Cryptothallus is unique among the bryophytes, since the thallus is a subterranean saprophyte with an abundance of endogenous fungi. The thallus lacks
chlorophyll and grows buried under several centimeters of raw humus or peat,
usually in somewhat shaded sites, especially in moss carpets.

Blasia and *Pellia* are especially frequent on mineral soil, and they stabilize
earth banks near streams and trails in shaded woodland. Some species of *Riccardia* occur on rotten wood, especially when the wood is sufficiently decom-

FIGURE 14-11
Ornamentation and anatomy of jacket cells in sporangia of Metzgeriales. (*A, K*) *Pellia endiviaefolia.* (*A*) Surface view of epidermal cells (×40). (*B, C*) *Calycularia crispula* (×68). (*B*) Surface view of epidermal cells. (*C*) Transverse section of a portion of the jacket, (*D, E*) *Hattorianthus erimonus* (×35). (*D*) Transverse section of a portion of the jacket. (*E*) Surface view of epidermal cells. (*F, G*) *Cavicularia densa* (×45). (*F*) Transverse section of a portion of the jacket. (*G*) Surface view of epidermal cells. (*H, I*) *Makinoa crispata* (×45). (*H*) Transverse section of a portion of the jacket. (*I*) Surface view of epidermal cells. (*J*) *Fossombronia cristula* (×30), inner cells of jacket. (*K*) Transverse section of a portion of the jacket of *Pellia endiviaefolia* (×40). (After Inoue, 1976.)

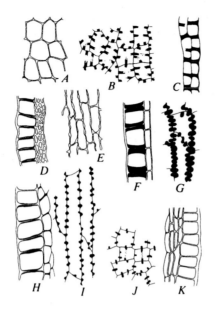

posed to retain moisture for long periods. Other species can remain submerged for extended periods. *Aneura pinguis* occurs in wetlands and on lake and stream margins, and tolerates long periods of submergence.

RELATIONSHIPS AND EVOLUTIONARY TRENDS

Evolutionary trends are extremely difficult to interpret in this order. The following appear to be some trends:

1. Simple irregular branching appears to be more generalized than pseudodichotomous or pinnate branching.
2. Exposed sex organs rather than those protected in a specialized structure or embedded within the thallus are the more generalized.
3. Numerous oil bodies in a chlorophyll-containing cell appears to be more generalized than the restriction of a single oil body to cells where chloroplasts are lacking. The absence of complex oil bodies may be a specialized feature.
4. Exogenous gemmae appear more generalized than endogenous gemmae.

FIGURE 14-12
Elaters of Metzgeriales. *(A, B) Metzgeria* sp. (×560), *(C) Pellia endivaefolia* (×525). *(D, E) Fossombronia* sp. (×490). *(F) Makinoa crispata* (×450). *(G) Calycularia crispa* (×300). (After Inoue, 1976.)

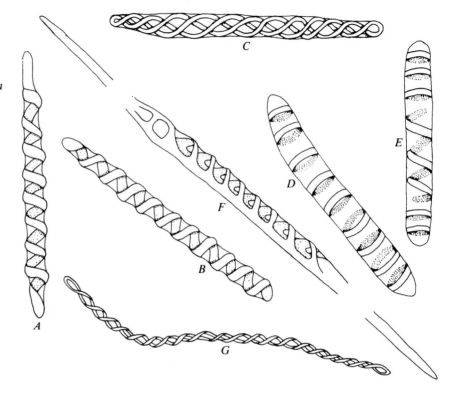

5. A distinctive stem with leaves, lobes, or "wings" appears to be less generalized than a thallus lacking a stem.
6. An elongate seta appears to be more generalized than a short one.
7. The saprophytic condition is specialized.

The closest relationships of the Metzgeriales appear to be with the Jungermanniales. They share most sporophytic features, even to the lobed spore mother cells, and cytological details are similar. The anacrogynous condition and the generally thallose gametophore separate the order as an independent evolutionary line. Even the thallose members of the Jungermanniales are decidely different from the Metzgeriales and appear to represent a series of specializations within that order rather than a line that could have been progenitor to the Metzgeriales.

FURTHER READING

Benson-Evans, K. 1960. Some aspects of spore formation and germination in *Cryptothallus mirabilis*. *Trans. Br. Bryol. Soc.* **3**:729–735.

Campbell, D. H. 1913. The morphology and systematic position in *Calycularia radiculosa* Steph. Stanford U. Publ. Dudley Mem. Vol., pp. 43–61, 12 figs.

—————— 1916. The morphology and systematic position of *Podomitrium. Am J. Bot.* **2**:199–210.

—————— and F. Williams. 1914. A morphological study of some members of the genus *Pallavicinia.* Leland Stanford Jr. Univ. Publ. Ser., pp. 1–144, 23 figs.

Cavers, F. 1910. The interrelationships of Bryophyta III. Anacrogynous Jungermanniales. *New Phytol.* **9**:197–234.

Clapp, G. L. 1912. The life history of *Aneura pinguis. Bot. Gaz.* **54**:177–192, plates IX–XII.

Clee, D. A. 1939. The morphology and anatomy of *Pellia epiphylla* considered in relation to the mechanism of absorbtion and conduction of water. *Ann. Bot.* **3**:106–111.

Evans, A. W. 1910. Vegetative reproduction in *Metzgeria. Ann. Bot.* **24**:271–303.

—————— 1921. The genus *Riccardia* in Chile. *Trans. Conn. Acad. Arts Sci.* **25**:93–209, figs. 1–13.

—————— 1925. A taxonomic study of *Hymenophytum. Bull. Torrey Bot. Club* **52**:491–506.

—————— 1925. The lobate species of *Symphyogyna. Trans. Conn. Acad. Arts Sci.* **27**:1–50, figs. 1–13.

—————— 1927. A further study of the American species of *Symphyogyna. Trans. Conn. Acad. Arts Sci.* **28**:295–354, figs. 1–12.

Farmer. J. B. 1894. Studies on Hepaticae: On *Pallavicina decipiens* Mitten. *Ann. Bot.* **8**:35–52.

Fulford, M. 1975. Young stages of some thalloid Hepaticae: a résumé of Anacrygynae. *Phytomorphology* **25**:176–193.

Greenwood, H. 1911. Some stages in the development of *Pellia epiphylla. Bryologist* **14**:59–70, 77–81, 93–100.

Grun, C. 1916. Monographische Studien an *Treubia insignis* Goebel. *Flora* **106**:331–392.

Haupt, A. W. 1918. A morphological study of *Pallavicinia lyellii. Bot. Gaz.* **66**:524–531.

—————— 1929. Studies in California Hepaticae II *Fossombronia longiseta. Bot. Gaz.* **88**:103–109.

Herzon, T. 1941. *Allisonia* Herzog, eine neue Gattung der Haplolaenaceae. *Hedwigia* **80**:77–83.

Hodgson, E. A. 1964. New Zealand Hepaticae XV. A new monotypic family of thalloid Hepaticae, Phyllothalliaceae Hodgson, *fam. nov. Trans. N. A. Roy. Soc. Bot.* **2**:247–250.

Humphrey, H. B. 1906. The development of *Fossombronia longiseta. Aust. Ann. Bot.* **20**:83–108.

Hutchinson, A. H. 1915. Gametophyte of *Pellia epiphylla. Bot. Gaz.* **60**:134–143.

Kuwahara, Y. 1966. The family Metzgeriaceae in North and South East Asia, Pacific Oceania, Australia, and New Zealand. *Rev. Bryol. Lichénol.* **34**:191–239.

Leitgeb, H. 1877. Untersuchungen über die Lebermoose III. Die frondosen Jungermannieen.

Malmborg, S. 1933. *Cryptothallus* nov gen., ein saprophytischen Lebermoos. *Ann. Bryol.* **6**:122–123.

McCormick, F. A. 1914. A study of *Symphyogyna aspera. Bot. Gaz.* **58**:401–418.

Proskauer, J. 1965. On the liverwort *Phyllothallia. Phytomorphology* **15**:375–379.

———— 1971. Notes on Hepaticae V Stalked gemmae in *Riccardia. Bryologist* **74**:1–7.

Renzaglia, K. S. 1982. A comparative developmental investigation of the gametophyte generation in the Metzgeriales (Hepatophyta). *Bryophyt. Bibl.* **24**:253 pp.

Rodgers, G. A. 1978. The effects of some external factors on nitrogenase activity in the free-living and endophytic *Nostoc* of the liverwort *Blasia pusilla. Physiol. Plant.* **44**:407–411.

Schuster, R. M. 1964. Studies on Antipodal Hepaticae IV. Metzgeriales. *J. Hattori Bot. Lab.* **27**:183–216.

———— 1966. Studies on Antipodal Hepaticae IX. Phyllothalliaceae *Trans. Br. Bryol. Soc.* **5**:283–288.

———— and G. A. M. Scott. 1969. A study of the family Treubiaceae (Hepaticae, Metzgeriales). *J. Hattori Bot. Lab.* **32**:219–268.

Showalter, A. M. 1923. Studies in the morphology of *Riccardia pinguis. Am. J. Bot.* **10**:148–166.

Smith, J. L. 1966. The liverworts *Pallavicinia* and *Symphyogyna* and their conducting system. *U. of Calif. Publ. Botany* **39**:1–46, 18 plates.

Stewart, W. D. P., and G. A. Rodgers. 1977. The cyanophyte-hepatic symbiosis II. Nitrogen fixation and the interchange of nitrogen and carbon. *New Phytol.* **78**:459–471.

15 The Bottle Hepatics — Order Sphaerocarpales

This order contains the Sphaerocarpaceae and Riellaceae, as well as the extinct family Naiaditaceae. Two genera, *Sphaerocarpos* (with 12 species) and *Geothallus* (with one species), constitute the Sphaerocarpaceae, while Riellaceae has only *Riella,* with 19 species, and Naiaditaceae has only *Naiadita* with a single fossil species.

Sphaerocarpos (Fig. 15-1) shows a wide distribution in both the Northern and Southern Hemispheres but is rarely common. *Riella* (Fig. 15-2), although exhibiting a similar geographic range, is restricted to a few localities. Both genera are best represented in climatic regions where the summer is dry and the winter is moderately wet and mild. *Geothallus* (Fig. 15-3) is extremely local in

FIGURE 15-1
Sphaerocarpales:
Sphaerocarpos texanus.
Antheridium-producing gametophores with flask-shaped "bottles" in center, surrounded by archegonium-producing gametophores (×15).

3 mm

FIGURE 15-2
Sphaerocarpales: *Riella affinis*. *(A)* Antheridium-producing gametophore (×15). *(B)* Detail of marginal chambers with antheridia (×150). *(C)* Sporophyte-bearing gametophore (×15). *(D)* Outline of transverse section of thallus, showing two gemmae borne laterally (×40). *(E)* Single gemma, surface view; the stalk is the dark area in the lower lobe (×70). *(F)* Germinating spore with short protonema, a single rhizoid and the sporeling (×900). *(A, B, after Wigglesworth, 1937; D, E, after Studhalter and Cox, 1940.)*

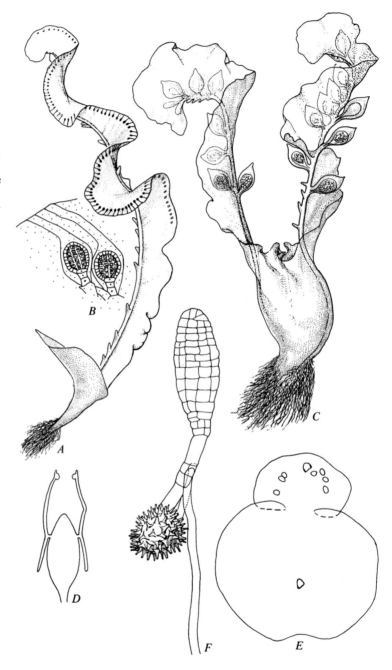

southern California, and *Naiadita* is known from a Triassic deposit in Great Britain (Fig. 15-4).

Sphaerocarpos and Geothallus occur on fine-textured mineral soil, especially in somewhat shaded sites. *Riella* is a submerged aquatic, usually on the

FIGURE 15-3
Sphaerocarpales: *Geothallus tuberosus.* (*A*) Antheridium-producing gametophore with "bottles" (×20). (*B*) Antheridium-producing gametophore, longitudinal view showing one "bottle" with a sporangium and another with an archegonium; the darkened area in the thallus represents a "tuber" (×20). (*C*) Longitudinal view of a sporophyte showing the unistratose sporangium jacket, the spores, small nurse cells, and the expanded foot (×54). (*A, B,* after Campbell, 1918.)

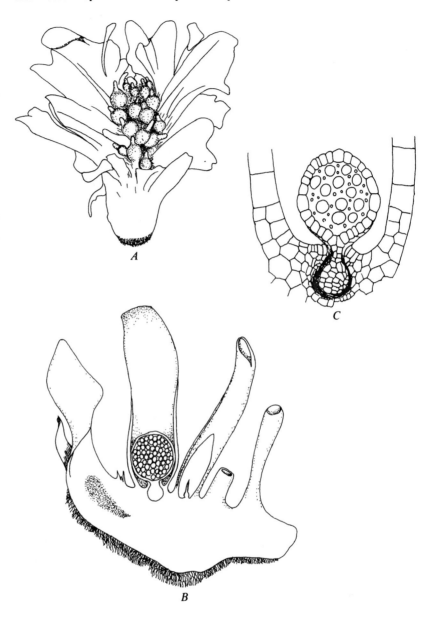

bottom muds in calcium-rich waters of shallow quiet pools. *Naiadita* presumably was also an aquatic and may have been tolerant of some salinity, as are some modern species of *Riella.*

Features that characterize the order include:

1. Unistratose involucres that surround each sex organ, (Figs. 15-1, 15-2, and 15-3)
2. A sporophyte with no seta (Fig. 15-3*B,C*)

FIGURE 15-4
Sphaerocarpales: *Naiadita lanceolata,* a fossil bryophte. *(A)* Reconstruction of a gametophore, bearing gemmae on lateral branch on left, and with lateral archengonia on stem (×7). *(B)* A single sporophyte on its branch (×7). *(C)* a longitudinal section through a sporangium (×7). *(D)* Reconstruction of a vegetative shoot (×7). *(E, F)* Surface views of leaves (×20). *(G)* Germinating gemma (×45). (After Harris, 1938.)

3. The sporangium with a unistratose unornamented jacket that opens irregularly by decomposition (Fig. 15-3C)
4. the absence of elaters in the sporangium but the presence of unicellular "nurse cells" (Fig. 15-3C)

SPORES AND SPORE GERMINATION

The spore wall ornamentation in *Sphaerocarpos* and *Riella* is important in distinguishing species, especially since features of the gametophore are similar among species or highly variable within a single species. Spores appear to possess a long period of viability, especially in *Riella*. In all genera the germinating spore ruptures the spore coat with a germ tube Figs. 15-2F, and 15-5A,B). This germ tube elongates considerably before undergoing any cell divisions. In *Riella* the tip of the germ tube first undergoes several transverse divisions to form a series of short broad cells, after which longitudinal divisions

FIGURE 15-5
Sphaerocarpales. *(A)* Germination in a tetrad of spores of *Sphaerocarpos stipitatus* (×100). *(B)* Spore germination in *Riella* sp. (×40). *(C, D)* *Sphaerocarpos donnellii* chromosomes of gametophore (×5500); *(C)* from an archegonium-bearing plant, the large chromosome is the X chromosome; *(D)* from an antheridium-bearing plant, the small chromosome is the Y chromosome. *(A,* after Proskauer, 1954; *B,* after Wigglesworth, 1937; *C, D,* after Allen, 1919.)

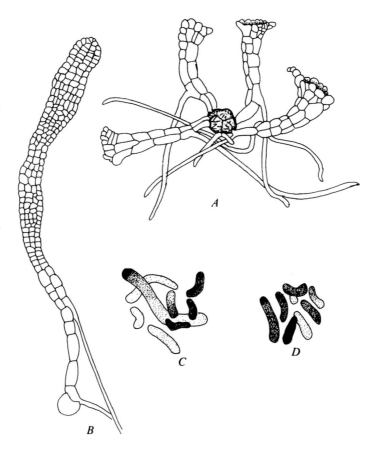

of succeeding cells produce a small flattened thallus, termed the sporeling (Figs. 15-2*F*, and 15-5*A*,*B*). A rhizoid is also formed in early cell divisions in the germ tube. In *Sphaerocarpos* and *Geothallus* the sporeling is initiated at the tip of the elongate unicellular germ tube. A cylindric multiseriate structure forms and produces rhizoids at its base, and an apical cell is initiated at the apex; this produces the gametophore.

The gametophore varies from a thallose to a leafy condition. All cells are thin-walled and chlorophyllose, with the exception of "scales" in the genus *Riella*, which have simple spherical oil bodies in scattered cells. Rhizoids are unicellular and smooth-walled. Most species are dioicous. The three families show considerable diversity in structure of the gametophore, but uniformity in sporophyte structure.

THE GAMETOPHORE

The gametophore of *Riella* is unlike that of any other bryophyte (Fig. 15-2). The mature plant consists of an erect cylindric stem with a single unistratose wing on one side or forms a helical spiral around the stem as it grows. The mature gametophore can reach lengths of 2–3 cm, but is generally smaller. Rhizoids are usually confined to the lower part of the stem. On the side of the stem opposite the wing are the scattered bottlelike involucres that contain the archegonia. These "bottles" are subspherical or tubular, and sometimes are fluted with parallel longitudinal wings. Frequently, on this same surface of the stem, are scattered unistratose scales and, sometimes, gemmae (Fig. 15-2*D*,*E*). Scattered cells of the scales contain simple spherical oil bodies. The gemmae, borne in two rows along the stem, are attached by a short stalk affixed to the equator of the gemma that is flattened parallel to the stem. Each gemma is unistratose and multicellular and consists of two rounded lobes with a constriction at the equator. Gemmae are sometimes abundant near the growing tip of the gametophore. Antheridia are in unistratose pockets on the outer edge of the thallus wing (Fig. 15-2*A*,*B*). Sometimes both antheridia and archegonia are on the same gametophore.

The thalli are always dioicous in *Sphaerocarpos* and *Geothallus* (Figs. 15-1 and 15-3). Antheridia are in flask-shaped or tubular "bottles" on the dorsal surface of the thallus, which is usually less than 2 mm in diameter. The thallus is thickened and multistratose in the center and has unistratose irregular lobes extending outward from this thickened portion. Rhizoids are confined to the undersurface of the thickened portion. Archegonia are in subspherical to cylindric "bottles" in the same relative position as the antheridia, but the thallus that bears them sometimes exceeds 8 mm in diameter. The form is similar to that of the antheridium-bearing thallus. No gemmae are produced in nature, but in culture, cells of the thallus surface sometimes bud off small, irregular masses of cells that can serve as vegetative diaspores. The wings of the thallus of *Geothallus* sometimes produce small, shallow, dorsal pockets. In this genus the thickened portion of the thallus can produce "tubers" (Fig. 15-3*B*). These consist of an apical cell surrounded by oil- and starch-rich cells enclosed in an envelope of dead cells. The tuber can tolerate drying and survive the unfavorable seasons. Tubers tend to appear later in the season of active growth of the gametophore.

THE SPOROPHYTE

The sporophyte in all genera consists of a spherical sporangium that is attached to the gametophore by a small bulging foot (Fig. 15-3*B,C*). The sporophyte is completely enclosed by the gametophytic "bottle." The sporangium jacket is unistratose and made up of parenchyma cells and is frequently sheathed by the calyptra, even when mature. Within the sporangium are spore tetrads (often enclosed in a single membrane and shed as a tetrad). There are also nurse cells, produced by mitosis from the sporogenous cell mass, while the spores are produced by meiosis. Spores are shed when the sporangium jacket disintegrates. The sporophyte appears to be nutritionally independent of the gametophore early in its growth.

NAIADITA

Naiadita is undoubtedly the best understood of the fossil bryophytes (Fig. 15-4). The elegant research of T. M. Harris (1938) has yielded interpretations of the beautifully preserved material of Triassic age. Most aspects of the life history of the organism are presented, except for the antheridial plants, which are unknown.

The gametophore of *Naiadita* was leafy, with unistratose elliptic to lanceolate leaves transversely attached and spirally arranged on the erect terete stem. Rhizoids were confined to the stem base. All cells of the gametophore were thin-walled. The whole gametophore was 1–3 cm high. Archegonia were lateral, and apparently supplanted the position of a leaf (Fig. 15-4*A*). The gametophore sometimes produced also gemmiferous branches terminated by a rosette of leaves forming a gemma cup enclosing several short elliptic stalked gemmae. Germinating gemmae produced an elongate rhizoid (Fig 15-4*G*).

The sporangium was spherical, with a unistratose jacket of thin-walled cells; it contained spores and no elaters. The enveloping structures resemble a stem-calyptra, and the stalk is similar to the pseudopodium in the Sphagnidae. Such a stalk is occasionally present in *Sphaerocarpos*. In *Naiadita*, however, the enlarging sporangium was enclosed by both expanding gametophytic tissue of the bases of the lobes of the involucre plus the calyptra itself, and the mature sporophyte was enclosed in an equatorially winged structure unlike anything in extant bryophytes (Fig. 15-4*B,C*).

The fact that *Naiadita* was leafy, supports the possibility that this condition is generalized while the thallose condition is specialized. Within the order it appears that both *Sphaerocarpos* and *Geothallus* are somewhat more specialized, but *Riella* shows the greatest specialization, both in morphology and habitat.

SPHAEROCARPOS AS A GENETIC TOOL

C. E. Allen devoted considerable research to the genetics and cytology of *Sphaerocarpos*. He discovered that in the chromosome complement there was one smaller chromosome in the nuclei of the antheridial plant while nuclei of archegonial plants showed one chromosome larger than the others (Fig. 15-5C,D). This led him to the hypothesis that these could be interpreted as sex chromosomes, the first discovered in the plant kingdom. Since two of the spores in a tetrad germinated to form antheridial plants while the other two produced archegonial plants, the genus served as a useful tool in his experiments. Besides this, the spore-ornamentation patterns, general morphology of "bottle" shape, and morphology of the thallus served as useful genetic markers to determine inheritance in the species.

His attempts at hybridization were successful in producing spores, but the spores were not viable. His experiments involving intraspecific hybridization of variant morphologic forms led him to the conclusion that simple mendelian inheritance was involved. In spore ornamentation patterns in hybridization between species, however, the ornamentation was always controlled by the maternal parent, and somatic control was indicated for the first time in organisms.

RELATIONSHIPS

The interpretations of relationships of this order are controversial. It is placed in the Marchantiae based on the unlobed spore mother cell, the nature of oil bodies (when present), the unistratose sporangial wall, and the absence of a seta. It shares with a few Marchantiales the absence of elaters and the presence of nurse cells. In general appearance, some gametophores superficially resemble Metzgeriales, including *Fossombronia*. The spirally arranged leaves of *Naiadita* are unlike any other hepatic, but the sporophyte is identical to that of the Sphaerocarpales, and the restriction of rhizoids, general nature of the stem, and position of archegonia make it most closely resemble *Riella* of the Sphaerocarpales.

FURTHER READING

Allen. C. E. 1919. The basis of sex inheritance in *Sphaerocarpos*. *Proc. Am. Phil. Soc.* **58**:289–316.
——— 1924. Inheritance by tetrad sibs in *Sphaerocarpos*. *Proc. Am. Phil. Soc.* **63**:222–235.

———— 1924–1930. Gametophytic inheritance in *Sphaerocarpos*. *Genetics* **9**:530–587. Ibid. **10**:1–16. Ibid. **11**:83–87. Ibid. **15**:150–188.

———— 1925. The inheritance of a pair of sporophytic characters in *Sphaerocarpos*. Genetics **10**:72–79.

———— 1930. Inheritance in a hepatic. *Science* **71**:197–204.

———— 1933. The occurrence of polyploidy in *Sphaerocarpos*. *Am. J. Bot.* **22**:664–680.

———— 1937. Fertility and compatibility in *Sphaerocarpos*. *Cytologia, Fujii Jubilee Vol.,* 494–501.

Campbell, D. H. 1896. The development of *Geothallus tuberosus* Campbell. *Ann. Bot.* **10**:489–510.

Doyle, W. T. 1962. The morphology and affinities of the liverwort *Geothallus*. *Univ. of Calif. Publ. Bot.* **33**:185–267.

Harris, T. M. 1938. *The British Rhaetic Flora*. London: Brit. Museum Nat. Hist.

———— 1939. *Naiadita*, a fossil bryophyte with reproductive organs. *Ann Bryol.* **12**:57–70.

Hassel de Menendez, G. G. 1959. Sobre el hallazgo del genero *Riella* en Sudamerica. *Rev. Bryol. Lichénol.* **28**:297–299.

Haynes, C. C. 1910. *Sphaerocarpos hians* sp. nov., with a revision of the genus and illustrations of the species. *Bull. Torrey Bot. Club* **37**:215–230, p. 25–32.

Howe, M. A., and L. M. Underwood. 1903. The genus *Riella*, with descriptions of new species from North America and the Canary Islands. *Bull. Torrey Bot. Club* **30**:214–224.

Jelenc, F. 1957. Les bryophytes nord-africains. IV Le genre *Riella* en Afrique mediterranéenne et au Sahara. *Rev. Bryol. Lichénol.* **26**:20–50.

Persson. H. 1960. The first find of *Riella* in Egypt and some words about the distribution of the genus in the world. *Rev. Bryol. Lichénol.* **29**:1–9.

Proskauer, J. 1954. On *Sphaerocarpos stipitatus* and the genus *Sphaerocarpos*. *J. Linn. Soc. (London) Bot.* **55**:143–157.

———— 1955. The Sphaerocarpales of South Africa. *J. S. Afr. Bot.* **21**:63–75.

Rickett, H. W. 1920. The development of the thallus of *Sphaerocarpos donnellii* Aust. *Am. J. Bot.* **7**:182–195.

Studhalter, R. A. 1931. Germination of spores and development of juvenile thallus of *Riella americana*. *Bot. Gaz.* **92**:172–191.

———— 1932. The elusive ruffle plant, *Riella*. *Sci. Monthly* **35**:303–311.

———— 1938. Independence of sporophyte in *Riella* and *Sphaerocarpos*. *Ann. Bryol.* **22**:153–154.

———— 1940. The gemma of *Riella americana*. *Bryologist* **43**:141–157.

———— 1942. The foot of *Riella americana* and its relation to nutrition of the sporophyte. *Bot. Gas.* **103**:633–650.

Studhalter, R. H., and M. E. Cox. 1941a. The lateral leaf scale of *Riella americana*. *Bryologist* **44**:19–27.

———— 1941b. The ventral scale of *Riella americana*. *Bryologist* **44**:29–40.

——— 1941–1942. The gemmaling of *Riella americana. Bryologist* **44**:77–93. Ibid. **45**:49–62.

Thomson, R. H. 1941. The morphology of *Riella affinis* I. *Am. J. Bot.* **28**:845–855.

Wigglesworth, G. 1937. South African species of *Riella,* including an account of the developmental stages of three of the species. *Linn. Soc. J. (Bot.)* **51**:309–332.

16

The Giant Thallose Hepatics — Order Monocleales

The order Monocleales contains a single family, Monocleaceae, with the genus *Monoclea*. There are two species, *M. forsteri* of New Zealand and South America and *M. Gottschei,* also of South America, but extending northward to Jamaica (Fig. 16-1).

THE GAMETOPHYTE

The spores are small, usually around 20 μm in diameter (Fig. 17-4A), and germinate soon after they are shed. The germinating spore forms a multicellular mass before a rhizoid is produced. From this multicellular mass an apical cell is differentiated, and it initiates formation of the thallus.

The gametophore is a remarkably large dark-green to olive-green succulent thallus up to 20 cm long and 5 cm wide. The plants commonly grow on mineral soil in damp to wet forest sites. Sometimes plants form floating mats in shallow pools. The thalli remain vegetative in deeply shaded sites, but in better illuminated areas produce sexual structures. The thalli are roughly dichotomous and branch repeatedly to form extensive colonies flattened against the substratum. Rhizoids are confined to the ventral surface and are most abundant near the thickened central portion of the thallus. Some rhizoids grow downward into the substratum, while others are oriented parallel to the length of the thallus and point backward from the thallus apex (Figs. 16-1C,D and 16-2C,D). Rhizoids are smooth and thin-walled or thick-walled, occasionally occur with rudimentary pegs, or may be sinuose (Fig. 16-2E).

The upper surface of the thallus is smooth and is composed of thin-walled cells (Fig. 16-1A,C). The uppermost cells contain numerous chloroplasts, while deeper in the thallus plastids are few and starch grains more numerous (Fig. 16-1C). Cells near the dorsal surface sometimes contain crystals of calcium oxalate. Scattered cells in the thallus contain single complex brown oil bodies; these cells lack chloroplasts, as in many of the Marchantiales. Within

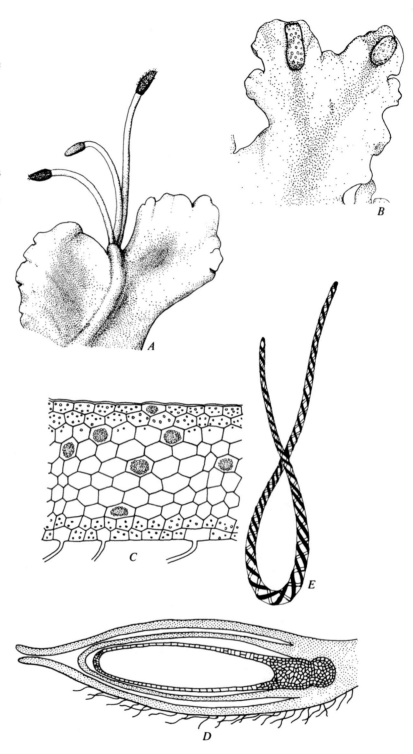

FIGURE 16-1

Monocleales: *Monoclea gottschei*. *(A)* View of sporophyte-producing gametophore (×1.5). *(B)* View of antheridium-producing gametophore showing antheridial pads near lobe apices (×1.5). *(C)* Transverse section through thallus, showing scattered oil bodies; chloroplasts are abundant in both surface cells and the cells near the undersurface (diagrammatic) (×65). *(D)* Longitudinal section through sporophyte within pouchlike chamber (diagrammatic) (×45). *(E)* A single elater (×800). *(D, E,* after Johnson, 1904.)

FIGURE 16-2
Monocleales: *Monoclea. (A)*
Longitudinal section of an
antheridium (×60). *(B)* Longitudinal section through jacket
wall of *M. forsteri,* showing
ornamentation, and spores
with fragments of elaters
(×600). *(C)* Longitudinal section through an antheridial
pad (diagrammatic); note the
youngest antheridia near the
apex of the lobe (×15). *(D)*
Longitudinal section through
pouchlike chamber of ar-
chegonium-bearing plant,
showing archegonia and mu-
cillage cells (diagrammatic
×15). *(E)* Segments of two
rhizoids (×200). *(F)* Chromosomes of a thallus cell
(×3900). *(A, C, D, G,* after
Johnson, 1904; E, F, after
Proskauer, 1951.)

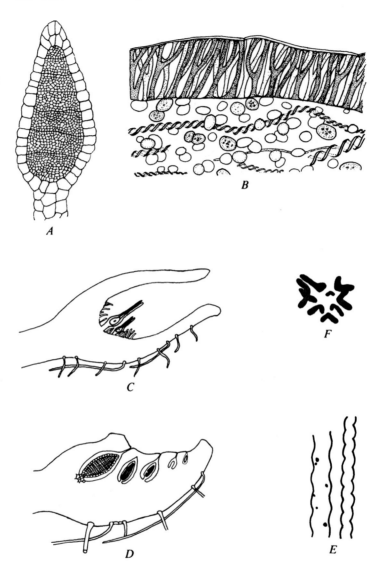

the thallus many cells have conspicuous slitlike pits in the walls. These slits
provide cytoplasmic interconnection among the cells and probably facilitate
water transport.

Thalli are dioicous. Antheridia are embedded in semicircular to elongate
cushionlike pads located behind the growing points of the thallus lobes (Fig.
16-1*B*). Each pad contains 25–50 or more antheridia, one in each chamber in
the pad. Antheridia are ovoid and pointed and attach by a short stalk to the
base of the chamber (Fig. 16-2*A,D*). Each antheridial chamber also contains
mucilage cells. The antheridial chambers open by a pore to the dorsal surface.

Within a single antheridial pad antheridia are often at various stages of maturity, those nearest the lobe apex being youngest (Fig. 16-2D).

Archegonia are within a tubular pouchlike cavity located behind the growing point of the lobes (Fig. 16-2C). This cavity is horizontal, and the flap enclosing it is fused to the thallus along its length with an opening near the lobe apex (Fig. 16-1A,D). At the distal end, opposite the cavity opening, within the cavity are many archegonia mingled with mucilage hairs. The archegonia are often in various stages of development, with the oldest farthest from the opening of the cavity (Fig. 16-2C). Indeed, within a chamber, young sporophytes are often developing while young archegonia are forming nearer the apex of the cavity. The neck of the archegonium is remarkably long, a feature usually considered generalized.

THE SPOROPHYTE

Early development of the zygote of *M. forsteri* is free-nuclear (i.e., successive nuclear divisions produce a number of nuclei before cell walls are formed), and cells begin to form at about the 26 nucleate stage, after which the foot and the rest of the sporophyte are differentiated.

Several sporophytes often emerge from a single involucral cavity (Fig. 16-1A). The mature sporophyte has a massive seta that sometimes reaches lengths of 2–4 cm. In *Monoclea,* as in many metzgerialean genera, a very long, pale calyptra enlarges to accommodate the growing sporophyte; it sometimes reaches a length of up to 1 cm before the sporophyte emerges through the tip of the calyptra and the opening of the archegonial pouch.

The sporangium is cylindric, dark-brown, and 6–8 mm long (Fig. 16-1A). It is of only slightly greater diameter than the seta (ca. 1.5 mm). The sporangium wall is unistratose and has forked transverse thickenings on all walls except the outer face, a feature unique to this genus (Fig. 16-2B). Spores are formed from an unlobed spore mother cell. Within the sporangium are longitudinally arranged, elongate, tapered elaters, each bearing two to three helical thickenings (Fig. 16-2B). The sporangium opens by a single longitudinal slit that exposes the contents and ejects the spores as the elaters uncoil violently and spring out.

RELATIONSHIPS

The relationships of the order Monocleales seem closest to the Marchantiales, with which it shares the orientation of rhizoids, specialized antheridial pads with embedded antheridia, large complex oil bodies (one per cell) in scattered cells of the thallus, and unlobed spore mother cells. It also produces the same general flavonoid type.

The elongate calyptra, sleevelike archegonial pouch, elongate elaters, massive elongate seta, sporangium dehiscence, ornamented jacket cells of the sporangium, and parenchymatous cells of the thallus are similar to the Metzgeriales, but these features may be coincidental. It seems best to treat this order as an independent evolutionary line closely allied to the Marchantiales.

FURTHER READING

Campbell, D. H. 1898. The systematic position of the genus *Monoclea*. *Bot. Gaz.* **25**:272–274.

Campbell, E. O. 1954. The structure and development of *Monoclea forsteri* Hook. *Trans. Roy. Soc. N. Z.* **82**:237–248.

——— 1963. New Zealand's largest liverwort, *Monoclea forsteri*. *Tuatara* **3**:16–19.

——— 1984. Looking at *Monoclea* again. *J. Hattori Bot. Lab.* **55**:315–319.

Cavers, F. 1904. On the structure and development of *Monoclea forsteri*. *Rev. Bryol.* **21**:69–80.

Gottsche, C. M. 1858. Über der Gattung *Monoclea*. *Bot. Z.* **16**:281–287, 289–292.

Hassel de Menendez, G. G. 1957. *Monoclea forsteri* en Argentina. *Bol. Soc. Argentina Bot.* **6**:248–250.

Johnson, D. S. 1904. The development and relationship of *Monoclea*. *Bot. Gaz.* **38**:185–205.

Markham, K. R. 1972. A novel flavone-polysaccharide compound from *Monoclea forsteri* (Monocleales). *Phytochemistry* **11**:2047–2053.

Proskauer, J. 1951. Notes on the Hepaticae II:4. On *Monoclea*. *Bryologist* **54**:258–266.

Schiffner, V. 1913. Phylogenetische Studien über die Gattung *Monoclea*. *Oest. Bot. Z.* **63**:29–33, 75–81, 113–121, 154–159.

Schuster, R. M. 1977. The evolution and early diversification of the Hepaticae and Anthocerotae, in Frey, W., H. Hurka and F. Overwinkler (eds.), *Beiträge zur Biologie der niederen Pflanzen*, pp. 107–115. Stuttgart: G. Fischer.

17 The Chambered Hepatics — Order Marchantiales

The order Marchantiales contains the liverworts that have given the class its common name. The widespread genus *Marchantia* (Fig. 17-1*C*) is probably the bryophyte most likely to be recognized by elementary botany students. Unfortunately, it is one of the least typical of the division. The order contains approximately 12 families with 27 genera and approximately 450 species, nearly half of which belong to the genus *Riccia* (Fig. 17-2*A*). The order is distributed throughout the world from arctic and alpine environments to tropical latitudes. The diversity in form is shown in Fig. 17-1, 17-2, and 17-3. A diversity of habitats are occupied, although most species grow on mineral soil; several are common on humid rock faces, and a few are aquatic. Most species thrive in well-illuminated sites, but several, including *Cyathodium,* grow in deep shade.

The order is characterized as follows:

1. All are thallose.
2. The thallus is generally differentiated into an upper photosynthetic region with dorsal pores and air chambers and a solid lower region of storage cells.
3. Rhizoids with pegged thickenings on the inner wall face are of frequent occurrence.
4. Scales are often present in two or more rows on the undersurface of the thallus.
5. Single large, complex oil bodies are usually found in scattered cells of the thallus.
6. Rhizoids are oriented both perpendicular to and parallel with the thallus surface.
7. Each specialized archegonial branch normally bears several sporophytes.
8. Sporophytes possess a short seta, or there is no seta.
9. The sporangium jacket is unistratose.
10. Gemmae are rare; when present they are usually borne in a gemma cup.
11. The sex organs are sometimes on specialized branches of the thallus.

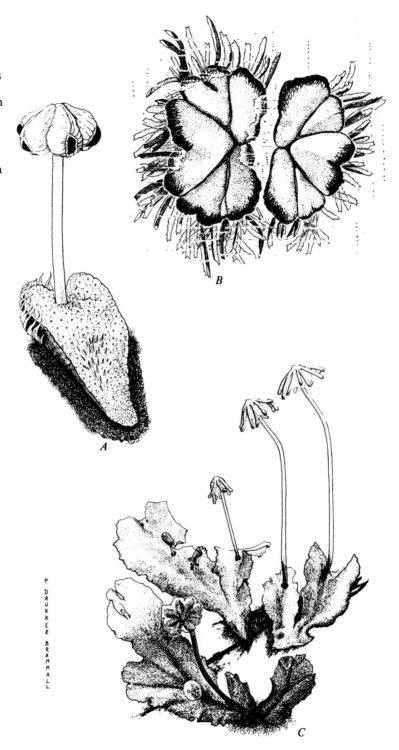

FIGURE 17-1
Diversity in Marchantiales.
(A) Sauteria alpina with carpocephalum; antheridia are
borne in a group on the thallus
(×4). *(B) Ricciocarpus
natans,* a species that floats on
the water surface (×4). *(C)
Marchantia polymorpha,*
showing gemma cup and an-
theridiophore on thallus below, and several carpocephala
on thalli above (×1).

P. DRUKKER BRAMMALL

FIGURE 17-2
Diversity in Marchantiales.
(A) Riccia sorocarpa on soil,
viewed from above (×10). *(B)
Asterella gracilis* showing
three carpocephala with
sporophytes (×6).

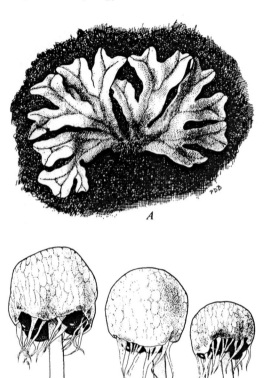

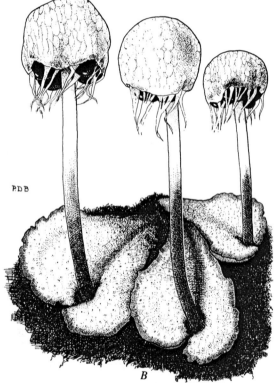

SPORES AND SPORE GERMINATION PATTERNS

The order shows considerable diversity in patterns of spore coat ornamentation and variation in spore size (Fig. 17-4*B–D*). In the species-rich genus *Riccia,* distinctive spore ornamentation patterns are useful in discriminating species. Spore size varies from an equatorial diameter of 10 μm to 140 μm, but

FIGURE 17-3
Diversity in Marchantiales.
(A) Mannia rupestris, with
carpocephalum (×10). *(B, C)*
Conocephalum conicum; (B)
showing carpocephala with
sporangia (×5). *(C)* showing
antheridial "pad" (×5). *(D)*
Targionia hypophylla, with
sporangia beneath lobes (×5).
(E) Preissia quadrata, with
carpocephala (×7).

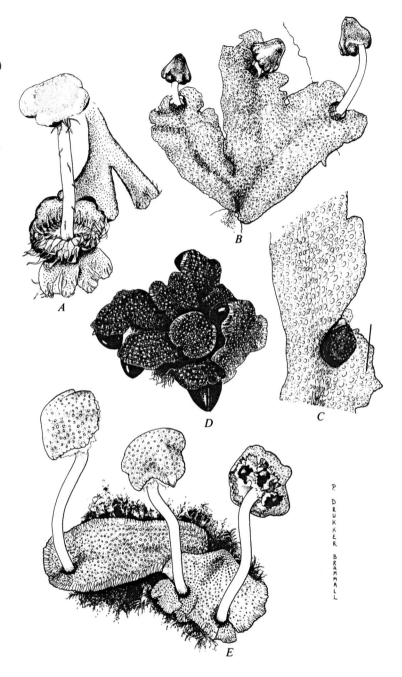

most spores are 40–60 μm in diameter. Most spores are unicellular when shed,
but in some species (e.g., *Conocephalum conicum*), they are multicellular.

Most spores have one hemispherical surface while three flattened faces form
a broad pyramid opposite this face. The hemispherical face is often elaborately

FIGURE 17-4
Spores of Monocleales and Marchantiales are revealed by scanning electron microscopy. *(A) Monoclea forsteri,* distal view (×1750). *(B) Targionia hypophylla,* distal view (×500). *(C) Marchantia berteroana,* distal view (×2000). *(D) Neohodgsonia mirabilis,* subequatorial view (×1500). (Provided by Jane Taylor.)

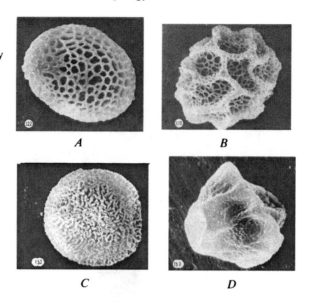

A

B

C

D

ornamented with reticulations, wings, papillae, spines, pustules, warts, or ridges. Scanning electron microscopy shows that complex ornamentation decorates both depressions and projections. Sometimes the surfaces are nearly smooth.

Spores usually form as a tetrad within a globose form; thus in most spores a triradiate scar is apparent. This scar is often important in determining where the germ tube emerges in a germinating spore. In *Conocephalum supradecompositum,* on the other hand, the tetrad is formed in a linear series.

Spore germination patterns in the order are treated in considerable detail by H. Inoue (1960). The basic pattern of spore germination (Fig. 17-5) and protonematal development is as follows. The contents of the spore differentiate a germ cell (Fig. 17-6) that produces a germ tube or filament. The apical cell of this filament divides to produce a quadrant of cells that, by further cell divisions, forms a plate of cells. This plate differentiates an apical cell that initiates the thallus (Fig. 17-7).

The manner in which the spore coat is ruptured, when and where the first rhizoid is formed, and the nature and persistence of the filament show various distinctive patterns (Fig. 17-5). In apolar spores the spore coat ruptures irregularly, while in cryptopolar spores the germ tube emerges either at the equator or through the apex of the triradiate face. In a polar spore, the germ tube emerges either through the triradiate face or through the hemispheric face opposite the triradiate scar. In most species the means of spore rupture and emergence of the germ tube is predictable.

The protonematal phase is also predictable and shows numerous patterns. Inoue recognized seven distinctly different patterns, shown in the accompanying figure (Fig. 17-6). Even the early stages of cell division following quadrant formation can be categorized into four distinctive patterns (Fig. 17-7). In

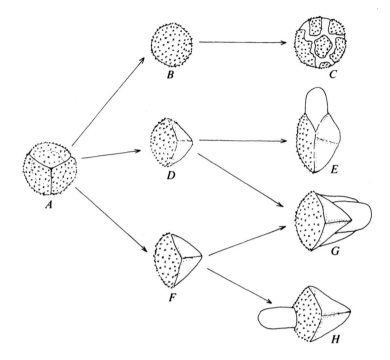

FIGURE 17-5
Schematic representation of the relation between spore morphology and spore-coat dehiscence in the Marchantiales. *(A)* Tetrad of spores. *(B, C)* Apolar spore with irregular dehiscence, *(D, E)* Cryptopolar spore with tangential dehiscence. *(F)* Polar spore. *(G)* Polar spore with proximal dehiscence. *(H)* Polar spore with distal dehiscence. *Note:* proximal face is shown facing right. (After Inoue, 1960.)

Conocephalum, where the spore is multicellular, there is neither a protonematal phase nor quadrant formation. The multicellular mass, upon rupturing the spore coat, produces rhizoids, differentiates an apical cell, and initiates the formation of a thallus. The germination patterns appear to show a certain consistency within genera or families, and when further detail is accumulated, it is possible that this will improve the understanding of interrelationships among genera in the order.

THALLUS STRUCTURE

In most genera, except *Riccia,* distinctive openings perforate the upper surface of the thallus. These pores fall into two categories. In simple pores, the epidermis bounding the opening is unistratose (Fig. 17-8*A*). Here, the aperture may be circular (Fig. 17-9*B,C*) or stellate (Fig. 17-9*A*), depending on the thickening of the walls of the cells that encircle it. In complex pores the aperture is surrounded by a barrel-shaped circle of cells, in which the "barrel" is three or four cells high, while the rest of the epidermis is unistratose (Figs. 17-8*F* and 17-10). These pores remain open at all times except in *Preissia* and some species of *Marchantia,* in which the lowermost cells of the "barrel," when turgid, keep the pore open, but when they lose their turgor, they collapse to nearly close the pore (Fig. 17-11).

Sometimes the chlorenchyma is confined to a single layer of air chambers,

FIGURE 17-6
Schematic representation of germ rhizoid formation in the Marchantiales. *(A) Targionia* type. *(B) Marchantia* type. *(C) Neohodgsonia* type. *(D) Stephensonialla* type. *(E) Mannia* type. *(F) Reboulia* type. *(G) Conocephalum* type. (After Inoue, 1960.)

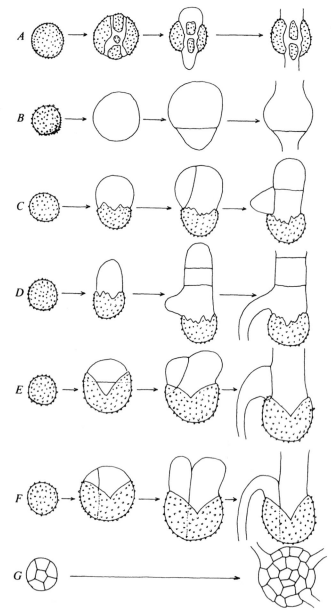

each of which has a dorsal pore. Within these chambers there are branched or unbranched chlorophyllose filaments (Fig. 17-12). Beneath the chlorenchyma layer, the thallus is made up of parenchymatous cells that contain starch, occasional cells with single complex oil bodies, and sometimes endogenous fungi. Occasionally blue-green algae (cyanobacteria) are found endogenously among these cells.

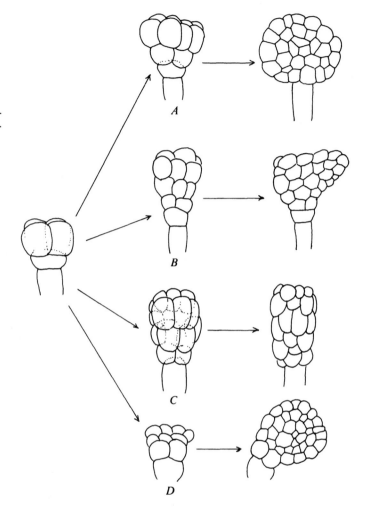

In some thalli the air chambers are in several layers. In such cases the walls
of the chambers are chlorophyllose, and filaments are usually lacking. Such
chambers are not interconnected with each other. In *Cyathodium* the thallus is
made up of chambers in which the cells of the upper surface are swollen and
lenslike. The swollen surfaces of these cells focus light upon the chloroplasts
within, enabling the plants to survive in very shaded sites and giving the thallus
a glowing appearance.

In *Riccia* the thallus is usually constructed of many slender perpendicular
chambers among columns of cells that constitute the chamber walls. These
open to the surface by a pore of very simple morphology (Fig. 17-8*C–E*). In
Dumortiera and *Monoselenium* the thallus has no air chambers; as in
Monoclea, the upper cells are chlorophyllose while those beneath lack
chlorophyll.

FIGURE 17-8
Anatomy of thalli of Marchantiales: transverse views. *(A) Asterella ludwigii*, showing simple pores and air chambers in several layers (×100). *(B, C) Riccia sorcarpa*: *(B)* showing diagrammatic view of thallus, containing one sporangium (×25); *(C)* detail of upper part of thallus with upper cells dissolved (×310). *(D) Riccia fluitans*, showing thallus made up largely of air chambers without pores (×80). *(E) Riccia frostii*, thallus mainly of air chambers with simple pores on upper surface (×50). *(F) Preissia quadrata*, showing complex barrel pores (×280). (After Müller, 1954.)

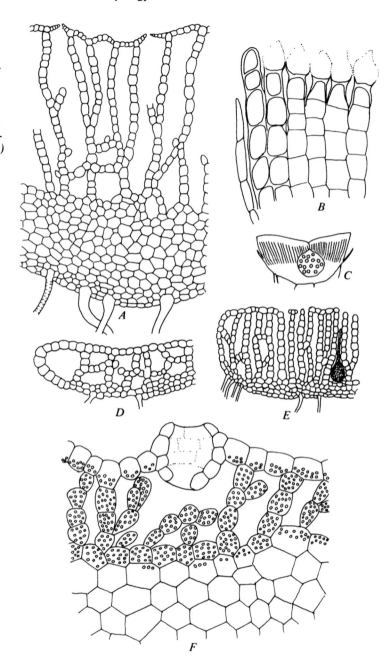

The parenchyma layer often shows groups of cells with numerous anastomosing thickenings in the walls and pits among these thickened bands. These pitted cells do not form a discrete continuous strand, but apparently facilitate longitudinal internal conduction of water and dissolved substances within the thallus.

FIGURE 17-9
Surface views of pores in thalli of Marchantiales. *(A) Peltolepis quadrata* (×230). *(B) Conocephalum conicum* (×230). *(C) Reboulia hemispherica* (×230). *(D) Marchantia polymorpha* (×230). *(E) Cyathodium smaragdinum* (×230). *(A,* after Hattori and Shimizu, 1954; *E,* after Hattori and Mizutani, 1959.)

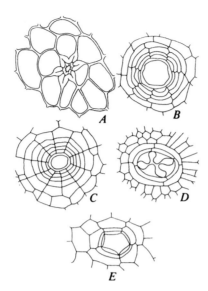

Rhizoids are usually abundant on the ventral surface of the thallus; some serve as anchoring and absorptive structures while others form an efficient capillary system for external conduction of water. Rhizoids are usually of two diameters: narrow and wide (Fig. 17-13). Within these categories are those with smooth walls that generally attach the thallus to the substratum, others with internal peglike or tuberclelike thickenings on the walls, and still others that are sinuous, with the indentations between the sinuosities marking the position of the inner pegs. These pegged and sinuous rhizoids are important in forming the capillary conducting system. The rhizoids arise mainly from beneath scales (when scales are present), and tend to be in clumps of fascicles. The smooth-walled rhizoids are mainly lateral to the midline of the thallus, while many of

FIGURE 17-10
Marchantiales: *Marchantia domingensis,* dorsal surface of thallus as revealed by scanning electron microscopy. *(A)* Apex of thallus, showing scales, rhizoids, and numerous pores (×37). *(B)* View into pore to the tips of photosynthetic filaments within (×250). (Provided by Dale M. J. Mueller.)

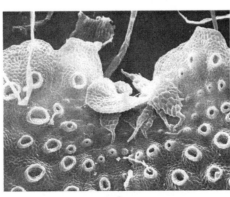
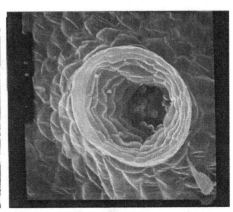

A *B*

FIGURE 17-11
Closing mechanism of the pore in *Preissia quadrata* (×400). *(A)* Open pore, longitudinal section on left, showing turgid cells bounding the pore; view from within the thallus on right. *(B)* Closed pore, longitudinal section on left, showing the collapsed lowermost cells bounding the pore; view from within the thallus on right. (After Walker and Pennington, 1939.)

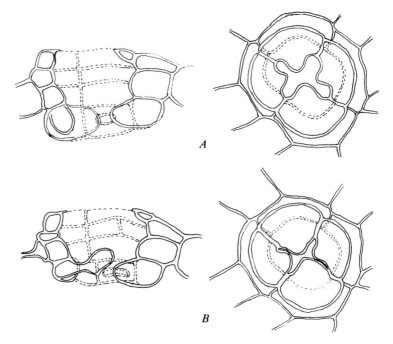

A

B

the pegged rhizoids arise lateral to the midline, but converge toward the midline of the thallus.

The undersurface of the thallus, in addition to bearing many rhizoids, often possesses several rows of scales (Fig. 17-14). M. McConaha (1941) notes that the rhizoids absorb water while the scales are passively wetted and form, with the rhizoids, a capillary network important in water conduction. The scales also ap-

FIGURE 17-12
Schmatic view of the thallus of *Marchantia* showing air chambers with pores and filaments, scattered oil bodies, and scattered cells with pits; scales (biseriate) and unicellular rhizoids shown on undersurface. (×25; modified from Parihar, 1965.)

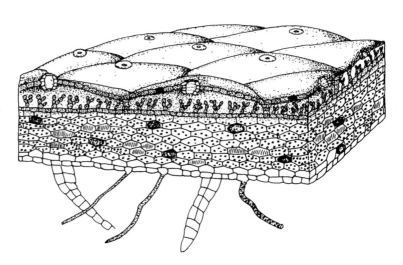

FIGURE 17-13
Rhizoids of Marchantiales.
(A) Pegged rhizoid of
Conocephalum conicum
(×250); *(B–E)* Rhizoids of
Marchantia polymorpha,
showing diversity in form, diameter, and ornamentation of
the inner walls (×250).

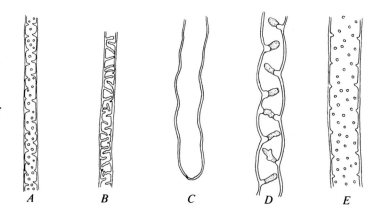

pear to affect the orientation of the pegged rhizoids, in particular. The scales
tend to be in two rows paralleling the thickened central band of the thallus
length, but sometimes, as in *Marchantia* and *Ricciocarpos,* the numerous
scales are scattered on the ventral surface.

ASEXUAL REPRODUCTION

Specialized means of asexual reproduction are uncommon in the Marchantiales. In only three genera, *Marchantia, Lunularia,* and *Neohodgsonia* are
gemma cups present (Fig. 17-15*A,C*). In these cups the gemmae are disclike
and with an indentation at the equator on opposite margins of the disc. The
gemmae terminate uniseriate short filaments and are accompanied by mucilage
filaments. The fact that the mucilage imbibes water assists in rupturing the
stalks of the gemmae so that, when raindrops splash into the cups, the loose
gemmae are dispersed. Rhizoids develop on the undersurface of the gemma,

FIGURE 17-14
Ventral scales in Marchantiales. *(A) Preissia quadrata*
(×4). *(B) Lunularia cruciata*
(×4). *(C) Reboulia hemis-
paerica* (×4). *(D) Marchantia
polymorpha* (×4). (After Mc-
Conaha, 1941.)

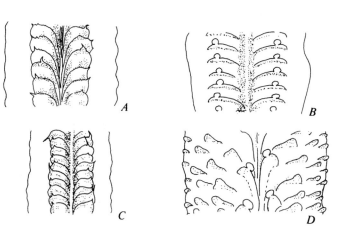

FIGURE 17-15
Vegetative diaspores of Marchantiales. *(A) Marchantia polymorpha,* thallus showing gemma cup on left lobe; antheridiophore also shown (×1). *(B) Conocephalum supradecompositum,* thallus lobe apex with apical gemmae (×6). *(C) Lunularia cruciata,* apex of thallus lobe with gemma cup and included gemmae (×6). *(B,* after Inoue, 1966.)

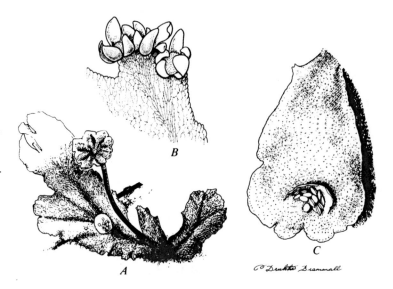

and the indentations serve as growing points to produce a rosettelike thallus that breaks apart to produce two separate thalli as the original gemma cells disintegrate. In *Marchantia planiloba* the gemmae are contained within a bottlelike structure, similar to that in *Blasia* (Metzgeriales). *Conocephalum supradecompositum* produces gemmae at the thallus apex, and they are covered by deformed appendages of ventral scales (Fig. 17-15A).

In some genera, including *Conocephalum,* older plants produce ovoid or spherical outgrowths within the superficial layer of the undersurface of the thickened portion of the thallus. When the thallus decays these outgrowths are released and serve as vegetative diaspores. Although tuberlike in structure, they appear to be intolerant of desiccation.

Adventitious branching is frequent, especially when the apex of the thallus is eliminated naturally (e.g., by formation of archegoniophores) or by adverse weather. Such branches are readily detached to serve as effective diaspores.

POSITION OF THE SEX ORGANS

Details of sex organ genesis differ somewhat from other bryophytes, as shown in Fig. 17-16. More than any other group of bryophytes, the Marchantiales show a great array of specialization in the position of the sex organs, especially the archegonia. Many genera produce an elaborate cylindric branch perpendicular to the reclining thallus. This branch is terminated by a somewhat symmetric disc that holds the sex organs. In internal structure this branch is like the vegetative thallus, with pores, chlorophyll, and oil bodies. It also sometimes possesses a relatively continuous system of pitted cells. If this branch bears archegonia it is termed an archegoniophore (Fig. 17-17A); when it

FIGURE 17-16
Ontogeny of the sex organs in Marchantiales. Note that the development of the antheridium (top row) is conspicuously different in earlier stages than in other bryophytes. (After Schuster, 1966.)

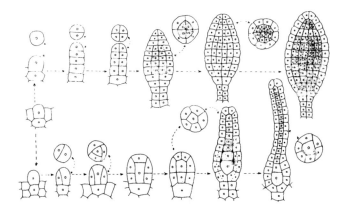

produces antheridia, it is an antheridiophore (Fig. 17-17*B*). Often these sexual branches are on separate thalli. In early stages of development the disclike portion of the branch is at the surface of the main thallus, and the stalk elongates later. The archegonia are exposed in radiating rows on the surface. Since the central cells of the disc remain meristematic while those of the rim grow more slowly, further growth of these central cells causes the top of the disc to expand over the archegonia so that the archegonium mouth is gradually oriented downward (Fig. 17-18). Coincident with this growth, the stalk of the

FIGURE 17-17
Marchantia polymorpha. (*A*) Archegonium-producing receptacle (archgeoniophore) of *Marchantia polymorpha* shown in longitudinal section. (Compare with the antheridial receptacle, *B*.) Archegonia are borne on the ventral surface of the receptacle as it matures (×40). (*B*) Antheridial receptacle (antheridiophore) shown in longitudinal section; note the pegged rhizoids and ventral scales on the main thallus, longitudinally on the stalk of the receptacle, and on the ventral surface of the receptacle. Pores and air chambers are on dorsal surfaces of the main thallus and in the disc; antheridia are shown embedded in the dorsal layers of the disc (×10). (After Smith, 1955.)

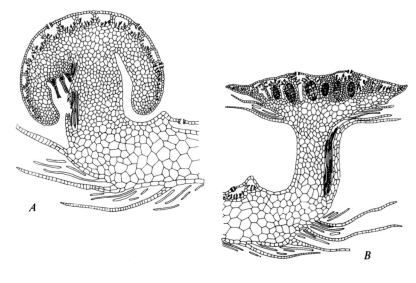

FIGURE 17-18
Diagrams of top view and vertical longitudinal sections of
stages in the development of
an archegonium-bearing receptacle of *Marchantia polymorpha*, showing the extensive growth of the upper cells
of the disc leading to the
change in the orientation of
the archegonia and their enclosed embryos (see text).
(After Smith, 1955.)

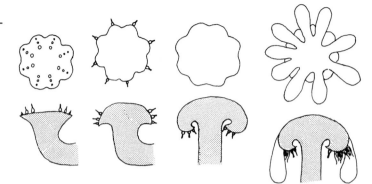

archegoniophore elongates. Such elongation may be very slow initially, but as the sporophytes mature, elongation is rapid, and the sporophytes are raised well above the thallus surface. The disc, when it bears sporophytes, is termed a carpocephalum.

The antheridia, although initially exposed, are soon embedded in chambers below the upper surface of the disc (Fig. 17-17*B*). This occurs by upgrowth of the surrounding tissue that encloses each antheridium in a separate chamber. A stalk often raises the antheridiophore above the thallus, enhancing wider dispersal of the sperms as raindrops splash them from this raised surface.

Archegoniophores show various degrees of complexity. The apex of the archegoniophore of *Neohodgsonia* is a dichotomizing system, with the tips of the dichotomies each producing a single sporophyte. Some species of *Marchantia*, on the other hand, bear many sporophytes on a radiate carpocephalum with nine elongate lobes. Other genera, including *Lunularia*, have lobes often reduced to four, and each lobe encloses a sporangium. In *Plagiochasma* these lobes are reduced to two or even one. The carpocephalum encloses a single sporangium in *Carrpos*, and this is sheathed by the overarching margins of the thallus. There is no specialized branch in *Corsinia*, but the sporangia are embedded in a bulge on the dorsal surface of the thallus. The archegonia, and ultimately the sporophytes, are embedded in the thallus in *Riccia*.

The antheridiophores show a similar reduction series. In *Marchantia*, for example, the antheridial disc surmounts an elongate stalk when mature (Fig. 17-1*C*). In other genera, including *Conocephalum*, the antheridia are embedded in a cushionlike pad near the apex of the flattened thallus lobes, much as in *Monoclea* (Fig. 17-3*C*). In *Riccia* the antheridia are deeply embedded in the thallus. Each of these morphological types strongly suggests a stage in the reduction in the structures that bear sex organs, with *Neohodgsonia* showing the most generalized branches and *Riccia* showing the greatest reduction in structure.

One unusual feature of the apical disc of the sexual branches is in the nature of the pores. Even in thalli where the pores are simple, those on the apical disc are sometimes barrel-shaped, as in *Conocephalum*.

THE SPOROPHYTE

Embryonic development follows two distinctive basic patterns: filamentous and octant. In the filamentous type the first divisions are transverse, resulting in a linear series of three to six cells; the succeeding division is longitudinal, and following cell divisions are basically similar in most bryophytes until endothecial and amphithecial cells are produced. The second type, termed octant, is usually considered more specialized. In this type the first cell division of the embryo is transverse, as in the filamentous type, but the third and fourth are longitudinal, and a quadrant of cells results; a transverse division of these cells results in an octant of cells; succeeding divisions are much as in the filamentous type.

The sporophyte of most Marchantiales has a short massive seta. Sometimes, as in *Riccia*, the seta is absent. The sporangium jacket is always unistratose and sometimes has transverse thickenings in the walls (Fig. 17-19), or the jacket cells are thin-walled and lack any ornamentation. The sporangium varies from spherical to elongate.

FIGURE 17-19
Ornamentation of epidermal cell walls in sporangial jackets of Marchantiales. (*A, B, D, E*) Surface views. (*C*) Transverse section. (*A*) *Asterella mitsumineneis* (×450). (*B, C*) *Peltolepis quadrata* var. *japonica* (×260). (*D*) *Conocephalum supradecompositum* (×215). (*E*) *Plagiochasma intermedium* (×120). (*A–C*, after Shimizu and Hattori, 1952, 1954; *D, E*, after Inoue, 1976.)

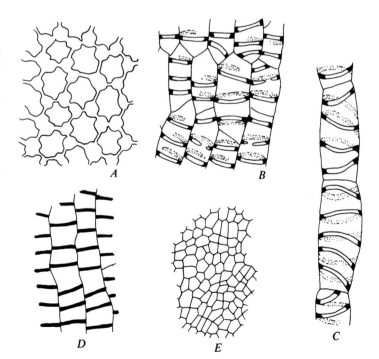

The spores (Fig. 17-4*B–E* and 17-20) are usually accompanied within the sporangium by elaters (Fig. 17-21). These elaters are often short and blunt and have one to three helical bands on the walls. Elaters are usually unbranched, but may sometimes be Y-shaped. They are usually smooth, but are papillose in a few species. In genera in which the sporophyte lacks a seta and in which dehiscence is by simple disintegration of the jacket (e.g., *Riccia*), elaters are absent.

Opening of the sporangium varies from four to six longitudinal lines to irregular lines of rupture. The upper part of the sporangium in *Cyathodium* produces toothlike processes terminated by a tiny disclike structure. When the disc is shed, the "teeth" open the apex of the sporangium and the spores and elaters are shed. In some genera the apex of the sporangium breaks off like an operculum, as in *Asterella*. Finally, in some genera, including *Corsinia* and *Riccia*, the spores are shed when the thallus enclosing the sporangia decomposes.

When a carpocephalum is present, it raises the sporophytes far above the thallus surface; the sporangia point downwards and setae are usually short. Habitats that dry up for long periods would tend to favor plants with sporophytes protected within the tissue of the dead gametophore. This is the situation in *Riccia* and some other genera.

RELATIONSHIPS AND EVOLUTIONARY TRENDS

Fossil material of apparent Marchantiales is found first in early Mesozoic sediments. Ventral scales, pegged rhizoids, pores, and spores are provided as convincing evidence of their identity. In Quaternary material even carpocephala and gemma cups are present, but sexual reproductive structures are extremely rare in any fossil material. Evolutionary trends in the Marchantiales

FIGURE 17-20
Spores of Marchantiales. (*A*) *Cyathodium smaragdinum* (×200). (*B*) *Preissia quadrata* (×450). (*C*) *Conocephalum conicum* (×200). (*D*) *Exormotheca pustulosa* (×360). (*E*) *Riccia nipponica* (×180). (*F*) *Sauteria alpina* (×180). (*G*) *Reboulia hemisphaerica* (×450). (*H*) *Athalamia glauco-virens* (×180). (*A, C–E*, after Miyoshi, 1966; *B, F, H*, after Shimizu and Hattori, 1953, 1954, 1959; *G*, after Hattori and Mizutani, 1959.)

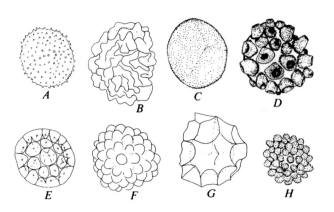

FIGURE 17-21
Elaters of Marchantiales. (*A*)
Sauteria alpina (×280). (*B*)
Athalamia glauco-virens
(×550); (*C*) *Sauteria alpina*
(×280). (After Shimizu and
Hattori, 1954.)

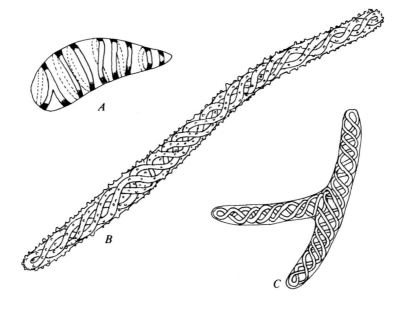

appear to be from the generalized condition to considerable reduction; they are summarized in Table 17-1.

The considerable variety in morphology within the order has undoubtedly enhanced its very wide distribution in the world. It is apparent that man has greatly expanded the distribution of some species, including *Marchantia polymorpha* and *Lunularia cruciata,* both of which have become common greenhouse and garden weeds. Most marchantialean genera (except *Riccia*) have few species and are local in distribution; these features strongly imply that these genera are relictual and may represent persistent and specialized remnants of a more diverse Marchantialian flora.

TABLE 17-1.
Generalized and Specialized
Features in Marchantiales

Generalized	Specialized
Thallus with elaborate air chambers containing filaments	Thallus with simple air chambers without filaments
Complex pores	No pores
Elaborate sex receptacles	Sex organs embedded in thallus
Numerous scales scattered	Scales restricted to two rows or absent
Gemma cups	No gemma cups
Sporangium dehiscence by preformed lines	Sporangium dehiscence by decomposition
Elaborate elaters	No elaters

FURTHER READING

Abeywikrama, B. A. 1945. The structure and life history of *Riccia crispatula* Mitt. *Ceylon J. Sci.* **12**:145–153.

Abrams, L. R. 1899. The structure and development of *Cryptomitrium tenerum. Bot Gaz.* **3**:110–121.

Ahmad, S. 1938. A study of *Aitchinsoniella himalayensis. Proc. Indian Acad. Sci. sect. B.* (7): 206–224, 8 figs.

—— 1940. Morphological study of *Exormotheca tuberifera. Bot. Gaz.* **104**:948–954.

Anthony, R. E. 1962. Greenhouse culture of *Marchantia polymorpha* and induction of sexual reproductive structures. *Turtox News* **40**:2–5.

Auret, T. B. 1930. Observations on the reproduction and fungal endophytism of *Lunularia cruciata* (L.) Dumortier. *Trans. Br. Mycol. Soc.* **15**:163–176, 8 figs.

Barnes, C. R., and W. J. G. Land. 1907. Bryological papers I. The origin of air chambers. *Bot. Gaz.* **44**:197–213, figs. 1–22.

—— 1908. Bryological papers II. The origin of the cupule of *Marchantia. Bot. Gaz.* **46**:401–409, figs. 1–14.

Benecke, W. 1903. Über die Keimung der Brutknospen von *Lunularia cruciata,* mit vergleichenden Ausblicken auf andere Pflanzen. *Bot. Z.* **61**:19–46.

Bergdolt, E. 1926. Untersuchungen über Marchantiaceen. *Goebel's Bot. Abhand.* **10**:1–86, figs. 1–121.

Black, C. A. 1913. The morphology of *Riccia frostii* Aust. *Ann. Bot.* **27**:511–532, pls. 37,38.

Bolleter, E. 1905. *Fegatella conica.* Eine morphologisch-physiologische Monographie. *Beih. Bot. Centralbl.* **18**:327–408, 16 figs., pls, XIII–XIII.

Bowen, E. J. 1935. A note on the conduction of water in *Fimbriaria blumeana. Ann. Bot.* **49**:844–848, 6 figs.

Burgeff, H. 1943. *Genetische Studien an Marchantia.* Jena: Gustav Fischer, 296 pp.

Campbell, D. H. 1918. Studies in some East Indian Hepaticae I. *Dumortiera* and *Wiesnerella. Ann. Bot.* **32**:319–338.

Campbell, E. O. 1954. The structure and development of *Marchasta areolata* Campb. *Trans. Roy. Soc. N. Z.* **82**:249–262, 45 figs.

—— 1965. *Lunularia* in New Zealand. *Tuatara* **13**:31–42.

—— 1965. *Marchantia* in New Zealand. *Tuatara* **13**:122–136.

Carr, D. J. 1956. Contribution to Australian Bryology I. The structure, development and systematic affinities of *Monocarpus sphaerocarpus* gen. et sp. n. (Marchantiales). *Austral. J. Bot.* **4**:175–191, 11 figs., 3 pls.

Carter, A. M. 1935. *Riccia fluitans* L.—a composite species. *Bull. Torrey Bot. Club* **62**:33–42, figs. 1–8, pls. 4,5.

Cavers, F. 1904. On the structure and biology of *Fegatella conica. Ann. Bot.* **18**:87–120, figs. 1–32, pls. 6,7.

————— 1904. *Contributions to the Biology of Hepaticae Part I–Targionia, Reboulia, Preissia, Monoclea,* 47 pp., 12 figs. London: Corley & Pickersgill.

————— 1910. The inter-relationships of the Bryophyta II. Marchantiales. *New Phytol.* 9:93–112, 157–186, 193–196.

Chavan, A. R. 1937. A morphological study of *Cyathodium barodae. Am. J. Bot.* 24:484–492.

Clee, D. A. 1943. The morphology and anatomy of *Fegatella conica* in relation to the mechanism of absorption and conduction of water. *Ann. Bot.* 7:185–193.

Deutsch, H. 1912. A study of *Targionia hypophylla. Bot. Gaz.* 53:492–503.

Donaghy, F. 1916. The morphology of *Riccia fluitans* L. *Proc. Indiana Acad. Sci.* **1915**:131–133.

Douin, R. 1920. Recherches sur les Marchantiées. *Rev. Gen. Bot.* 33:1–99.

————— 1922. Le sporophyte des Marchantiées. *Rev. Gen. Bot.* 34:1–15.

Douin, C., and R. Douin. 1918. Le *Reboulia* Raddi. *Rev. Gen. Bot.* 30:129–145.

Dupler, A. W. 1921. The air chambers of *Reboulia hemispherica. Bull. Torrey Bot. Club* 48:241–252.

————— 1922. The male receptacle and antheridium of *Reboulia hemispherica. Am. J. Bot.* 9:285–295.

Duthie, A. V., and J. Garside. 1936, 1939. Studies in South African Ricciaceae I & II. *Trans. Roy. Soc. S. Africa* 24:93–133. Ibid. 27:17–28.

Ernst, A. 1908. Untersuchungen über Entwicklung, Bau und Verteilung der Infloreszenzen von *Dumortiera. Ann. Jard. Bot. Buitenzorg (Ser. 2)* 7:153–223, pls. 18–24.

Evans, A. W. 1915. The genus *Plagiochasma* and its North American species. *Bull. Torrey Bot. Club* 42:259–308.

————— 1917. The American species of *Marchantia. Trans. Conn. Acad. Arts Sci.* 21:201–313.

————— 1918. The air chambers of *Grimaldia fragrans. Bull. Torrey Bot. Club* 45:235–251.

————— 1919. A taxonomic study of *Dumortiera. Bull. Torrey Bot. Club* 46:167–182.

————— 1920. The North American species of *Asterella. Contrib. U. S. Natl. Herb.* 20:247–312.

Fitting, H. 1939. Untersuchungen über den Einfluss von Licht und Dunkelheit auf die Entwicklung von Moosen I. Die Brutkörper von Marchantieen. *Jahrb. Wiss. Bot.* 88:633–722.

Forster, K. 1927. Die Wirkung ausserer Factoren auf Entwicklung und Gestaltbildung bei *Marchantia polymorpha. Planta* 3:325–390.

Garber, J. F. 1904. The life history of *Ricciocarpus natans. Bot. Gaz.* 37:161–177.

Goebel, K. V. 1882. Zur vergleichenden Anatomie der Marchantieen. *Arbeit. Bot. Inst. Wurzburg* 2:529–535.

—— 1910. Archegoniatenstudien 13 *Monoselenium tenerum* Griffith. *Flora* **101**:43–97.

Golenkin, M. 1902. Die Mycorrhiza-ähnlichen Bildungen der Marchantiaceen. *Flora* **90**:209–220, pl. x1.

Graft, P. W. 1936. Invasion of *Marchantia polymorpha* following forest fires. *Bull. Torrey Bot. Club* **63**:67–74.

Hassel de Menendez, G. G. 1963. Studio de las Anthocerotales y Marchantiales de la Argentina. *Opera Lilloana* **7**:1–297.

Haupt, A. W. 1921. Gametophyte and sex organs of *Reboulia hemisphaerica*. *Bot. Gaz.* **71**:61–74.

—— 1926. Morphology of *Preissia quadrata*. *Bot. Gaz.* **82**:30–54.

—— 1929. Studies in Californian Hepaticae I. *Asterella californica*. *Bot. Gaz.* **87**:302–319.

Haynes, C. C. 1920. Illustrations of six species of *Riccia,* with the original descriptions. *Bull. Torrey Bot. Club* **47**:279–287, pls. 10–13.

Hirsch, P. 1910. The development of the air chambers in the Ricciaceae. *Bull. Torrey Bot. Club* **37**:73–77.

Inoue, H. 1960. Studies in spore germination and the earlier stages of gametophyte development in the Marchantiales. *J. Hattori Bot. Lab.* **23**:148–191.

—— and T. Asakawa. 1966. Making a motion picture of the life history of *Marchantia polymorpha. Bryologist* **69**:369–373.

Jovet-Ast, S. 1964. Essai sur le genre *Funicularis* Trev. *Rev. Bryol. Lichénol.* **32**:193–211.

—— 1964. Essai sur le genre *Cronisia* Berkeley. *Rev. Bryol. Lichénol.* **33**:180–184.

Kachroo, P. 1954. Morphology of Rebouliaceae II. On some species of *Mannia* Corda, *Asterella* Beauv., and *Plagiochasma*. L & L. *J. Hattori Bot. Lab.* **12**:34–52.

—— 1955. Morphology of *Dumortiera,* with particular reference to D. *hirsuta* Nees. *J. Univ. Gauhati* **6**:127–137.

—— 1958. Morphology of Rebouliaceae III. Development of sex organs, sporangium and inter-relationships of the various genera. *J. Hattori Bot. Lab.* **19**:1–24.

Kammerling, Z. 1897. Zur Biologie und Physiologie der Marchantiaceen. *Flora* **84**:1–173.

Kashyap, S. R. 1914. Morphological and biological notes on new and little known West-Himalayan liverworts I & II. *New Phytol.* **13**:206–226, 308–323.

—— 1915. Morphological and biological notes on new and little known West-Himalayan liverworts III. *New Phytol.* **14**:1–18.

Khanna, L. P. 1929. The morphology of *Cyathodium kashyapii. J. Indian Bot. Soc.* **8**:118–122.

Klingsmüller, W. 1959. Zur Entwicklungsphysiologie der Ricciaceen. *Flora* **147**:76–122.

Kny, L. 1864. Beiträge zur Entwicklungsgeschichte der laubigen Lebermoose. *Jahrb. Wiss. Bot.* **4**:64–100.

———— 1867. Über Bau und Entwicklung der Riccieen. *Jahrb. Wiss. Bot.* **5**:264–386, pls. 44–46.

Lampa, E. 1909. Über die Beziehung zwischen dem Lebermoosthallus und dem Farnprothallium. *Osterr. Bot. Z.* **59**:405–414.

Land, W. H. 1905. On the morphology of *Cyathodium. Ann. Bot.* **19**:411–426.

Leitgeb. H. 1879. *Untersuchungen über die Lebermoose IV. Die Riccieen.* 101 pp., 9 pls.

———— 1880. Die Atemöffnungen der Marchantiaceen. *Sitzb. Akad. Wiss. Wein I. Abtheil* **81**:40–54.

———— 1880. Die Inflorescenzen der Marchantiaceen. *Sitzb. Akad. Wiss. Wien I. Abtheil* **81**:123–143.

Lindenberg, J. B. W. 1833. Über die Lebermoos-Gattungen *Corsinia* und *Grimaldia. Allgem. Bot. Z. (Flora)* **16**:161–176.

———— 1836. Monographie der Riccien. *Allgem. Bot. Z. (Flora)* **18**:361–504, pls. 19–37.

Lundblad, B. 1954. Contributions to the geological history of the Hepaticae. Fossil Marchantiales from Rhaetic-Liassic coal mines of Skromberga (Prov. of Scania), Sweden. *Sv. Bot. Tidskr.* **48**:401–417.

———— 1955. Contributions to the geological history of the Hepaticae. II. On a fossil member of the Marchantiinae from the Mesozoic plant-bearing deposits near Lago San Martin, Patagonia. *Bot. Notiser* **108**:22–39.

———— 1959. On *Ricciosporites tuberculatus* and its occurrence in certain strata of the "Hollviken II" Boring in S. W. Scania. *Grana Palynol.* **2**:1–10.

Mashabale, I. S., and P. D. Bhate. 1945. The structure of life-history of *Asterella (Fimbriaria) angusta. J. Univ. Bombay* **13**:5–15.

Mashabale, I. S., and S. R. Dashpande. 1947. Life history of *Plagiochasma articulatum* Kask. *J. Univ. Bombay* **15**:23–36.

Massalongo, C. 1916. Le Marchantiaceae della flora europa. *Atti Reale Inst. Veneto Sci. Lett. Arti.* **75**:669–817, pls. 4–30.

McConaha, M. 1939. Ventral surface specializations of *Conocephalum conicum. Am. J. Bot.* **26**:353–355.

———— 1941. Ventral structures effecting capillarity in the Marchantiales. *Am. J. Bot.* **28**:301–306.

Mehra, P. N. 1957. A new suggestion on the origin of thallus in the Marchantiales I & II. *Am. J. Bot.* **44**:505–513, 573–581.

———— and H. L. Mehra. 1939. Life history of *Stephensoniella brevipedunculata* Kash. *Proc. Indian Acad. Sci.* **99**:287–315.

———— and B. R. Vashisht. 1950. Embryology of *Petalophyllum indicum* Kash. and a new suggestion of the evolution of thalloid habit from foliose forms. *Bryologist* **53**:89–114.

Menge, F. 1930. Die Entwicklung der Keimpflanzen von *Marchantia polymor-*

pha L. und *Plagiochasma rupestre* (Forst.) Steph. *Flora* **124**:423−478.

Meyer, K. 1931. Die Sporophytenentwicklung und die Phylogenie bei den Marchantiales. *Planta* **13**:210−220.

Miller, M. W., and P. O. Voth. 1962. Geotropic responses of *Marchantia*. *Bryologist* **65**:146−154.

Miyoshi, N. 1966. Spore morphology of Hepaticae in Japan. *Bull. Okayama Coll. Sci.* **2**:1−46.

O'Hanlon, M. E. 1927. A study of *Pressia quadrata*. *Bot. Gaz.* **89**:208−218.

———— 1930. Gametophyte development in *Reboulia hemispherica*. *Am. J. Bot.* **17**:765−769.

———— 1934. Comparative morphology of *Dumortiera hirsuta*. *Bot. Gaz.* **96**:154−164.

O'Keefe, L. 1915. Structure and development of *Targionia hypophylla*. *New Phytol* **14**:105−116.

Orth, R. 1931. Vergleichende Untersuchungen über die Luftkammerentwicklung der Ricciacen. *Flora* **25**:232−259.

———— 1943. Zur Luftkammerentwicklung von *Marchantia*. *Bot. Arch.* **44**:544−550.

Pande, S. K. 1924. Notes on the morphology and biology of *Riccia sanguinea* Kash. *J. Indian Bot. Soc.* **4**:117−128.

———— 1933. On the morphology of *Riccia robusta* Kash. *J. Indian Bot. Soc.* **12**:110−121.

Patterson, P. M. 1933. A developmental study of *Dumortiera hirsuta* (Sw.) Nees. *J. Elisha Mitchell Sci. Soc.* **49**:122−150, pls. 4−8.

Pickett, F. L. 1925. The life-history of *Ricciocarpus natans*. *Bryologist* **28**:1−3.

Proskauer, J. 1951. Notes on the Hepaticae II. *Bryologist* **54**:243−258.

———— 1962. On *Carrpos* I. *Phytomorphology* **65**:213−233.

Register, T. E., and W. R. West 1971. *Marchantia* life cycle. *Carolina Tips.* **34**:17−19.

Saxton, W. T. 1931. The life-history of *Lunularia cruciata* (L) Dum., with special reference to the archegoniophore and sporophyte. *Trans. Roy. Soc. S. Africa* **19**:259−268.

Schiffner, V. 1908. Morphologische und biologische Untersuchungen über die Gattungen *Grimaldia* und *Neesiella*. *Hedwigia* **47**:306−320, pl. 8.

———— 1908. Untersuchungen über die Marchantiaceen-Gattung *Bucegia*. *Beih. Bot. Centralbl.* **23**:273−290.

———— 1909. Studien über die Rhizoiden der Marchantiales. *Ann. Jard. Bot. Buitenzorg 2nd ser. Suppl. III*:473−492.

———— 1938−39. Monographie der Gattung *Cyathodium*. *Ann. Bryol.* **11**:131−140. Ibid. **12**:123−142.

Seth, M. L. 1931. On *Sauchia spongiosa* Kash. *J. Indian Bot. Soc.* **10**:175−183.

Shimizu, D., and S. Hattori. 1953−1955. Marchantiales of Japan I−IV. *J. Hattori Bot. Lab.* **9**:32−44. Ibid. **10**:49−55. Ibid. **12**:53−75. Ibid. **14**:91−107.

Solms-Laubach, H. 1899. Die Marchantiaceae Cleveideae und ihre Verbreitung. *Bot. Zt.* **57**:15–37.

Strasburger, E. 1870. Die Geschlechstorgane und die Befruchtung bei *Marchantia polymorpha* L. *Jahrb. Wiss. Bot.* **7**:409–422.

Udar, R. 1960. Studies in Indian Sauteriaceae II. On the morphology of *Athalamia pinguis* Falc. *J. Indian Bot. Soc.* **39**:56–77.

Underwood, L. M. 1895. Notes on our Hepaticae III. The distribution of the North American Marchantiaceae. *Bot. Gaz.* **20**:59–71.

Voth, P., and K. Hammer. 1940. Responses of *Marchantia polymorpha* to nutrient supply and photoperiod. *Bot. Gaz.* **102**:169–205.

Walker, R., and W. Pennington. 1939. The movement of air pores of *Preissia quadrata* (Scop.) Nees. *New Phytol.* **38**:62–68.

Wenzl, H. 1934. Untersuchungen über den Wasseraushalt von *Marchantia polymorpha*. *Jahrb. Wiss. Bot.* **79**:311–352.

18 Evolutionary Trends and Interrelationships among the Hepatics

The relationships in the hepatics are inferred mainly from gametophytic features. This contrasts to the situation in the mosses, in which features of the sporophyte are given priority in distinguishing evolutionary lines. Although the fossil record is sketchy for the hepatics, many useful specimens assist in speculations concerning the time of origin of the major groups and suggest possible evolutionary trends. Because the record is so incomplete, attempts to construct phylogenetic sequences are necessarily tentative.

GAMETOPHORE DIVERSIFICATION

Two contrasting evolutionary trends are represented in the gametophores of the hepatics. On the one hand, leafy gametophores have given rise to thallose growth forms. This is exemplified by some genera and families in the Jungermanniales and Sphaerocarpales. On the other hand, thallose gametophores, as in the Metzgeriales, have apparently served as progenitors for leafy members in the same order. The ancestry of some orders, as in the Marchantiales and Sphaerocarpales, remains controversial, and each of the available hypotheses has serious flaws either in morphogenetic assumptions or in the fundamental data upon which the hypothesis was based.

Fossil material attributed to the Metzgeriales appears in the Devonian. The fact that gametophytes of extant lignified archegoniates ("pteridophytes") also resemble Metzgerialean gametophores should not be overlooked. It is possible that these fossilized gametophytes attributed to the Metzgeriales could belong to some of the early "pteridophytes." Regrettably sporophytes are unknown for these fossil specimens, and the gametophytes possess no sex organs, thus their implied relationships are based on circumstantial evidence.

These early fossils, including *Hepaticites (Pallaviciniites) devonicus*, *H. (Treubiites) kidstonii*, *H. (Blasiites) lobatus*, *H. langii*, and *H. metzgerioides* represent most of the main trends in morphological variety in the order. Except for *H. kidstonii*, these are essentially thallose. This evidence, plus the existence of such a wide diversity of gametophore morphology in the extant Metzgeriales, strongly suggests that this order represents the closest approximation to the earliest hepatics. This should not be interpreted as an indication that this group served as a progenitor for all hepatics. There is too little evidence to warrant such an assumption.

A hypothetical gametophore of most generalized type for the hepatics might be expected to be anacrogynous and have exposed sex organs scattered on the upper surface of a flattened thallus. The thallus would be composed of thin-walled cells, have a thickened central portion, and become thin toward the margins; rhizoids would be smooth and abundant on the undersurface of this thickened portion. Branching was probably irregular. It is possible that a mycorrhizal relationship existed in such a thallus. Such a thallus is extremely similar to some gametophytes of extant fern families of ancient lineage. This similarity in form suggests that such a structure is adapted to the microenvironment in which these gametophytic phases grow. In the hepatics, selection has resulted in the production of diverse morphologies tolerant of a wide variety of microenvironments, while in the ferns the gametophytes are more restricted in their habitat requirements. Within and among the evolutionary lines of the hepatics, apparent trends in morphological change permit either a changeable morphology that permits diverse habitat tolerance or a very specific morphology that adapts the gametophore to a very restricted site.

It is the Metzgeriales that most closely approach this possible ancestral type, but no extant genus possesses all archaic features. In a genus like *Pellia* the gross morphology is similar to this ancestral type, but the antheridia are embedded in the thallus, a specialized feature. The restriction of archegonia to a lateral group within a protective involucre is also derivative.

Other divergent trends in the Metzgeriales are shown by *Treubia,* in which the sex organs are exposed on the dorsal surface of the gametophore, but the thallus shows considerable specialization in the production of lateral lobes and dorsal scales. The presence of complex oil bodies in scattered cells and rhizoids which show a tendency to produce helically arranged tubercles on their inner walls are also specialized features.

Another line of specialization is illustrated by the genus *Moerckia,* where there are elongate pitted internal conducting cells in the thallus. In *Metzgeria,* although no pitted conducting cells have been reported, the "midrib" is formed of elongate cells, and the midrib itself is a very discrete structure. In *Metzgeria* the sex organs are enclosed in reduced postical branches, a specialized feature.

Finally, *Fossombronia* also shows strong tendencies toward a leafy condition but retains the generalized feature of exposed antheridia; the archegonia are initially exposed, but become surrounded by a perianthlike involucre with a wide-flaring mouth, especially after development of the sporophyte.

In the Calobryales the evolutionary lines are remarkably distinctive. Indeed, it is possible that *Takakia* should be separated into its own evolutionary line, Takakiales. *Haplomitrium* can be compared to the ancestral hepatic gametophore as follows: it shares the thin-walled cells of the entire gametophore, and the sex organs are lateral and exposed. Specialization is evident in the lack of rhizoids, the possession of a subterranean creeping, much-branched, nonchlorophyllose system, the existence of an erect axis in which leaves are arranged in three ranks, and sometimes a stem with a central conducting strand. The nature of spore germination also can be interpreted as specialized.

Takakia possesses, as far as is known, similar features, except that the leaves are cylindric and sometimes bifid or bisbifid. The mode of origin of archegonia to supplant the position of a leaf suggests a generalized condition, while the beaked mucilage cells imply specialization.

The Jungermanniales stand out as an independent evolutionary line in various fundamental features. The possession of an apical cell with three cutting faces is, in itself, a specialized feature, and this structure determines the leaf arrangement on the stems and also the nature of the stem itself. Another remarkable feature is acrogyny. This is hinted at in *Haplomitrium,* especially in those species where fertilization of the archegonia prevents further growth of the shoot. Further features of gametophore specialization in the Jungermanniales are shown in the elaborate lobing of the leaves, the complexity of ornamentation on leaf-cell surfaces, the structurally modified perianths, diversity in origin and patterns of branches, and localization of the rhizoids.

There are some unusual trends in the Jungermanniales toward the leafless condition (except for the sexual branches) in the genus *Phycolepidozia,* and to the production of a thallus, as in *Metzgeriopsis, Schiffneria,* and *Pteropsiella,* among other genera. These are considered specialized, rather than generalized types, "regressing" to a generalized morphology while retaining essential features of their specialized ancestors. They cannot be considered as ancestral to the Metzgeriales; their acrogyny, leafy sexual branches, perianths, and basic form of the gametophore remain jungermannialean features and do not point to the thallose orders of hepatics. The many features shared with the Metzgeriales, however, suggest that these two orders had a common ancestor.

In the Sphaerocarpales the thallus morphology again shows features reminiscent of the ancestral type, especially in the genera *Sphaerocarpos* and *Geothallus,* but even in these genera there is a tendency toward the leafy condition, and the production of "tubers" is a derivative feature, as is the presence of specialized unistratose involucres around each sex organ. In *Riella* the thallus shows remarkable specialization in the stem and its production of sex organs, its production of gemmae, and its pleated involucres surrounding the archegonia. Even *Naiadita* shows considerable specialization in the production of gemmiferous branches and in the nature of the protective leaves associated with the archegonia.

The Monocleales show several gametophytic features in common with the hypothesized ancestral hepatic, but these appear to represent trends in specialization that reach their zenith in the Marchantiales. In *Monoclea* the presence of differentiated structures protecting the sex organs is specialized as is the nature of the rhizoid orientation and the possession of complex scattered oil bodies in a thallus in which there are clusters of pitted cells.

In the Marchantiales the thallus has reached its greatest morphological complexity and diversity, possibly associated with adaptive radiation into drier habitats. The radiation was accompanied by selection favoring individuals possessing a chlorenchyma with air chambers and pores, and a growth form tolerant of drier sites than the metzgerialean thallus morphology usually permits. Other devices adapted to a drier environment are the ventral scales and the capillary external conducting system. Further specialization exists in the position of the sex organs and restriction of the oil bodies.

Gametophytic features of the orders of hepatics are compared in Table 18-1.

SPOROPHYTE DIVERSIFICATION

Patterns of sporophytic diversification are not entirely consistent with the trends in gametophore specialization. The following features are generally considered to characterize a generalized sporophyte: a massive elongate seta surmounted by a sporangium with a multistratose jacket and containing elongate elaters and numerous small spores. It is probable that a stem calyptra protected the developing sporophyte preceding the rapid elongation of the seta. Dehiscence of the sporangium was probably by four longitudinal lines.

Monoclea, Haplomitrium, and many genera of the Metzgeriales exhibit this type of sporophyte morphology. Within the Metzgeriales, however, a degree of specialization is shown: production of elaterophores in such genera as *Riccardia* and *Pellia,* the reduced number of dehiscence lines in the sporangium, as in *Moerckia,* and the irregular dehiscence of the sporangium, as in *Fossombronia,* accompanied by such specializations as shorter elaters and abbreviated seta. In the Calobryales and Monocleales, the morphology of the sporophyte is superficially similar. The jacket is unistratose, the seta is massive, and the elaters are elongate, thus presenting a mixture of generalized and derivative features. In dehiscence there is a tendency toward specialization, in *Monoclea* reduced to a single suture and in *Haplomitrium* varying from four lines to one line of dehiscence.

In the Jungermanniales there is also considerable specialization in the dehiscence of the sporangium: helical dehiscence in such genera as *Gyrothyra* and *Calypogeia* and the remarkably specialized dehiscence in *Frullania.* Considerable specialization exists in the structure of an elongate seta, reaching its extreme in such genera as *Cephaloziella,* where the seta is four cells in circumference. The reduction of the length of the seta appears to be common in genera that tolerate desiccation, including *Radula, Frullania,* and many genera of the Lejeuneaceae.

In the Marchantiales the sporophyte shows considerable specialization: the seta is relatively short or essentially absent, the jacket is unistratose, and dehiscence varies from the generalized conidtion of four longitudinal lines through the production of an operculum to dehiscence by irregular rupture or decomposition of the jacket. There is also a trend toward reduction in length and effectiveness of the elaters, which proceeds to their complete absence.

In the Sphaerocarpales the sporophyte shows the same ultimate reduction exhibited by the most specialized sporophytes of the Marchantiales: unistra-

TABLE 18-1.
Comparison of Gametophytic
Features of the Hepaticae

Calobryales	Jungermanniales	Metzgeriales	Sphaerocarpales	Monocleales	Marchantiales
Erect from nonchlorophyllose reclining shoots	Erect or creeping	Creeping or erect	Creeping or erect	Creeping	Creeping
Radial symmetry	Radial or (usually) bilateral symmetry	Bilateral symmetry	Asymmetrical	Bilateral symmetry	Bilateral symmetry except for sexual branches
No rhizoids	Smooth rhizoids	Smooth rhizoids	Smooth rhizoids	Smooth rhizoids	Smooth and "pegged" rhizoids
No scales	No scales	Sometimes scales on dorsal surface	Scales in *Riella*	No scales	Usually scales on ventral surface
Leafy	Leafy (rarely thallose)	Thallose or somewhat leafy	Thallose to almost leafy	Thallose	Thallose
No true amphigastria	Amphigastria in most	Never amphigastria	Never amphigastria	Thallose	Thallose
Cells thin-walled	Thin-walled or trigonous cells	Thin-walled cells	Thin-walled cells	Thin-walled cells	Mostly thin-walled cells
Many oil bodies and chlorophyll in each cell	Many oil bodies and chlorophyll in each cell	Many oil bodies and chlorophyll in each cell (usually)	Single oil bodies in scattered cells (*Riella*)	Scattered single oil bodies in cells without chloroplasts	Scattered single oil bodies in cells without chloroplasts
No air chambers	No air chambers	No air chambers	No air chambers	No air chambers	Air chambers commonly
No gemmae, but caducous lvs. (*Takakia*)	Gemmae and caducous lvs.	Gemmae	Gemmae in *Riella*	No gemmae	Gemmae in cup (rare in order)
No tubers	No tubers	Tubers occasional	Tuber present (*Geothallus*)	No tubers	No tubers
Anacrogynous	Acrogynous	Anacrogynous	Anacrogynous	Anacrogynous	Anacrogynous
Sex organs exposed	Sex organs usually enclosed in protective structure	Sex organs usually enclosed, or sometimes exposed	Sex organs enclosed	Sex organs enclosed	Sex organs enclosed when mature
Stem calyptra	Perianth in most	Involucre	Involucre	Involucre	Involucre

tose jacket, dehiscence by irregular rupture or decomposition, absence of elaters (but presence of nurse cells supplanting the elaters), and absence of seta. The sporophytic features of the hepatics are compared in Table 18-2.

TABLE 18-2.
Comparison of Sporophytic
Features of the Hepaticae

Calobryales	*Jungermanniales*	*Metzgeriales*	*Sphaerocarpales*	*Monocleales*	*Marchantiales*
Seta elongate, massive	Seta generally elongate, usually slender	Seta elongate, sometimes massive	Seta essentially none	Seta elongate, massive	Seta usually short, or none
Sporangium cylindric	Sporangium spherical to cylindric	Sporangium spherical to cylindric	Sporangium spherical	Sporangium cylindric	Sporangium spherical to cylindric
Sporangium wall unistratose	Sporangium wall 2–10 stratose	Sporangium wall 2–5 stratose	Sporangium wall unistratose	Sporangium wall unistratose	Sporangium wall unistratose
Sporangium with longitudinal band thickenings on transverse wall	Sporangium wall with nodular or annular thickenings	As in Jungermanniales	No thickenings on jacket of sporangium	Forked thickenings on sporangium jacket on all walls except the outer face	Thickenings various on jacket cells, or none
Dehiscence 1–4 lines	Dehiscence usually 4 lines	Dehiscence 1–4 lines	Dehiscence by decay of jacket	Dehiscence by 1 line	Dehiscence 1–4 lines, or decay, or operculum
Elaters long	Elaters long	Elaters long (usually)	No elaters, nurse cells present	Elaters long	Elaters short, long, or absent, nurse cells sometimes
Particular Features Characterizing Orders					
Flask-shaped mucilage cells on "stolons" (*Takakia*); internal conducting strand in stem	Perianth; amphigastria; variously lobed lvs.; trigones (essentially restricted); *Frullania* dehiscence; marsupium	Lamellae on thallus; (*Petalophyllum*); Elaterophore "vascular system"; scales on upper surface; Gemma flasks or cups	"Bottles"; tubers; ruffle; completely aquatic life cycle (*Riella*)	Jacket cells with forked thickenings on all but outer surface; multiple sporophytes from involucre	Archegoniophore; antheridiophore; gemma cups; air chambers; pores

In the hepatics, then, there are several very distinctive and independent evolutionary lines in gametophore structure, accompanied only in part by equivalent changes in sporophytic specialization. The specialization within each of the orders appears to be a result of adaptive radiation into diverse microenvironments, but details of these adaptive advantages remain very incomplete. The extant Marchantiales and Metzgeriales appear to represent persistent fragments of groups that may have been much more diverse. The restricted distribution of many taxa and the representation of many genera by a single species also strongly implies remnants of an evolutionary line, particularly since many of these taxa are taxonomically isolated.

FURTHER READING

Berrie, G. K. 1963. Cytology and phylogeny of liverworts. *Evolution* **17**:347–357.

Campbell, D. H. 1936. The relationships of the Hepaticae. *Bot. Rev.* **2**:53–64.

Campbell, E. O. 1971. Problems in the origin and classification of bryophytes with particular reference to liverworts. *N. Z. J. Bot.* **9**:678–688.

Chopra, R. S. 1967. Relationship between liverworts and mosses. *Phytomorphology* **17**:70–78.

Crandall-Stotler, B. 1984. Musci, Hepatics and Anthocerotes—an essay on analogies, in Schuster, R. M. (ed.), *New Manual of Bryology* pp. 1093–1129. Nichinan, Japan: Hattori Bot. Lab.

Fulford, M. 1964. Contemporary thoughts in plant morphology. Hepaticae and Anthocerotae. *Phytomorphology* **14**:103–119.

——— 1965. Evolutionary trends and convergence in the Hepaticae. *Bryologist* **68**:1–30.

Jeffrey, C. 1962. The origin and differentiation of the archegoniate landplants. *Bot. Not.* **115**:446–454.

Kashyap, S. R. 1919. The relationship of liverworts, especially in the light of some recently discovered Himalayan forms. *Proc. Asiat. Soc. Bengal* **15**:152–166.

Mehra, P. N. 1957. A new suggestion on the origin of thallus in Marchantiales. *Am. J. Bot.* **44**:505–513, 573–581.

——— 1967. Phyletic evolution in the Hepaticae. *Phytomorphology* **17**:47–58.

——— 1969. Evolutionary trends in Hepaticae with particular reference to the Marchantiales. *Phytomorphology* **19**:203–218.

Miller, H. A. 1974. Rhyniophytina, alternation of generations, and the evolution of bryophytes. *J. Hattori Bot. Lab.* **38**:161–168.

——— 1974. Hepaticae through the ages. *Rev. Fac. Cienc. Lisboa* **17**:763–775.

Proskauer, J. 1961. Hepaticae, in Gray, P. (ed.), *The Encyclopedia of the Biological Sciences,* pp. 472–474. New York: Reinhold.

Richards, P. W. 1978. The taxonomy of bryophytes, in Street, H. E. (ed.), *Essays in Plant Taxonomy,* pp. 177–209. New York: Academic Press.

Schuster, R. M. 1966. *The Hepaticae and Anthocerotae of North America East of the Hundredth Meridian,* Vol. I, 802 pp. New York: Columbia University Press.

———— 1977. The evolution and early diversification of the Hepaticae and Anthroceratae, in Frey, W., et al. (eds.), pp. 107–115. *Beiträge zur Biologie der Niederen Pflanzen.* Stuttgart: G. Fischer.

———— 1979. The phylogeny of the Hepaticae, in Clarke, G. C. S., and J. G. Duckett (eds.), *Bryophyte Systematics.* Syst. Assoc. Spec. Vol. 14, pp. 41–82.

———— 1981. Paleoecology, origin, distribution through time, and evolution of Hepaticae and Anthocerotae, in Niklas, K. J. (ed.), *Paleobotany, Paleoecology and Evolution,* Vol. 2, pp. 129–191. New York: Praeger.

———— 1984. Evolution, phylogeny and classification of the Hepaticae, in Schuster, R. M. (ed.), *New Manual of Bryology,* pp. 892–1070. Nichinan, Japan: Hattori Bot. Lab.

Sharp, A. J. 1974. Hepatopsida, in *Encyclopedia Brittanica* (ed. 15). Macropaedia, Vol. 8, pp. 779–781.

Steere, W. C. 1958. Evolution and phylogeny in bryophytes. in *Current Topics in Plant Science,* pp. 134–143. New York: Academic Press.

19 The Hornworts — Class Anthocerotae

The hornworts form an isolated evolutionary line. Indeed some researchers consider them to be entirely independent from the bryophytes and place them in the division Anthocerotophyta. The "horn" describes the sporophyte, a tapered cylinder that shows indeterminate growth in most species. The class consists of at least four genera, although some researchers recognize six (Fig. 19-1). These are in a single order Anthocerotales and are sometimes included in a single family, Anthocerotaceae, but some authors place *Notothylas* in its own family, Notothylaceae (Fig. 19-1*B*). Commonly recognized genera, besides *Notothylas* (ca. 13 species), are *Anthoceros* (ca. 250 species), *Megaceros* (ca. 46 species), and *Dendroceros* (ca. 51 species). *Phaeoceros* (ca. 30 species) and *Folioceros* (ca. 19 species) are sometimes segregated from *Anthoceros*. The number of species is undoubtedly greatly exaggerated; detailed study of living populations of plants is seriously needed. The careful researches of Johannes Proskauer, in particular, suggest that there is considerable synonymy among the species.

The class is widely distributed in temperate and tropical latitudes throughout the world. All genera are in the tropics and subtropics. *Anthoceros, Phaeoceros,* and *Notothylas* are also widespread in temperate climates. *A. punctatus* occurs at subalpine sites in some mountains. Most species are found on moist mineral soil of banks and cliffs, among grasses, and along streams, but *Megaceros* often occurs on splashed rocks of streams while *Dendroceros* is commonly found on tree trunks in humid forests. In temperate areas the plants are seasonal, flourishing when moisture and temperature are favorable and dying in the unfavorable season. In the tropics, however, and to a certain degree in milder temperate climates, the gametophores are sometimes perennial. Features that characterize the Anthocerotae are as follows:

1. A dorsiventrally flattened thallus commonly forms a rosette.
2. The thallus, composed of thin-walled cells, is attached to the substratum by smooth rhizoids.

FIGURE 19-1
Diversity in the Anthocerotae. (*A*) *Dendroceros japonicus* (×10). (*B*) *Notothylas orbicularis* (×8). (*C*) *Phaeoceros laevis* (×12). (*A*, after Inoue, 1976; *B*, after Schuster, 1953.)

3. Each of the cells of the thallus usually contains a single large disc-shaped chloroplast, which frequently has an included pyrenoid.
4. The thallus often has mucilage-filled cavities formed by breakdown of

groups of cells; these cavities are often invaded by the blue-green alga, *Nostoc.*

5. The thallus sometimes has ventral pores, resembling stomata in form (sometimes termed as slime pores).

6. The sex organs, although embedded in the upper layers of the thallus when mature, are formed from superficial cells.

7. Some thalli produce internal "tubers" that survive the unfavorable season.

8. Numerous antheridia often originate within a single antheridial chamber. Antheridia are discrete organs.

9. Archegonia are not discrete organs, but are represented by neck canal cells and an egg surrounded by essentially undifferentiated cells of the thallus.

10. The first division of the zygote is by a longitudinal line, thus differing from other bryophytes.

11. The sporophyte is always a tapered horn with no seta.

12. The sporophyte possesses a basal intercalary meristem and has indeterminate growth in most genera (except *Notothylas*).

13. The sporophyte usually grows throughout the favorable season, shedding spores at the apex and differentiating new spores from the intercalary meristem above the foot.

14. The young sporophyte is often protected by a thallus-calyptra (sometimes termed an involucre) that elongates as the sporophyte elongates, and encloses it in a sleeve. Sometimes the apex is ruptured to form an apical calyptra while the remainder sheathes the base of the sporophyte.

15. The sporangium usually opens by one or two longitudinal lines; initially the split does not extend to the apex of the sporangium; the jacket walls are often hygroscopic.

16. The sporangium jacket is multistratose and frequently possesses stomata.

17. There is usually a cylindric columella in the sporangium.

18. The sporogenous layer overarches the columella.

19. The spores usually have multicellular, somewhat hygroscopic elaters among them.

As apparent from the preceding list of features, the class is highly distinctive. Since it is an archegoniate plant, its basic design in gametophyte and sporophyte is extremely similar to many hepatics, and it shows numerous other bryophytic features, it is logical to include it with the Bryophyta.

SPORE MORPHOLOGY AND GERMINATION PATTERNS

The spores of hornworts are always somewhat rounded and possess a conspicuous pyramidal triradiate scar on one face. They tend to adhere in tetrads until they are shed, and sometimes are shed as tetrads. Spore size varies from 27 to 70 μm in unicellular spores to more than 80 μm in multicellular spores. In *Dendroceros* the spores are always multicellular (Fig. 19-2*A*), and in some species of *Anthoceros* spores are either unicellular or multicellular.

FIGURE 19-2
Anthocerotae: spores, sporelings, and elaters. (*A*) Multicellular spore of *Dendroceros japonicus* (×845).
(*B*) Spore of *Dendroceros endiviaefolius* (×1000), (*C*) Spores of *Anthoceros fuciformis* (×1000). (*D*) Spore of *Anthocerpos punctatus* (×1000).
(*E*) Elater of *Dendroceros endiviaefolius* (×1000).
(*F*) Elater of *Anthoceros fuciformis* (×1000).
(*G*) Elater of *Notothylas orbiculars* (×525). (*H–M*) *Notothylas*, spore germination and formation of sporeling (×120). (*N–P*) *Anthoceros punctatus*, spore germination and formation of sporeling (×65). (*A, G*, after Inoue, 1976; *B, C, E, F*, after Proskauer, 1953; *H–P*, after Mehra and Kachroo, 1962.)

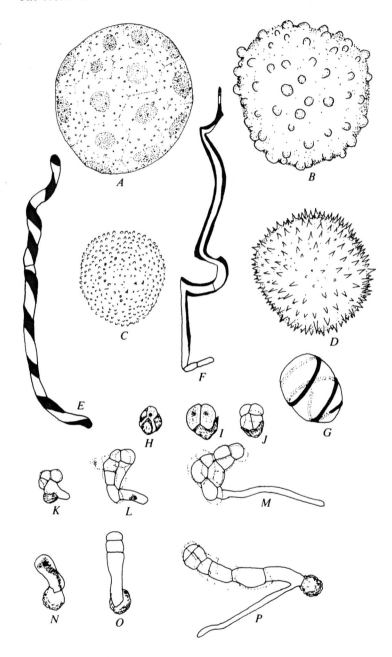

Spores are green in *Megaceros* and *Dendroceros,* but all the other genera have golden-yellow, dark-brown, or black spores. Ornamentation of the spore coat varies from almost smooth in *Notothylas* to vermicular ridged or spiny in several genera. Ornamentation can vary to a certain degree, even within a single sporangium. Spores are viable for more than 10 years in most species and tend to germinate relatively rapidly when under suitable moisture conditions.

Three main patterns of spore germination appear to be represented in the hornworts. In many species of *Anthoceros* the germinating spore produces a short uniseriate germ tube that becomes two to four cells in length. The apical cell soon divides to produce a sporeling that initiates the thallus (Fig. 19-2*N–P*). In *Notothylas,* the spore, when it germinates, usually ruptures the spore wall irregularly and forms a multicellular mass that produces the sporeling directly (Fig. 19-2*H–M*). In *Dendroceros* the spores are chlorophyllose and multicellular when they are shed and initiate the sporeling directly after rupture of the spore coat.

GAMETOPHORE MORPHOLOGY

The thalli of *Anthoceros* and *Notothylas* form dark-green orbicular to semiorbicular rosettes. Sometimes many gametophores are aggregated, and the growth form is irregular with semierect lobes. In these two genera, the thallus is usually much dissected with radiating lobes. The thallus surface is sometimes smooth (Fig. 19-1*A,B*), but some species (e.g., *A. punctatus*), form leaflike lamellae on the dorsal surface. Branching in the thallus is irregular or dichotomous. Thin ruffles occur on either side of a straplike midrib in *Dendroceros* (Fig. 19-1*A*). Branching in *Dendroceros* varies from irregular to somewhat pinnate. *Megaceros* often produces broad, elongate, flattened thallus lobes. Thalli of *Anthoceros* and *Notothylas* are usually 1–2 cm in diameter, but in *Megaceros* the lobes are sometimes more than 5 cm long and more than 1 cm in width. *Dendroceros* thalli are 2–3 cm long and 4–5 mm in width.

The thallus is always multistratose, at least near the center (Fig. 19-3*A,D*). Only in *Dendroceros* is there always a well-defined, ribbonlike, multistratose midrib from which the unistratose lobes or ruffles extend. Thalli tend to be somewhat succulent and brittle. In all thalli most of the cells are uniformly thin-walled, but in *Dendroceros* and *Megaceros,* elongated cells with slitlike pits resemble the clusters of pitted cells in the Marchantiales. These appear to be important in internal conduction. Stomatelike slime pores occur on the undersurface of the thallus (Fig. 19-3*B*). Rhizoids are all unicellular, smooth, colorless, and perpendicular, and usually are most crowded on the thickened portions of the thallus.

In most hornworts of temperate regions the thallus has many internal mucilage cavities that are formed by the breakdown of cells and their replacement by mucilage (Fig. 19-3*D*). These form very early in the development of the thallus. As the thallus matures the mucilage in these cavities often dries out, and air-filled chambers result. Such cavities occur in many species of *Anthoceros, Dendroceros,* and *Notothylas,* but they are absent in *Megaceros* and *Phaeoceros.* In all thalli the cyanobacterium *Nostoc* enters through the slime pores (Fig. 19-3*B*), and colonies live in the thallus. The presence of these

FIGURE 19-3
Gametophytic structure in
Anthocerotae. (*A*) Longitudinal section through thallus
of *Anthoceros punctatus,*
showing mucilage cavities
(stippled) and a single archegonium (×300). (*B*) Pores
from ventral surface of
Phaeoceros laevis game-
tophore (×300). (*C*) Antheridia of *Anthoceros punctatus,*
the left one shedding sperms
(×15). (*D*) Longitudinal sections through a thallus of
Anthoceros punctatus, with
developing embryo in
chamber at left, antheridia in
chamber at right, and enlarged
mucilage cavities (×120).
(*A, D,* after Renzaglia, 1978;
C, after Proskauer, 1951.)

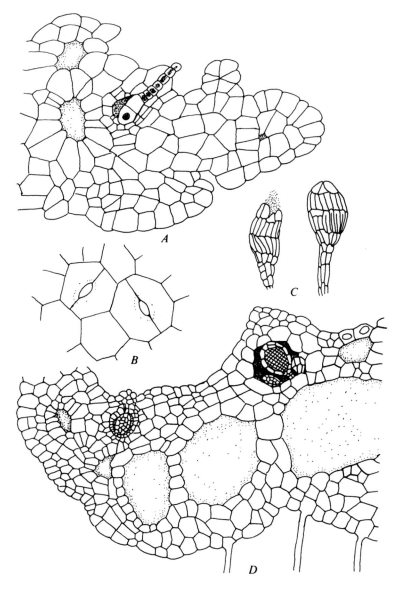

colonies is useful to the hornwort since *Nostoc* fixes nitrogen. The hornwort
supplies a carbohydrate source to the *Nostoc*. Endogenous fungi are also often
present, especially in central regions of the thallus.

Cells of the thallus usually contain a single lenticular chloroplast (Fig.
19-4*B*), except in *Megaceros*, where there are up to 12 chloroplasts per cell.
The chloroplast often encloses the nucleus and, in many cases, encloses a
pyrenoid. All genera except *Dendroceros* possess a wedge-shaped apical cell
with four cutting faces. In *Dendroceros* there are three cutting faces.

FIGURE 19-4
Anthoceros punctatus
(diagrammatic). *(A)* Transverse section through
sporangium, showing angular
columella in center, spore tetrads and elaters, and multistratose jacket (×60). *(B)*
Cells of thallus showing chloroplasts invested with
starch grains (×100).

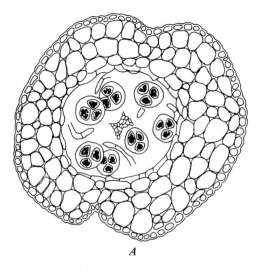

A

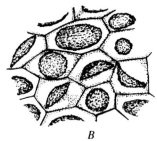

B

ASEXUAL REPRODUCTION

Some hornworts, especially *Anthoceros,* produce perennating "tubers" that resemble those in the Metzgeriales and in *Geothallus* of the Sphaerocarpales. "Tubers" tolerate unfavorable conditions, especially desiccation, to which most of the thallus is intolerant, and thus permit the thallus to be perennial. Since many "tubers" are produced by a single thallus, especially late in the season, "tubers" also serve as vegetative diaspores. Some species of *Anthoceros* produce gemmae on the thallus margin. These are small spherical clusters of cells which can produce a new gametophore. In humid sites, as the thallus ages and the older portions decay, isolated lobes become independent gametophores and ultimately produce an expanding clone.

DEVELOPMENT AND POSITION OF THE SEX ORGANS

The development of sex organs is the same in all the hornworts. Most thalli are monoicous, but dioicous thalli exist, even within a single species. In *Phaeoceros laevis,* for example, some geographic races are monoicous, while others are dioicous. In a monoicous thallus antheridium production precedes archegonium production, and usually only after antheridia are mature does archegonium production begin. This tends to prevent self-fertilization, but it is not entirely effective, at least in timing.

Antheridium production begins with transverse division in an epidermal cell on the upper surface of the thallus. This cell never becomes papillate. A periclinal division of the outer of the two cells produces the cells that become the bistratose roof; the cell beneath becomes a mucilage cavity which enlarges to become the antheridial chamber. On the floor of this cavity an antheridial initial forms in some genera, including *Dendroceros* and *Megaceros,* and this initial gives rise to a single antheridium by continued divisions. In other genera several initials are formed, each of which produces an antheridium (Fig. 19-5). As antheridium development proceeds, further cell divisions occur in the cells that bound the antheridial chamber, enlarging it and the rest of the thallus. There may be as many as 15 antheridia (and sometimes more) within a single chamber in *Anthoceros,* and each of these may be at a different stage of maturity. Mature antheridia are ovoid to subspherical and have a short stout stalk. As the antheridia mature and elongate, they push upward in the enclosing chamber and rupture the roof irregularly. The antheridia open at the apex and expel the sperms (Fig. 19-3C). Antheridia occur in scattered patches on the thicker portion of the thallus, usually behind the growing point.

Initiation of the archegonia is similar to that of the antheridia (Fig. 19-6A–D). A transverse wall divides an epidermal cell into an inner and outer cell. The outer cell undergoes three longitudinal divisions, resulting in an inner cell enclosed by an outer three, and triangular in cross-sectional view. This axial cell, by further transverse divisions, forms the neck-canal cells and the egg. Further divisions of the surrounding cells of the thallus enclose these internal cells. The outer cells are indistinguishable from the rest of the cells of the thallus, and the archegonium is not a discrete organ. The surface cells above the neck-canal cells become a mucilage papilla, and when the contents of the

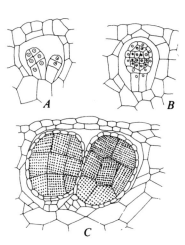

FIGURE 19-5
Development of the antheridium in *Anthoceros fusiformis* in vertical section. (*A*) Formation of two antheridia from floor of chamber (×240). (*B*) Developing single antheridium (×240). (*C*) Two nearly mature antheridia (×110). (After Smith, 1955.)

FIGURE 19-6
Development of archegonium and sporophyte in *Phaeoceros laevis* (×275). (*A–D*) Early stages in differentiation of neck canal cells and egg. (E, F) Early cell divisions in embryo. (*G*) Young embryo removed from thallus, showing development of foot and development of amphithecial and endothecial layers. (*H*) Sporophyte within thallus, showing development of columella and sporogenous layer (stippled). (After Smith, 1955.)

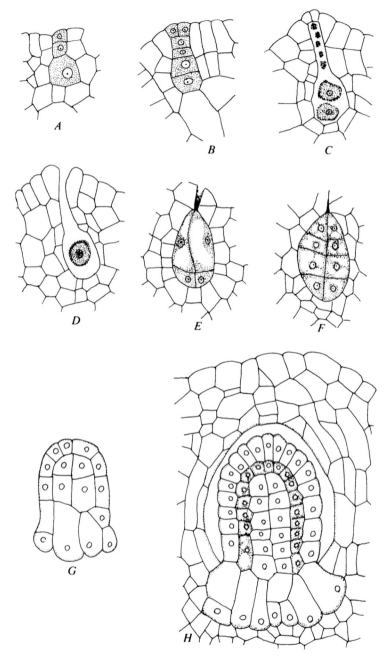

archegonium are mature, these cells disintegrate, and the mucilaginous remains of the neck-canal cells extrude to the surface of the thallus and expose the canal to entrance by the sperms. Archegonia tend to form in clusters of one to three on the thicker portions of the thallus, usually behind the growing point. In monoicous thalli, they occupy the same relative position as the antheridia that developed before them.

SPOROPHYTE DEVELOPMENT AND MORPHOLOGY

In the hornworts the first division of the zygote is longitudinal, in this way differing from other bryophytes. Succeeding divisions are transverse and longitudinal (Fig. 19-6*E,F*). The lower cells initiate the formation of the foot, while the upper ones form the remainder of the sporophyte. An intercalary meristem is initiated early. Early in embryonic development there is a differentiation of amphithecium and endothecium (Fig. 19-6*H*). The endothecium, as in *Sphagnum,* produces the columella, while the amphithecium produces the sporogenous layer and jacket.

The sporophyte is a tapered cylinder with a bulging foot that penetrates the dorsal surface of the thallus. The foot often produces short haustorial projections that push among the cells of the thallus. The sporophyte, although chlorophyllose, is dependent on the gametophore. If removed from the thallus it ceases to undergo further differentiation of spores and vegetative cells although those already differentiated before isolation will mature after isolation. Sporophytes sometimes reach lengths of 12 cm, but are usually much smaller. In *Notothylas* the entire sporophyte is often less than 5 mm long.

In all genera except *Notothylas,* the intercalary meristem at the base of the sporangium permits continued growth and differentiation of cells in the sporophyte as long as conditions permit (Fig. 19-7). These conditions may involve seasonal changes and sometimes, especially in the tropics, age of the thallus. As the thallus becomes old it tends to decompose; any sporophytes on this part of the thallus are isolated, and death follows.

The jacket in most sporophytes is multistratose and has stomata, but in *Dendroceros* and *Notothylas* stomata are absent, and they are rudimentary or absent in *Megaceros* as well as in a few species of *Anthoceros* and its segregate genera. The jacket in *Megaceros* sometimes is up to 16 cells thick, while in other genera it is usually 4 or 5 cells thick. The epidermal cells of the jacket sometimes have thick walls, as in *Dendroceros*. In *Dendroceros* too, the inner cells of the jacket (the lining layer) sometimes possess weakly defined transverse thickened bands.

Rupturing of the sporangium is usually by two longitudinal lines, but occasionally by a single line and rarely by four lines. Some species of *Notothylas* dehisce by decay of the jacket walls. In most genera, as the apex of the sporangium matures and dries out, the slits gape open and expose the spores and elaters. The elaters, as they dry, tend to twist, loosening the spores and releasing them. This process continues downward in the sporangium as new spores mature. In most genera the jacket walls are also hygroscopic and their coiling and uncoiling assists in spore release.

The sporangium usually contains a central columella of elongate cells, which

FIGURE 19-7
Sporophyte structure in Anthocerotae. Schematic longitudinal view through segments of the sporophyte of *Anthoceros* (×80). (*A*) Apex, showing spores and elaters being released; columella is in middle. (*B*) Mature segment showing tetrads of spores and elaters. (*C*) Segment showing earlier stage with sporogenous cells with nuclei. (*D*) Basal portion showing expanded foot, intercalary meristem, and differentiating sporogenous layer and columella. (Modified after Parihar, 1965.)

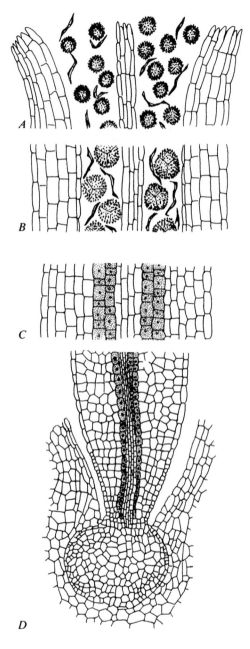

in *Dendroceros* may have spiral and annular thickenings. The columella is important in internal conduction. In some species of *Notothylas* a columella is absent.

Multicellular or unicellular elaters are among the spores (Fig. 19-2*E–G*). In *Anthoceros* (including its segregate genera) the elater walls have irregularly arranged thick and thin areas. In *Notothylas* the walls are uniformly thin-walled, but in *Dendroceros* and *Megaceros* the elater walls have a single spiral thickening, much as in the hepatics. Elaters are somewhat hygroscopic and are important in spore dispersal. In *Megaceros* and *Dendroceros* they probably also assist in internal conduction. Elaters vary considerably in length, from as little as 50 μm to as much as 700 μm. Considerable variation in elater size exists within a single sporangium. The elaters tend to be longitudinally arranged, but in *Notothylas* they are transversely arranged.

As the embryo enlarges, within the thallus there is cell division and elongation of the adjacent cells of the thallus to form a sleeve to accommodate the elongating sporangium. This is the thallus-calyptra, so called because it is derived from cells of the thallus that occupy the same position as the archegonial cells leading to calyptra formation in other bryophytes. The thallus-calyptra is initiated in response to embryonic growth. As the sporophyte elongates it ruptures the summit of the thallus-calyptra and sometimes carries this summit as a cap at the tip of the sporophyte, but leaves the rest as a basal protective sleeve that may reach a length of a centimeter or more.

EVOLUTIONARY TRENDS AND RELATIONSHIPS

Within the hornworts it appears that *Notothylas* could have been derived from *Phaeoceros*. This was achieved presumably through reduction of some features of the sporophyte. The relationships of *Dendroceros* and *Megaceros* to *Anthoceros* are less easy to interpret. These genera appear to be generalized, especially in sporophytic structure, but in gametophytic morphology *Dendroceros* shows some specialization; such specialization is represented also in the multicellular spores and in sporeling development.

Trends in development in the gametophore appear to be from a large thallus lacking mucilage cavities to a small thallus possessing these cavities. The ability to produce "tubers" also appears to be specialized.

The generalized sporophyte can be visualized as one with indeterminate growth, a thick multistratose stomata-containing jacket, a massive columella with some cells possessing spiral thickenings, and with helically thickened-wall elaters. Dehiscence would be by two longitudinal lines (as in *Dendroceros*). The specialized condition would be represented by a short sporangium of determinate growth, a thin-walled, multistratose jacket without stomata, a rudimentary columella or none, and small thin-walled elaters. Dehiscence would be by decomposition of the jacket. This is best represented in *Notothylas*.

The relationship of the hornworts to other hepatics is very distant, and it seems probable that they did not have the same immediate ancestor. Some authors suggest that the hornworts may have arisen through reduction from the tracheophyte class Rhyniophytina. Gross structure of the sporangia is similar, and even spore size and ornamentation are almost identical to some hornworts. The gametophore is extremely similar to that of many "pteridophytes," and the early embryonic development is like many vascular plants. J. Proskauer has also drawn attention to the similarity of the columellar cells and elaters of *Dendroceros* to the tracheids in the stele of *Horneophyton* in the Class Rhyniophytina. Based on these final observations, it would support the placement of the hornworts in their own division.

FURTHER READING

Bartlett, E. M. 1928. A comparative study of the development of the sporophyte in the Anthocerotae, with special reference to the genus *Anthoceros*. *Ann. Bot* **42**:409–430.

Bharadwaj, D. C. 1950. Studies in Indian Anthocerotae. I. The morphology of *Anthoceros crispulus* (Mont.) Douin. *J. Indian Bot. Soc.* **29**:145–163.

——— 1965. Studies in Indian Anthocerotae. VI. Some aspects of morphology of *Phaeoceros* Prosk. *Phytomorphology* **15**:140–150.

——— 1971. On *Folioceros*, a new genus of Anthocerotales. *Geophytology* **1**:6–15.

——— 1972. On some Asian species of *Folioceros* Bharad. *Geophytology* **2**:74–89.

——— 1973. Taxonomy of some Indo-Pacific species of *Folioceros* Bharad. *Geophytology* **3**:215–221.

Burr, F. A. 1970. Phylogenetic transitions in the chloroplasts of the Anthocerotales. I. The number and ultrastructure of the mature plastids. *Am. J. Bot.* **57**:97–110.

Campbell, D. H. 1898. On the structure and development of *Dendroceros* Nees. *J. Linn. Soc. (Bot.)* **33**:467–478.

——— 1907. Studies on some Javanese Anthocerotaceae. I. *Ann. Bot.* **21**:467–486.

——— 1908. Studies on some Javanese Anthocerotaceae. II. *Ann. Bot.* **22**:91–102.

——— 1924. A remarkable development of the sporophyte in *Anthoceros fusiformis*. *Ann. Bot.* **38**:473–483.

Crandall-Stotler, B. 1984. Musci, Hepatics and Anthocerotes—an essay on analogues, in Schuster, R. M. (ed.), *New Manual of Bryology*, pp. 1093–1129. Nichinan, Japan: Hattori Bot. Lab.

Hassel de Menendez, G. G. 1976. Taxonomic problems and progress in the study of the Hepaticae. *J. Hattori Bot. Lab.* **41**:19–36.

Howe, M. A. 1898. The Anthocerotaceae of North America. *Bull Torrey Bot. Club* **25**:1–24, plates 321–326.

Inoue, H. 1978. Memoir of *Dendroceros tubercularis* Hatt., an endemic hornwort in the Bonin Islands. *Mem. Nat. Sci. Museum* (Tokyo) **11**:31–36.

Isaac, J. 1941. The structure of *Anthoceros laevis* in relation to its water supply. *Ann. Bot.* **5**:339–351.

Leitgeb, H. 1879. Untersuchungen über die Lebermoose III. Die Frondosen Jungermannien V. Die Anthoceroteen. Jena, pp. 1–59, v plates.

McAllister, F. 1914. The pyrenoid of *Anthoceros*. *Am. J. Bot.* **1**:79–93.

Mehra, P. N., and P. Kachroo. 1962. Sporeling germination studies in Anthocerotales, *J. Hattori Bot. Lab.* **25**:145–153.

Meier, W. 1957. Notes on some Malayan species of *Anthoceros* II. *J. Hattori Bot. Lab.* **18**:1–13.

Mottier, D. M. 1894. Contributions to the life-history of *Notothylas*. *Ann. Bot.* **8**:391–402.

Pande, S. K. 1932. On the morphology of *Notothylas indica* Kashyap. *J. Indian Bot. Soc.* **11**:169–177.

Pierce, G. J. 1906. *Anthoceros* and its *Nostoc* colonies. *Bot. Gax.* **42**:55–58.

Proskauer, J. 1948. Studies on the morphology of *Anthoceros* I–II. *Ann. Bot.* **12**:237–265, 427–439.

——— 1951. Studies on Anthocerotales III. The genera *Anthoceros* and *Phaeoceros*. *Bull. Torrey Bot. Club* **78**:331–349.

——— 1953. Studies on Anthocerotales IV. *Bull. Torrey Bot. Club* **80**:113–135.

——— 1957. Studies on Anthocerotales V. *Phytomorphology* **7**:113–135.

——— 1960. Studies on Anthocerotales VI. *Phytomorphology* **10**:1–19.

——— 1967. Studies on Anthocerotales VII. *Phytomorphology* **17**:61–70.

——— 1969. Studies on Anthocerotales VIII. *Phytomorphology* **19**:52–66.

Renzaglia, K. S. 1978. A comparative morphology and developmental anatomy of the Anthocerotophyta. *J. Hattori Bot. Lab.* **44**:31–90.

Rink, W. 1935. Zur Entwicklungsgeschichte, Physiologie und Genetik der Lebermoosgattungen *Anthoceros* and *Aspiromitus*. *Flora* **130**:87–130.

Schuster, R. M. 1984. Morphology, phylogeny and classification of the Anthocerotae, in Schuster, R. M. (ed.), *New Manual of Bryology*, pp. 1071–1092. Nichinan, Japan: Hattori Bot. Lab.

Wilsenach, R. 1963. Differentiation of the chloroplast of *Anthoceros*. *J. Cell Biol.* **18**:419–428.

20 History of Bryology

The study of bryophytes is of relatively recent origin. The contribution of the ancient Greeks and Romans is negligible. Indeed, their erroneous concepts of bryophytes show the same confusion that persists to the present day, where the term "moss" is used to include any organism that shows what is interpreted as a mosslike appearance, independent of its relationships; this includes lichens, vascular plants, algae, and even some invertebrate animals. The medieval herbalists refined the concepts only slightly. The term "liverwort" is derived from the herbalists, who attributed curative properties for liver ailments to any organism thought to resemble a liver. Lobate thalli of both lichens and liverworts satisfied this basic structure, and the term was, in consequence, extremely general in its application. Several sixteenth century herbals illustrate thallose liverworts which can be determined confidently as *Marchantia polymorpha*.

EIGHTEENTH CENTURY

P. A. Micheli (1679–1737) presented, in 1729, the names and accurate figures for several liverworts, among them *Riccia*, *Lunularia*, *Blasia*, and *Marchantia*. He also recognized and illustrated *Anthoceros*. He possessed a vague understanding of the structures of bryophytes but misinterpreted the function of both sporophytic and gametophytic features. Like his successors, he was trapped by the notion that bryophyte structures of form similar to those in vascular plants served the same functions, but he did correctly observe that the spores of mosses were not equivalent to pollen.

The study of bryology actually began with J. J. Dillenius (1648–1747), who published his *Historia Muscorum* in 1741 (Fig. 20-1A). In this he illustrated and presented short descriptions of numerous bryophytes. The name *Lichenastrum* was applied to a diversity of thallose liverworts, but a number of lichens were also treated under that category. The name *Mnium* was applied to such

FIGURE 20-1
Bryologists. (*A*) J. J. Dillenius
(1684–1747). (*B*) J. Hedwig
(1730–1799). (*C*) W. S. Sullivant (1803–1873). (*D*) M.
Fleischer (1861–1930). (Redrawn from portraits.)

unrelated mosses as *Tetraphis* and *Aulacomnium*. The name *Polytrichum*, although including that genus, also included a diversity of unrelated moss genera.
Many of the plates illustrate the gross form of the bryophytes very accurately,
and his publication served as a means for identifying these plants for many
years. Like Micheli, Dillenius endeavored to seek homologies for structures in

bryophytes to match morphologically similar structures in seed plants. He interpreted the sporangium as an anther and the spores as pollen in spite of the contrary opinion of Micheli.

It was C. C. Schmidel (1718–1792), who recognized, in 1747, the function of the liverwort sporangium and also interpreted the antheridium as the male organ, although he did not recognize sperms. He appears also to have recognized the function of the elaters in the liverwort sporangium.

The great taxonomist C. Linnaeus (1707–1778) relied heavily on the work of Dillenius and utilized names that Dillenius had proposed. In consequence, and because no microscopic characters were used, he considered mosses and vascular plants closely related, and treated *Lycopodium* as related to the mosses; some plants now treated as algae he classified together with some plants now treated as hepatics.

J. Hedwig (1730–1799) is the first researcher who interpreted byrophytes in much the concept that is maintained today (Fig. 20-1*B*). He used the microscope extensively and achieved a magnification near 300×. He correctly interpreted the structure and function of antheridia, archegonia, and the sporophyte. He studied the development of the moss sporophyte and understood the structure and function of the peristome. He made the first clear interpretation of the differences between mosses and liverworts and presented, for the first time, an essentially complete and correct account of the bryophyte life cycle, which he accompanied with accurate, elegant illustrations. He described and illustrated many genera and species of mosses, and these, with his beautiful hand colored plates, were published posthumously in his *Species Muscorum* (1801). This publication serves as the starting point of nomenclature of most mosses. C. F. Schwaegrichen (1775–1853), who acquired the Hedwig herbarium and unpublished material, published, between 1811 and 1842, numerous further moss names. Unfortunately, Hedwig's ideas were rejected by many contemporary workers.

NINETEENTH CENTURY

N. J. Necker (1730–1793) first interpreted gemmae correctly, but unfortunately he drew the conclusion that bryophytes did not reproduce sexually. It was not until 1851 that W. Hofmeister (1824–1877) described and properly interpreted the egg cell and clearly outlined the concept of alternation of sporophytic and gametophytic generations. He noted also the basic affinities of the bryophytes and vascular archegoniates.

Although preceding researchers had seen sperms in bryophytes, it F. J. A. N. Unger (1800–1870), who, in 1834, first correctly suggested their function, in the moss genus *Sphagnum*. C. von Nägeli (1817–1891) in 1845 described the apical cell in bryophytes and noted segments cut off from it. W. Hofmeister in 1870 interpreted the contribution of these segments to the axis of the hepatic shoot.

W. P. Schimper (1808–1880) greatly influenced the study of mosses during much of the 19th century. In 1848 he published studies which contained graceful and accurate figures showing details of the morphology and anatomy of mosses. His study of *Sphagnum* is especially significant. In collaboration with P. Bruch (1781–1847) and W. T. von Gümbel (1812–1858) he produced the monumental *Bryologia Europaea*. Serially published from 1836 to 1855, this work contains beautiful etchings of all European taxa known at that time as well as careful and detailed descriptions and discussions of each. These authors refined the generic concepts from those of Hedwig and Schwaegrichen and greatly improved the definition of species as applied to mosses.

P. G. Lorentz (1835–1881) published between 1864 and 1869 an assessment of the developmental anatomy of mosses. The meticulous morphologic studies of H. Leitgeb (1835–1888) established, from 1874 to 1881, an understanding of the fundamental branching patterns in the Hepaticae; his studies have served as one of the foundations for modern understanding of the interrelationships of the hepatics. He noted also the distinction between acrogyny and anacrogyny in the Hepaticae.

In 1878 N. Pringsheim (1823–1894) demonstrated the aposporous production of a gametophyte from sporophytic tissue in a moss. This discovery somewhat weakened the basic concept of the life cycle, since there appeared to be no definitive factors that controlled the basic morphology of the gametophyte. E. A. Strasburger (1844–1912) recognized in 1894 the haploid and diploid phases in the life cycle of plants, thus completing the general foundation to its understanding.

K. I. E. Goebel (1855–1932) began a careful assessment of developmental anatomy in bryophytes, summarized in *Organographie der Pflanzen,* the final edition of which appeared in 1930. His research was enhanced by the invention of the microtome and the accurate figuring of anatomical details; combined with his assessment of the developmental stages, this set a firm foundation for subsequent research. D. H. Campbell (1859–1953) began his researches in bryophyte structure and development at approximately the same time. As with Goebel, his observations were based on careful preparation of serial sections and were accompanied by accurate drawings. A synthesis of his understanding was presented in the final edition of *The Structure and Development of Mosses and Ferns* (1913).

The nineteenth century marked a conspicuous increase in discovery and description of the bryophytes of the world. Since accurate identification is fundamental to understanding of plants, this activity was vital to succeeding researchers. Unfortunately, a number of researchers were hasty in their judgments and presented new names for thousands of species, a high proportion of which have proven to be based on faulty concepts or erroneous judgments. The hepaticologist F. Stephani (1842–1927) and the muscologists K. A. F. W. Müller (1818–1899) and N. C. Kindberg (1832–1910) are notorious for creating species that have been relegated to synonymy by succeeding researchers. Since all of these species were legitimately described, it means that a great deal of time has been expended to prove which species names represent distinctive species, and which are based on poor judgment and must be discarded.

Many researchers have produced manuals that have been of immense importance in improving the accurate understanding of the diversity, interrelationships, and distribution of the bryophytes. A number of these are mentioned because of the impact they have made on an accurate assessment of the bryophytes and because the sound scholarship of their researches has positively influenced all succeeding research.

S. E. Bridel-Brideri (1761–1828) described all mosses known at the time in his *Bryologia universa* (1826–1827). This has been a valuable source of information, but possibly stimulated careless researchers to add to the number of species without careful assessment of taxa before they were proposed. C. G. D. Nees von Esenbeck (1776–1858) elaborated an extensive system of classification for the Hepaticae (1833–1838), and, with his collaborators, produced between 1844 and 1847, the classical *Synopsis Hepaticarum*.

W. J. Hooker (1785–1865) in his *British Jungermanniae* (1816), *Plantae cryptogamicae* (1816), and *Musci exotici* (1818–1820), presented descriptions and sometimes hand-colored figures of numerous bryophytes, not only of Britain, but also of South America, North America, and other parts of the world, treating specimens presented to him for study by A. von Humboldt and other naturalists.

W. S. Sullivant (1803–1873) was the first resident scholar in North America to make major contributions to bryology (Fig. 20-1C). He issued exsiccati of bryophytes from the eastern United States and published careful studies from his own collections, as well as those from exploring expeditions. His major work was *Icones Muscorum* (1864) and its supplement (1874) published after his death. Works of equal caliber were those of F. Dozy (1807–1856) and J. H. Molkenboer (1816–1854), whose beautifully illustrated treatments of the mosses of the Indomalayan and Indonesian area and adjacent regions were published between 1844 and 1870. At approximately the same time, R. Spruce (1817–1893) and W. Mitten (1819–1906) produced their important treatments of bryophytes of tropical America. Spruce's *Hepaticae* (1884–1885) of the Amazon and of the Andes of Peru and Ecuador (1885) remains a fundamental treatment for that region, and Mitten's *Musci austro-americani* (1869) is of equal importance.

Late in the nineteenth century, and in the period preceding the First World War, a number of researchers in Europe presented comprehensive studies that are basic to an understanding of the North Temperate bryoflora, in particular. F. Renauld (1837–1910), besides publishing many important discussions of the mosses of this broad region, also published a comprehensive moss flora of Madagascar and adjacent islands (1898, 1909). In 1915, J. Cardot (1860–1934) produced a collaborative flora of the same area, which he completed after Renauld's death. Cardot also presented the first detailed treatment of Antarctic mosses (1908, 1913), a monograph of the moss family Fontinalaceae (1892), and numerous papers concerning the moss flora of many parts of the world. R.

Braithwaite (1824–1817) published important volumes on the mosses of Great Britain (1880–1905), and W. H. Pearson (1849–1923), on the Hepaticae of Great Britain (1902). For many years these served as vital references for the bryoflora of the North Temperate region. Since they were fully illustrated, sometimes with hand-colored plates, and had extensive descriptions, their importance was considerable. These floras were supplanted by the valuable Student's Handbooks written by H. N. Dixon (1861–1944) on British mosses (the first edition of which appeared in 1896, and the third, and final, edition in 1924), and by S. M. Macvicar (1857–1932) on British Hepaticae and Anthocerotae (first edition in 1912 and second, final edition in 1926). These works were inexpensive, written with the enthusiastic amateur in mind, and were fully illustrated. Besides containing full descriptions of all taxa, useful comments were given that enhanced a precise understanding. Dixon published numerous papers on the mosses from many parts of the world, based on collections submitted to him. Possibly his most important extra-European contribution was his studies of the New Zealand moss flora (1913–1929), which greatly clarified the comprehension of the moss flora of the Southern Hemisphere.

TWENTIETH CENTURY

In continental Europe there was a similar outburst of vital floristic research resulting in several comprehensive manuals of profound importance and serving as a stimulus to accurate detailed research on many taxa. The volumes of K. G. Limpricht (1834–1902) for the central European moss flora (1885–1903) remain the most comprehensive treatment for that region. The descriptions of species are models of detailed and accurate observation. The working manual of W. Moenkemeyer (1862–1938) published in 1927, with extensive illustrations, is especially useful in revealing the polymorphism of many taxa. L. Loeske (1865–1935) produced monographs of the European Funariaceae (1929) and Grimmiaceae (1930), revealing the considerable insight to be gained in studying one group in detail. His *Studien zur vergleichenden Morphologie und phylogenetischen Systematik der Laubmoose* (1910) was a major attempt to synthesize the details of comparative morphology with interpretations of interrelationships of the mosses. Unfortunately, the book is not illustrated. For the Hepaticae and Anthocerotae of Europe, the publications of K. Müller (1881–1955) (Fig. 20-2*C*) and V. F. Schiffner (1862–1944), in particular, are of basic importance, supplying rich detail that has determined the direction of all succeeding research on these bryophytes. The final edition of *Die Lebermoose Europas* of K. Müller was published serially from 1951 to 1958. Schiffner in 1893–1895 also contributed the most recent attempt to assess all of the hepatics of the world. In 1917 he provided a discussion of his systematic and phylogenetic investigations in Hepaticae.

V. F. Brotherus (1849–1929) was the most important researcher of his time in floristics of mosses (Fig. 20-2*A*). He acquired a rich collection of material

FIGURE 20-2
Bryologists. (*A*) V. F. Brotherus (1849–1929). (*B*) H. Buch (1883–1964). (*C*) K. Mueller (1881–1955). (*D*) A. Evans (1868–1959). (Redrawn from portraits.)

from throughout the world and published numerous papers detailing his discoveries. His major contributions were *Die Laubmoose Fennoskandias* (1923) and his contribution *Musci (Laubmoose)* to Engler and Prantl's *Die Natürlichen Pflanzenfamilien,* the first edition of which appeared in 1909, and the second and final edition in 1924–1925. This remains as the only attempt to treat all known families, genera, and species of the mosses of the world. The work is extensively illustrated and has full descriptions of all families and genera. Although now somewhat dated, it is a vital component of the working library of any serious bryologist.

Outside Europe, the researches on mosses by M. Fleischer (1861–1930) of the Indonesian area (especially 1904–1924) with his proposal of a phylogenetic system of classification of mosses, based on sporophytic characters, is of particular significance (Fig. 20-1*D*); this system was elaborated and improved by Brotherus (Fig. 20-2*A*). The work of G. O. K. Sainsbury (1880–1957), which culminated in his *Handbook of New Zealand Mosses* (1955), stimulated renewed research in New Zealand mosses. The researches of Y. Horikawa (1902–1976) activated research in the bryophytes of Japan and adjacent areas. The publications of P.-C. Chen (1907–1970) led to a resurgence in the understanding of the bryophytes of the People's Republic of China. The careful researches of S. R. Kashyap (1882–1934) revealed the richness of the hepatic flora of the Indian subcontinent. The studies of R. A. L. Potier de La Varde (1878–1961) and T. R. Sim (1858–1938) revealed the considerable significance of the bryophyta of Africa. The inspiration of L. I. Savicz-Ljubitzkaja (1886–1978) initiated a comprehensive study of the bryophytes of the Soviet Union.

In North America, E. B. Bartram (1878–1964) made major contributions to an understanding of the moss flora of North America, and also produced detailed manuals of the mosses of the Hawaiian Islands (1933), the Philippines (1939), and Guatemala (1949) as well as innumerable papers concerning the mosses of nearly all parts of the world. He encouraged numerous students to proceed in the study of mosses and was a major influence in modern bryology. In the hepatics, A. W. Evans (1868–1959) made major contributions by production of beautifully researched studies of a diversity of hepatics, especially of the Americas (Fig. 20-2*D*). His studies contributed significantly to an understanding of the phylogeny of the hepatics. T. C. Frye (1869–1962) and L. Clark (1884–1967) presented a treatment of all North American Hepaticae and Anthocerotae (1937–1947); their work suffers seriously in its uncritical approach.

It was A. J. Grout (1867–1947) who stimulated much of the recent research in the taxonomy of mosses in North America. He accomplished this through his private publication of manuals to the mosses, which culminated in his *Moss Flora of North America* (1828–1940) which, in spite of its shortcomings, is likely to remain the only treatment of all of North American mosses for a considerable period of time.

Although floristic researches are of great value, other aspects of their biology provide insight into the bryophytes as living plants and greatly enhance an understanding of basic biological processes. C. F. E. Correns (1864–1933) published a comprehensive treatise in which he described, illustrated, and discussed the modes of vegetative reproduction in the mosses. F. Cavers (1876–1936), in a series of papers (1910–1911), presented a general discussion of the interrelationships of the bryophytes. He considered thallose hepatics to be primitive, and the other bryophytes to have been derived from thallose forms. This contrasted markedly with Goebel's interpretation; he considered the *Riccia* type of thallus to represent the most specialized morphology, reaching its relatively simple structure through reduction.

The artificial production of polyploids in plants was first demonstrated in 1911 by E. Marchal (1839–1923) and E. J. J. Marchal (1871–1954). They did this by culturing diploid tissue of mosses that differentiated to form protonemata and ultimately gametophores. Fertilization resulted in a tetraploid sporophyte, and this tetraploid material was cultured to produce tetraploid gametophores. C. E. Allen (1872–1954) in 1919 noted sex-correlated chromosomes for the first time in plants; this was demonstrated in *Sphaerocarpos,* in which one chromosome (suggested to be the X chromosome) in the archegonium-bearing gametophore was larger than the rest of the chromosome complement, while the cells of the antheridium-producing gametophore showed one chromosome (denoted as the Y chromosome) to be much smaller than the others. Allen also carried out extensive genetic studies in the genus *Sphaerocarpos* and was the first researcher to attempt genetic studies with haploid organisms. E. Heitz (1892–1982) discovered, described and named heterochromatin in the nucleus of plants, using mosses to demonstrate it. He figured heterochromosomal bivalents and suggested that they might be sex chromosomes (1928). His discovery of heterochromatin is of considerable importance in cytological research, since heterochromatic bands of chromosomes have served as valuable markers to distinguish the different chromosomes within sets.

Beginning in 1919 and 1920 H. R. V. Buch (1883–1964) published on experimental researches showing the considerable phenotypic plasticity of leafy hepatics (Fig. 20-2*B*). These researches led him to clarify the taxonomy of the genus *Scapania*. It marked the beginning of field-oriented taxonomic studies and attempted to direct researchers to living plants rather than exclusive examination of dead herbarium materal.

T. K. G. Herzog (1880–1961) produced the first comprehensive analysis of bryogeography with his *Geographie der Moose* (1926). His wide field experience in bryophytes and careful analysis of the available literature makes this work the foundation for all succeeding bryogeographic studies.

The careful genetic studies in *Marchantia* in 1943 by H. E. H. Burgeff (1883–1976) serve as the only example of a comprehensive genetic analysis of a single bryophyte genus. He was able to determine the apparent origin of a very common species, *M. polymorpha,* from what had been interpreted as varieties of that species.

Between 1949 and 1968 A. Vaarama (1912–1975) produced his meticulous studies on the cytology of mosses. His work stimulated a considerable body of cytological research, and his interpretations and methodology have been vital in succeeding studies.

Between 1948 and 1969 J. M. Proskauer (1923–1970) published a series of papers concerning the Anthocerotae. He applied cytological, anatomical, and culture studies to the elucidation of species and generic concepts. Like Buch, he discovered that environment sometimes altered the phenotype, and these morphotypes had been named as distinctive species.

M. F. Neuberg (1894–1962) published in 1960 her elegant studies on fossils of the Permian from Angaraland (U.S.S.R.). These fossils occupy a critical position in the understanding of the interrelationships of mosses, and their meticulous study and figuring have served as a model for all succeeding studies of fossil mosses. The publication of the *Index Muscorum* (1959–1969), with R. van der Wijk (1895–1981) as senior editor, has greatly enhanced a more comprehensive understanding of the mosses.

Anatomy of bryophytes and its relationship to conducting systems had been relatively neglected until C. Hébant (1941–1982) began to publish a series of papers, culminating in his book *The Conducting Tissues of Bryophytes* (1977). His application of the techniques of chemistry and light and electron microscopy have helped to stimulate further studies to clarify the understanding of possible routes that evolution has taken in the production of functionally similar conducting systems in entirely unrelated plant groups.

CURRENT RESEARCH

The preceding discussion has dealt with only deceased researchers, who can be treated within an historic context. Current bryological research is remarkably healthy. Taxonomic studies have tended to be concentrated on monographic studies, but attempts to clarify nomenclature also have been significant. Major manuals have been published for bryophytes of various parts of North America, Europe, Africa, Asia, Australia, South America, and the Pacific Oceanic Islands. The utilization of biochemistry in taxonomic understanding has increased considerably with the refinement of analytical techniques. Identification of substances unique to specific bryophyte groups has assisted considerably in making taxonomic interpretations. Similar improvements in the available technology have opened up an entirely new means to evaluate microscopic and submicroscopic structures and their significance in the biology of bryophytes. These two fields still remain little exploited in bryology; results are highly preliminary and based on a very limited sample.

Research in physiological ecology has been renewed in the past few years, also in response to the diversity of new technology to assist in acquiring detailed information concerning specific bryophytes. Cytological research has also been renewed within the past few years, and the information concerning mosses, in particular, is attracting active researchers in areas outside temperate floras, where most of the preceding analyses were made.

Several valuable summaries of the status of knowledge of fossil and subfossil bryophytes have appeared, mainly concerning bryophytes of the last 12,000 years.

Researches in bryogeography have attempted to assess the genesis of distributional patterns as related to speculations concerning continental positions through time. Such interpretations have aided in evolutionary and phylogenetic speculation.

Research in genetics has lagged behind most other bryological studies, although significant contributions were made in the 1930s. Part of the problem lies in the culturing of the material, but problems in clarity of taxonomy, in the difficulties in dealing with haploid material, and simply with the fact that genetic research appears to concentrate its analyses to very few kinds of organisms, have hampered advance in genetics of bryophytes. Induction of genetic mutation and somatic hybridization in mosses promises to yield new insight into bryophyte genetics.

PERIODICALS DEVOTED TO BRYOLOGY

The first journal devoted to bryology, *Revue Bryologique,* was established by T. Husnot in 1874. This journal continued until 1926. A. J. Grout and E. G. Britton began the journal *The Bryologist* in 1898, which continues to the present. P. Allorge revived *Revue Bryologique* as *Revue Bryologique et Lichénologique* in 1928. This publication continued under this name until 1980, when it was incorporated into *Cryptogamie: Bryologie et Lichenologie,* which contuinues to the present. Frans Verdoorn began the *Annales Bryologici* in 1929; this journal appeared annually until 1939, when it was discontinued. In 1947 three bryological periodicals were born: *Transactions of the British Bryological Society* in Great Britain, *Buxbaumia* in the Netherlands, and *The Journal of Hattori Botanical Laboratory* in Japan. In 1971 *Buxbaumia* was discontinued, and *Lindbergia* appeared as a journal published by the Dutch and Nordic Bryological Societies. The two other journals continue to the present day, but the name of the British journal was changed to *Journal of Bryology* in 1972. The small publication *Miscellanea Bryologica et Lichenologica* appeared in 1955 with notes of local interest published in Japanese; this was published until 1983 with the contributions increasingly international in scope, and in several languages. The journal *Herzogia* was established in Germany in 1968 and continues to the present. The most recent periodical to appear is *Bryologische Beiträge,* in 1982.

The ephemeral *Bryologische Zeitschrift* appeared as a single volume in 1916–1917 and disappeared. Numerous other periodicals have been important in presenting the results of bryological research; *Hedwigia* (1852–1944) and its successor *Nova Hedwigia* (1959–present) have been especially significant.

FURTHER READING

Allen, C. E. 1919. The basis of sex inheritance in *Sphaerocarpos. Proc. Am. Phil. Soc.* **58**:289–316.

———— 1937. Fertility and compatibility in *Sphaerocarpos*. *Cytologia, Fuji Jubilee Volume,* 494–501.

Andrews, A. L. 1911. Notes on North American *Sphagnum* I The groups. *Bryologist* **14**:72–75.

———— 1913. Order Sphagnales, Family 1. Sphagnaceae, 1. *Sphagnum. North Am. Flora* **15**(1):1–31.

———— 1960. Notes on North American Sphagnum XII. *Sphagnum cyclophyllum. Bryologist* **63**:229–234.

Bartram, E. B. 1933. *Manual of Hawaiian Mosses. Bishop Mus. Bull.* **101**:1–275.

———— 1939. *Mosses of the Philippines. Phillip. J. Sci.* **68**:1–437.

———— 1949. *Mosses of Guatemala. Fieldiana.* **25**:1–442.

Braithwaite, R. 1888–1905. *The British Moss Flora,* 3 vols. London: L. Reeve.

Bridel-Brideri, S. E. de. 1826–1827. *Bryologia universa seu systematica ad novam methodum disposito, historia et descripto omnium muscorum frondosorum hucusque cognitorum cum synonymia ex auctoribus probatissimus.* 2 vols. Leipzig.

Brotherus, V. F. 1923. *Die Laubmoose Fennoskandias. Flora Fennica I.* Helsingfors: Akadem. Buchhandlung.

———— 1924–1925. Musci, in Engler A. and K. Prantl (eds.), *Die natürlichen Pflanzenfamilien* (2nd ed.), Vols. 10, 11. Leipzig: W. Engelmann.

Bruch, P., W. P. Schimper, and W. T. von Gümbel. 1936–1855. *Bryologia Europaea seu genera muscorum Europaeorum monographice illustrata.* 6 vols. Stuttgart.

Buch, H. 1919. Über den Einfluss von Licht und Feuchtigkeit auf der Wachstumsrichtung der Lebermoosgametophyten. *Översikt. Finsk Vetensk. Soc. Förhandl.* **61**(10):1–8.

———— 1920. Physiologische und experimentell-morphologische Studien an beblätterten Lebermoosen I-II. *Översikt. Finsk. Vetensk Soc. Förhandl.* **62**(Afd. A, Nr. 6):1–46.

Burgeff, H. 1933. *Genetische Studien an Marchantia.* Jena: Gustav Fischer.

Campbell, D. H. 1913. *The Structure and Development of Mosses and Ferns* (3rd ed.). New York: Macmillan.

Cardot, J. 1892. *Monographie des Fontinalacées. Mem. Soc. Natl. Sci. Nat. Cherbourg* **28**:1–152.

———— 1908. *La flore bryologique des Terres Magellaniques, de la Géorgie du Sud et l'Antarctide. Wiss. Ergeb. Schwed. Südpolar-Exped. 1901–1903,* **4**(pt. 8): 1–298.

———— 1912. *Mousses. Deuxième Expédition Antarctique francaise.* Paris: Masson.

Cavers, F. 1910–1911. *The Interrelationships of the Bryophyta. New Phytologist Reprint No. 4,* 200 pp. + 15 unnumbered pages.

Chen, P-C. 1963. *Genera Muscorum Sinicorum (Pars Prima).* Peking: Chinese Acad. Sci.

———— 1978. *Genera Muscorum Sinicorum (Pars Secunda)*. Peking: Chinese Acd. Sci.

Correns, C. 1899. *Untersuchungen über die Vermchrung der Laubmoose durch Brutorgane und Stecklinge*. Jena: Gustav Fischer.

Dillenius, J. J. 1741. *Historia Muscorum in qua circiter sexcentae species veteres et novae ud sua genera relatae describuntur, et iconibus genuinis illustrantur: cum appendice et indice synonymorum*, 576 pp., 85 pls. Oxford.

Dixon, H. N. 1913–1929. *Studies in the bryology of New Zealand with special reference to the herbarium of Robert Brown, of Christchurch, New Zealand. N. Z. Inst. Bull.* 3:1–372.

———— 1924. *The Student's Handbook of British Mosses* (3rd ed.). Eastbourne: V. V. Sumfield.

Dozy, F., and J. H. Molkenboer. 1844. *Muscorum frondosorum novae species ex archipelago indico et Japonica*. Leiden.

———— 1845–1854. *Musci frondosi inediti archipelagi indici*. Leiden.

———— 1855–1870. *Bryologia javanica*. Leiden.

Fleischer, M. 1904–1924. *Die Musci der Flora von Buitenzorg (zugleich Laubmoosflora von Java)*, 4 vols. Leiden: E. J. Brill.

Frye, T. C., and L. Clark. 1937–1947. *Hepaticae of North America*. Univ. Wash. Publ. Biol., Vol. 6, 1018 pp.

Goebel, K. 1930. *Organographie der Pflanzen* (3rd ed.), *Part 2: Bryophyten-Pteridophyten*. Jena: Gustav Fischer.

Grout, A. J. 1903–1908. *Mosses with a Hand-lens and Microscope*. New York: Mount Pleasant Press.

———— 1928–1940. *Moss Flora of North America*, 3 vols. Newfane, Vermont.

———— 1947. *Mosses with Hand-lens* (4th ed.). Newfane, Vermont.

Hébant, C. 1977. *The Conducting Tissues of Bryophytes*. Vaduz: J. Cramer.

Hedwig, J. 1801. *Species Muscorum frondosorum descriptae et tabulis aeneis lxxvii*. Leipzig.

Heitz, E. 1928. Das heterochromatin der Moose I. *Jahrb. Wiss. Bot.* **69**:762–818.

Herzog, T. 1926. *Geographie der Moose*. Jena: Gustav Fischer.

Hofmeister, W. 1851. *Vergleichende Untersuchungen der Keimung. Entfaltung und Fruchtbildung höher Kryptogamen*. Leipzig.

Hooker, W. J. 1816. *British Jungermanniae*. London.

———— 1816. *Plantae cryptogamicae*. London.

———— 1818–1820. *Musci Exotici*, 2 vols. London.

Horikawa, Y. 1929–1933. Studies on the Hepaticae of Japan I–VIII. Sci. Reports Tokoku Imper. Univ. Ser. 4. Biol **4**(1):37–72. Ibid. **4**(2):395–429. Ibid. **5**:623–650. *J. Sci. Hiroshima Univ., Ser. B., Div. 2* **1**:13–94, 121–134, 197–204.

Kashyap, S. R. 1929–1932. *Liverworts of the Western Himalayas and the Punjab Plains*. 2 parts. Lahore: Univ. of Punjab.

Leitgeb, H. 1874–1881. *Untersuchungen über die Lebermoose I–VI.* Jena: (reprinted, J. Cramer, 1968).

Limpricht, K. G. 1885–1903. *Die Laubmoose Deutschlands, Oesterreichs und der Schweiz. Unter Berücksichtigung der übrigen Länder Europas und Sibiriens,* 3 vols. Leipzig: E. Kummer.

Loeske, L. 1910. *Studien zur vergleichenden Morphologie und phylogenetischen Systematik der Laubmoose.* Berlin: M. Lande.

———— 1929. *Die Laubmoose Europas II. Funariaceae.* Berlin: M. Lande.

———— 1930. *Monographie der europäischen Grimmiaceen. Bibl. Bot.* **101**:1–236.

Lorch, W. 1931. Anatomie der Laubmoose, in Linsbauer, K. (ed.), *Handbuch der Pflanzenanatomie 7.* Berlin: Gebrüder Borntraeger.

Lorentz, P. G. 1864. Studien zur vergleichenden Anatomie der Laubmoose. *Flora* **50**:241–248, 257–264, 287–297, 305–313, 526–540, 544–558.

———— 1867–1868. Grundlinien zu einer vergleichenden Anatomie der Laubmoose. *Jahrb. Wiss. Bot.* **6**:363–466.

———— 1869. Studien zur Anatomie des Querschnittes der Laubmoose. *Flora* **52**:161–173, 193–208, 209–219, 225–233, 241–250.

Macvicar, S. M. 1926. *The Student's Handbook of British Hepatics* (2nd ed.). Eastbourne: V. V. Sumfield.

Marchal, É., and É. Marchal. 1911. Aposporie et sexualité chez les mousses III. *Belg. Acad. Belg. Cl. Sci.* **1911**:750–778.

Margadant, W. D. 1968. *Early Bryological Literature.* Pittsburgh: Hunt. Bot. Library.

Micheli, P. A. 1729. *Nova Plantarum Genera Juxta Tournefortii Methodum Disposita.* Florence.

Mitten, W. 1869. *Musci austro-americani. J. Linn. Soc. Bot.* **12**:1–659.

Moenkemeyer, W. 1927. *Die Laubmoose Europas IV. Ergänzungsband. Andreaeales-Bryales.* Leipzig: Akadem. Verlagsges.

Müller, K. 1951–1958. Die Lebermoose Europas, in *Rabenhorst's Kryptogamen-Flora* (3rd ed.). Leipzig: Geest and Portig K.-G.

Müller, K. A. F. 1849–1851. Synopsis muscorum frondosorum omnium hucusque cognitorum. 2 parts. Berlin.

Nägeli, C. W. von. 1845. Wachsthumsgeschichte der Laub- und Lebermoose I–IV. *Z. Wiss. Bot.* **2**:138–150. Ibid. **3**:150–158, 158–164, 164–166.

Nees von Esenbeck, C. G. 1833–1838. *Naturgeschichte der europäischen Lebermoose,* 4 vols. Berlin: Breslau.

Neuberg, M. F. 1960. Leafy-stemmed mosses from Permian deposits of Angarid. (in Russian). *Akad. Nauk SSSR, Trudy Inst. Geol.* **19**:104 pp.

Pearson, W. H. 1902. *The Hepaticae of the British Isles,* 2 vols. London: L. Reeve.

Potier de la Varde, R. 1928. Mousses de l'Oubangi. *Arch. Bot.* **1**(Mem 3):1–152.

———— 1936. Mousses du Gabon. *Mem. Soc. Natl. Sci. Nat. Math. Cherbourg* **42**:1–270.

Pringsheim, N. 1878. Über Sprossung der Moosfrüchte und den Generationswechsel der Thallophyten. *Jahrb. Wiss. Bot.* **11**:1–46.

Proskauer, J. 1948. Studies on the morphology of Anthoceros I. *Ann. Bot. (n.s.)* **12**:237–265.

―――― 1969. Studies on Anthocerotales VIII. *Phytomorphology* **19**:52–66.

Renauld, F. 1897–1909. *Prodrome de la flore bryologique de Madagascar*, 2 vols. Monaco.

Renauld, F., and J. Cardot. 1915. Histoire naturelle des plantes: Mousses, in Grandidier, A., and G. Grandidier, (eds.), *Histoire physique, naturelle et politique de Madagascar*, Vols. 29, 30. Paris: Imprimerie Nationale.

Sainsbury, G. O. K. 1955. *A Handbook of the New Zealand Mosses. Roy. Soc. N. Z. Bull. No. 5*, 490 pp.

Savicz-Ljubitzkaja, L., and Z. N. Smirnova 1970. *The Handbook of the Mosses of the U.S.S.R.: The Mosses Acrocarpous* (in Russian), 824 pp. Leningrad: Acad. Sci. U.S.S.R., Komarov. Bot. Inst.

Schiffner, V. 1893–1895. Hepaticae, in Engler and Prantl (eds.), *Natürlichen Pflanzenfamilien*, Vol. 1, no. 3. Leipzig: Wm. Engelmann.

Schmidel, C. C. 1747. *Icones Plantarum et Analyses Partim, etc.* Nuremberg.

Schwaegrichen C. F. 1811–1842. *Species Muscorum frondosorum descriptae et tabulis-aeneis colorates illustratae. Opus posthumum* [*J. Hedwig*]. *Supplementum*, 4 vols. Leipzig.

Sim, T. R. 1926. *The Bryophyta of South Africa. Trans. Roy Soc. South Africa* **15**:1–475.

Spruce, R. 1884–1885. Hepaticae amazonicae et andinae. *Trans. Proc. Bot. Soc. Edinburgh* **15**:1–590.

Stephani, F. 1898–1924. *Species Hepaticarum sive Enumeratio Monographica Hepaticarum Orbis Terrarum Hucusque cognitarum*, 6 vols. Geneva: Univ. Genève et Herbier Boissier.

Strasburger, E. 1894. The periodic reduction of the number of chromosomes in the life-history of living organisms. *Ann. Bot.* **8**:281–316.

Sullivant, W. S. 1864, 1874. *Icones Muscorum and Supplement.* Cambridge, Mass.: Sever and Francis.

Vaarama, A. 1949. Meiosis in moss species of the family Grimmiaceae. *Port. Acta Biol. (A)*, Vols. 47–78. R. B. Goldschmidt.

――――1969. Structurally and deviating chromosome types in Bryophyta. *Nucleus* **12**:285–294.

Wijk, R. van der, W. D. Margadant, and P. A. Florschütz (eds.). 1959–1969. *Index Muscorum*, Volume I (*Regnum Vegetabile*, Vol. 17), 548 pp. Volume II (*Regnum Vegetabile*, Vol. 26), 535 pp. Volume III (*Regnum Vegetabile*, Vol. 33), 529 pp. Volume IV (*Regnum Vegetabile*, Vol. 48), 604 pp. Volume V (*Regnum Vegetabile*, Vol. 65), 922 pp.

21 Cytology and Genetics

GENERAL STRUCTURE OF CELLS

Bryophyte cell walls are relatively simple and constituted of cellulose. Often, especially in jungermannialean hepatics, the walls are of two to three layers. They are sometimes pigmented, conspicuously so in the subclass Andreaeidae and in some genera of the order Jungermanniales, but are usually colorless. The pigments are apparently tannins. Between adjacent cells these walls are sometimes pitted and are usually perforated by plasmodesmata. The middle lamella is apparent with the light microscope in many of the genera of bryophytes. Exposed surfaces of cells are sometimes ornamented (sometimes cutinized), with thickenings that form papillae of varying complexity. Such cellular ornamentation is especially apparent in leafy bryophytes of drier habitats. The inner faces of cells are sometimes ornamented with papillae, striations, and fibrils, as in some species of *Sphagnum*. There are peglike thickenings on the inner surface of the cell in rhizoids of most of the Marchantiales. Sometimes rhizoids in the subclass Bryidae have warts and papillae on the cell surface (e.g., Bartramiaceae). Large pores perforate the cell walls of some mosses (e.g., *Sphagnum;* and some genera of the subclass Bryidae, e.g., *Leucobryum* and *Calymperes*). Markedly pitted cells often occur in many thallose hepatics, especially in the orders Metzgeriales, Marchantiales, and Monocleales, as well as in many mosses.

Younger living cells contain a single nucleus, vacuoles of various sizes, chloroplasts (usually lenticular), starch grains, mitochondria, and other organelles that characterize green plants. Complex oil bodies are conspicuous in the hepatics, and each is enclosed in a membrane. Simple oil spherules are usually present in the mosses. Such spherules are especially apparent in the spores of bryophytes. Fungal hyphae often occur in bryophyte cells and are frequent in rhizoids and in cells of the ventral layer of hepatic gametophores.

In immature cells the walls are uniformly thin and sometimes remain thin, even when mature. Some bryophytes have additional layers of cellulose, often at the corners of the cells, resulting in the trigonous condition frequent in leaf cells of many of the jungermannialean hepatics.

Cells also exhibit considerable diversity of shapes, with elongate cells frequent in the mosses and isodiametric cells characterizing most hepatics. In leafy bryophytes the diversity and predictability of the cell shape and arrangement in the leaves is a valuable feature in the identification of species.

CELL DIVISION

Mitosis follows the same pattern as in all green land plants and in most bryophytes is confined mainly to the cells adjacent to the apical cell. An obvious exception is in the sporophyte of the Anthocerotae, where mitotic divisions occur in the intercalary meristem between the foot and sporangium as long as growth conditions favor it. Recent electron microscopic study shows that there are differences in spindle arrangement of this subclass, compared to other bryophytes.

Meiotic divisions occur only in the spore mother cells of the sporogenous layer. In mosses and hepatics many spore mother cells within a sporangium undergo meiotic divisions synchronously. This greatly enriches the available information concerning meiotic behavior within a single sporangium, which forces one to recognize the gross features that represent various stages of maturation in a sporangium in order to obtain suitable chromosomal preparations. Sporangial maturation appears to be strongly controlled by season, reflecting the time of induction and maturation of the sex organs and, consequently, the time of fertilization of the egg. Favorable temperature and humidity are also important.

Chromosomes in bryophytes are considerably smaller than those in seed plants ($0.5-9.0$ μm): they tend to clump at metaphase (especially in the mosses) and are difficult to spread apart. Cell inclusions often become stained and obscure the chromosomes.

CHROMOSOME NUMBER AND BEHAVIOR

There is considerable uniformity in the chromosome number of hepatics (n = 8, 9, or 10) and hornworts (n = 4, 5, or 6), with secondary polyploidy occurring relatively infrequently. The chromosome number of the hepatic *Takakia* is exceptional for the hepatics, being n = 4 or 5. In mosses basic numbers vary from n = 4, 5, 6, 7 and multiples of these with n = 9 being infrequent; n = 11 is extremely well represented, especially among pleurocarpous Bryidae. Polyploidy is relatively frequent in the mosses and has occurred in at least 10% of the species studied. In bryophytes the same degree of complexity of chromosome number is represented as in the lignified plants, as detailed by R. Fritsch (1982). Indeed, the mosses, as A. Vaarama (1968), has noted, exhibit the frequent existence of exceptional or deviant chromosome types, the number of which far surpasses that present in other plant groups. A widespread species, common as a greenhouse weed in temperate areas, *Physomitrium pyriforme,* shows a polyploid series of n = 9, 18, 27, 36, and 54.

Most of the research concerning chromosome numbers and behavior in the bryophytes has been utilized to aid in determining relationships among taxa, especially at the species level. Nearly all of this research has used very simple techniques of preparation and staining of the material, based on examination through the light microscope.

In the hepatics and hornworts most chromosome studies have been on mitotic divisions in the gametophore. In the mosses, many chromosome studies have been based on only meiotic divisions in the spore-mother cells. Squash techniques have been utilized for most studies.

In hepatics the spore mother cell is four-lobed before meiosis in the subclass Jungermanniidae and unlobed in the Marchantiidae. In the Anthocerotae and most of the mosses, the spore mother cell is also unlobed, but in the Polytrichidae it has been reported as four-lobed. In mosses premeiotic cells contain a single plastid, but in hepatics the number of plastids varies.

E. Heitz (1928) discovered, described, and named heterochromatin, noting it first in the interphase and prophase nuclei during mitosis in the hepatic *Pellia.* These deeper staining areas have been valuable markers in distinguishing particular chromosomes and determining their behavior. Heterochromatin is defined as regions of chromosomes that remain condensed during interphase, in contrast to the rest of the chromosome material, or euchromatin, which uncoils and becomes thin and more or less threadlike during the same period.

Utilizing mosses, artificial induction of polyploidy was first carried out by E. and Em. Marchals in 1911. By very simple techniques they were able to induce polyploidy through apospory. This involves direct production of gametophytes from sporophytic tissue, without meiosis and spore production. Tissue of young sporophytes (usually injured), was placed in culture; the cells dedifferentiated to produce protonemata which, in time, produced diploid leafy gametophores with sex organs. When 2n eggs were fertilized with 2n sperms, a polyploid sporophyte was produced (4n or tetraploid). This polyploid showed somewhat larger cells than the haploid but was otherwise structurally similar to the haploid gametophore. In some mosses in which the haploid gametophore was dioicous, the diploid gametopore produced gametes that failed to function.

F. von Wettstein in 1923, and for more than a decade following, with his students carried out many experiments that induced polyploidy in mosses. He was able to produce sporophytes up to 32n! Furthermore, Wettstein was able to create what was interpreted as a new polyploid species of *Bryum* (*B. corrensii,* n = 20) from a haploid parent (*B caespiticium,* n = 10). This "new species" had double the chromosome complement of the parent and was monoicous, while the parental species was dioicous. Initially the polyploid produced embryonic sporophytes, but no sporangia formed. After the passage of 11 years normal sporophytes were formed with viable spores. Although this polyploid differed in sexual condition and chromosome number from its parent, it showed no other differences. Most bryologists would consider it as a race within *B. caespiticium*, although some would treat it, and other intraspecific chromosome races, as an independent species. Such a procedure considerably complicates identification of bryophytes, demanding chromosome counts for all specimens, which is obviously impractical.

Induced polyploidy tends to result initially in increased vigor and size of the gametophores, but after a number of generations this size difference tends to stabilize in the perennial gametophore, and new growth attains the same size as the original parent. Polyploids occur within and among species in nature. Indeed, H. P. Ramsay (1967) has shown that several different polyploids can exist within a small population on a single log, as in *Hypopterygium rotulatum* in Australia.

In mosses polyploidy presumably arises in several ways: by unreduced spores (= diplospory), by chromosome doubling by the combination of two nuclei within a single cell during mitosis but without formation of a new cell wall, and by apospory. Certainly all of these phenomena have been observed in laboratory material. Polyploids arise therefore mainly through autopolyploidy. Rare examples of allopolyploids have been reported.

Polyploids are usually considered to be rather uncommon among the hepatics but, as S. Tatuno (1941) has shown, in gemma production in *Calypogeia neesiana*, some gemma cells on a single leaf possess the expected haploid chromosome number of the parent gametophore while others are clearly diploid. It has been suggested however, that all hepatics may be polyploids, having arisen from $x = 4$ or 5. Polyploidy provides conspicuous advantages to a basically haploid organism, as L. E. Anderson (1981) has stated permitting it "greater variability and protection against deleterious mutants whose effects would otherwise be instantly expressed."

TAXONOMIC IMPLICATIONS OF CHROMOSOME NUMBER

Approximately 75% of the species of hepatics have the chromosome number $n = 9$. *Takakia* differs with $n = 4$ or 5, which provides further evidence of its isolation among the hepatics, and *Porella*, a clearly jungermannialean hepatic, has $n = 8$. Since $n = 9$ is rare among the mosses, occurring mainly in specialized groups, it suggests that these two evolutionary lines of bryophytes (the mosses and hepatics) are probably very distantly related. In the Anthocerotae, constantly $n = 5$ or 6, cytological isolation is again demonstrated.

The mosses show a remarkable diversity in chromosome number, with the lowest numbers $n = 4$ found only in *Hypnodendron*, while $n = 5$ occurs in a number of unrelated groups (Fig. 21-1). The highest reported numbers, e. g., $n > 60$, are extremely rare. There are some clear trends, with the Sphagnidae cytologically isolated with a basic number $n = 19$ plus a variable number of m-chromosomes. The Polytrichidae show a constant basic number of n = 7, and Tetraphidae with n = 7 or 8, and Buxbaumiidae with n = 8 or 9. Andreaeidae, on the other hand, has n = 10 or 11. The number is n = 13 in the Archidiidae. Since there is such a vast diversity of chromosome numbers in the Bryidae, undoubtedly it will be necessary to acquire greater detail concerning chromosome structure and behavior before any reliable assumptions can be made concern-

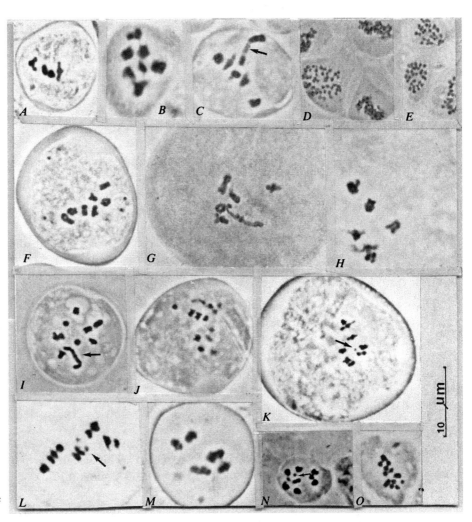

FIGURE 21-1
Photomicrographs of moss sporocytes at meiosis. Note the varying size of cells, chromosome numbers, and chromosome types present. (*A*) *Hypnodendron colensoi*, n = 4 (smallest chromosome number in mosses). Other species have n = 4, 5, or 9. (*B*) *Dawsonia longiseta*, n = 7. *Dawsonia* is closely related to *Polytrichum*, which has the same chromosome number and similar large chromosomes. (*C*) *Pyrrhobryum* (*Rhizogonium*) *parramattense*, n = 6. One large bivalent is present; compare (*G*) and (*H*). (*D, E*) *Tortula muralis*, n = 26, 52. These two numbers, n = 26 (*E*) and n = 52 (*D*), were found in British populations. Many other chromosome numbers have been found in this species, e.g., n = 48, U.S.A. and Australia. (*F*) *Dicranoloma robustum*, n = 7. Chromosomes depicted here are from an Australian population. (*G*) *Mnium hornum*, n = 6. The six chromosomes of this species are similar to *Rhizomnium glabrescens* (*H*) but differ markedly from those of *Rhizogonium* thought at one time to be closely related. This is from a British collection. (*H*) *Rhizomnium glabrescens*, n = 6. Compare with (*G*). This photograph is from a British Columbian collection. (Provided by H. Ramsay.)

ing the taxonomic implications of cytological information in the mosses. Analysis of karyotypes, rather than chromosome number alone, is more likely to be useful in assessing evolutionary trends and relationships among the bryophytes. The acquisition of this detail is somewhat complicated since most chromosomal information from the mosses is derived from meiotic counts. Moss chromosomes, on the average, are much smaller than those of flowering plants. Furthermore non-disjunction of meiosis and the tendency of chromosomes to aggregate at metaphase of meiosis, as well as difficulties in spreading apart in mitosis, place cytological work in the hands of a specialist. The usual presence of non-nuclear inclusions in the cell that take up the nuclear stains also contributes to the difficulties in obtaining assessment of chromosome number and behavior.

SEX CHROMOSOMES

Many dioicous hepatics possess a heteromorphic bivalent (i.e., a pair of chromosomes of unequal size), in which the smaller is associated with the male gametophore and the larger with the female. This has been interpreted as forming an X-Y mechanism in *Sphaerocarpos*, for example, which provided first evidence of sex chromosomes in plants.

In the mosses the evidence for sex chromosomes is not clearly established, although there are examples of mosses in which size differences have been noted in a bivalent at meiosis, and these are often suggested to be sex chromosomes. The most satisfactory evidence has been obtained from the bryidean genus *Macromitrium*, in which a dimorphic bivalent is present at meiosis, and cells of female gametophores were found to have a large chromosome corresponding to the larger member of the bivalent. However, unquestionable sex chromosomes in the mosses have not been positively demonstrated.

GENETICS

Hybridization has been achieved experimentally and has been noted in natural populations of bryophytes. The researches of C. E. Allen, beginning in 1919 and continuing to 1945, provide most of the detailed information available for hepatics and are based entirely on *Sphaerocarpos*. The researches of F. von Wettstein between 1924 and 1942 on mosses demonstrated intergeneric, inter- and intraspecific crosses in the bryidean families Funariaceae (several genera) and Bryaceae (*Bryum*). In more recent researches (1954–1969), J. Moutschen (1954–1969) used induced mutant races to study genetic mechanisms in mosses.

C. E. Allen, in his studies of *Sphaerocarpos*, found that most interspecific crosses failed. In *S. donnellii* (female) × *S. texanus* (male), he found that some spores resulted, and the ornamentation was controlled by the female parent, indicating basic somatic control of this feature. Some of the spores of this cross germinated, and the gametophores were interfertile. He also performed numerous crosses between induced and natural mutant strains within *S. donnellii*. The results of his intra-tetrad analyses concerning morphologic variants in gametophores show the situation to be very complex, and no satisfactory explanation could be given for the results. Since these researches, there has been a comparative neglect of hepatic genetics.

The discovery of moss hybrids in nature is most readily recognized in taxa in which the sporophyte maintains the basic features of the paternal parent, and in which the sporophyte is structurally markedly different from that of the maternal parent. In the bryidean families Funariaceae, Pottiaceae, and Ditrichaceae, such natural intergeneric hybrids have been described and, in some cases, induced experimentally. In many cases the paternal parent possessed sporangia that lacked an operculum and/or a peristome, while the maternal parent possessed both. In all cases, of course, the gametophore of the maternal parent bore the hybrid sporophyte. It appears that no analysis of the gametophores derived from the spores of such hybrids has been carried out, even when the spores proved to be viable. Crossing experiments among the progeny of such a hybrid sporangium would be of considerable interest. It is probable that such hybrids also exist among other families, but since the sporophytes are extremely similar among many genera in the pleurocarpous Bryidae, for example, their discovery is extremely difficult.

No natural hybrids have been noted in the hepatics, but a number of natural interspecific hybrids have been reported in the mosses. It is possible, for example, that the sporophyte variability that has been noted in *Pogonatum nanum* and *P. aloides* is the result of their interbreeding.

The induction of genetic variants has been accomplished in a small number of moss and hepatic species. The earliest mutants were morphological, but intriguing insights into the genetics of haploid organisms have also been gained through somatic hybridization of nutritionally deficient mutant strains. Research with the hepatic *Sphaerocarpos donnellii* and the moss *Physcomitrella patens* has contributed some understanding concerning polyploid origin of taxa as well as the parthenogenetic origin of sporophytes. It is possible that recombinant DNA technology applied to bryophytes also will provide a new and fruitful direction for bryophyte genetics.

FURTHER READING

Allen, C. E. 1930. Inheritance in a hepatic. *Science* **71**:197–204.
———— 1935. Genetics of bryophytes I. *Bot. Rev.* **1**:269–291.
———— 1945. Genetics of bryophytes II. *Bot. Rev.* **2**:260–287.
Anderson, L. E. 1963. Heteropycnosis and sex chromosomes in mosses. *Proc. XI Int. Congr. Genet.* **1**:101.
———— 1964. Biosystematic evaluations in the Musci. *Phytomorphology* **14**:27–91.

—— 1981. Cytology and reproductive biology of mosses, in Taylor, R. J., and A. E. Leviton (eds.), *The Mosses of North America*, pp. 37–76. San Francisco: AAAS.

—— and B. E. Lemmon. 1972. Cytological studies of natural intergeneric hybrids and their parental species in the moss genera *Astomum* and *Weissia. Ann. Missouri Bot. Gard.* **59**:382–416.

—— 1974. Gene flow distances in the moss *Weissia controversa* Hedw. *J. Hattori Bot. Lab* **38**:67–90.

—— and J. A. Snider. 1982. Cytological and genetic barriers in mosses. *J. Hattori Bot. Lab.* **52**:241–254.

Berrie, G. K. 1960. The chromosome number of liverworts. *Trans. Br. Bryol. Soc.* **3**:688–705.

—— 1963. Cytology and phylogeny of liverworts. *Evolution* **17**:347–357.

Bowers, M. C. 1980. A cytotaxonomic classification of the Mniaceae (Bryophyta). *Lindbergia* **6**:22–31.

Brown, R. C., and B. E. Lemmon. 1980. Ultrastructure of sporogenesis in a moss *Ditrichum pallidum. Bryologist* **83**:137–160.

—— 1982. Ultrastructural aspects of moss meiosis; review of nuclear and cytoplasmic events during prophase. *J. Hattori Bot. Lab.* **53**:29–39.

Bryan, V. S. 1955. Chromosome studies in the genus *Sphagnum. Bryologist* **58**:16–39.

Cove, D. J. 1983. Genetics of Bryophyta, in Schuster, R. M. (ed.), *New Manual of Bryology*, Vol. I, pp. 222–231. Nichinan: Hattori Botanical Laboratory.

Crosby, M. R. 1980. Polyploidy in bryophytes with special emphasis on mosses, in Lewis, W. H. (ed.), *Polyploidy: Biological Relevance. Basic Life Sciences*, Vol. 13, pp. 193–198.

Cummins, H., and R. Wyatt. 1981. Genetic variability in natural populations of the moss *Atrichum angustatum. Bryologist* **84**:30–38.

Dill, F. J. 1964. Dictyotene stage of meiosis in mosses. *Science* **144**:541–543.

Farmer, J. B. 1895. On spore-formation and nuclear division in the Hepaticae. *Ann. Bot.* **9**:469–523.

Fritsch, R. 1982. *Index to Plant Chromosome Numbers—Bryophyta. Regnum Vegetabile*, Vol. 108, xiv + 268 pp.

Heitz, E. 1928. Das heterochromatin der Moose I. *Jahrb. Wiss. Bot.* **69**:628–818.

Inoue, S. 1973. Karyological studies on *Takakia ceratophylla* and *T. lepidozioides. J. Hattori Bot. Lab.* **37**:275–286.

Jensen, K. G., and R. L. Hulbary. 1978. Chloroplast development during sporogenesis in six species of mosses. *Am. J. Bot.* **65**:823–833.

Khanna, K. R. 1960. The haploid and spontaneous diploid race of *Octoblepharum albidum. Cytologia* **25**:334–341.

—— 1965. Differential evolutionary activity in bryophytes. *Evolution* **18**:652–670.

Lazarenko, A. S. 1967. Polyploidy in the evolution of Musci. *Tsitologiya in Genetika* (Kiev) **1**(2):15–26.

Lewis, K. R. 1957. Squash techniques in the cytological investigation of mosses. *Trans. Br. Bryol. Soc.* **3**:279–284.

—— 1961. The genetics of bryophytes. *Trans. Br. Bryol. Soc.* **4**:111–130.

Longton, R. E. 1982. The biosystematic approach to bryology. *J. Hattori Bot. Lab.* **53**:1–19.

Lowry, R. J. 1954. The number and morphology of moss chromosomes. *Stain Tech.* **29**:17–20.

Mehra, P. N., and K. R. Khanna. 1961. Recent cytological investigations in mosses. *Res. Bull. Punjab Univ. (N.S.)* **12**:1–29.

Moutschen, J. 1954. Quelques considerations sur l'aposporie chez les mousses. *Congrès Int. Bot., Paris 1954*, pp. 114–121.

—— 1962. Quelques tendances de la génétique des mousses. *Bull. Soc. Roy. Bot. Belgique* **95**:61–71.

—— 1969. L'hérédité des caractères gamétophytiques chez les mousses. *Rev. Bryol. Lichénol.* **36**:617–624.

Neidhardt, H. V. 1979. Comparative studies of sporogenesis in bryophytes, in Clarke, G. C. S., and J. G. Duckett (eds.), *Bryophyte Systematics*, pp. 251–280. London: Academic Press.

Newton, M. E. 1979. Chromosome morphology and bryophyte systematics, in Clarke, G. C. S., and J. G. Duckett (eds.), *Bryophyte Systematics*, pp. 207–229. London: Academic Press.

—— 1983. Cytology of the Hepaticae and Anthocerotae, in Schuster, R. M. (ed.), *New Manual of Bryology*, Vol. I, pp. 117–148. Nichinan: Hattori Botanical Laboratory.

Ramsay, H. P. 1966. Sex chromosomes in *Macromitrium*. *Bryologist* **69**:293–311.

—— 1967. Intraspecific polyploidy in *Hypopterygium rotulatum* (Hedw.) Brid. *Proc. Linn. Soc. N.S.W.* **91**:220–230.

—— 1979. Anisospory and sexual dimorphism in the Musci, in Clarke, G. C. S., and J. G. Duckett (eds.), *Bryophyte Systematics*, pp. 281–316. London: Academic Press.

—— 1982. The value of karyotype analysis in the study of mosses. *J. Hattori Bot. Lab.* **53**:51–71.

—— 1983. Cytology of mosses, in Schuster, R. M. (ed.), *New Manual of Bryology*, Vol. I, pp. 149–221. Nichinan: Hattori Botanical Laboratory.

—— and G. K. Berrie. 1982. Sex determination in bryophytes. *J. Hattori Bot. Lab.* **52**:255–274.

Segawa, M. 1965. Karyological studies in liverworts, with special reference to sex chromosomes I. *J. Sci. Hiroshima Univ. ser. B., div. 2.* **10**:69–80.

Sinoir, Y. 1952. Génétique et cytotaxonomie des bryophytes. *Rev. Bryol. Licéénol.* **21**:32–45.

Smith, A. J. E. 1978. Cytogenetics, biosystematics and evolution in the bryophyta. *Adv. Bot. Res.* **6**:196–277.

—— and H. P. Ramsay. 1982. Sex, cytology and frequency of bryophytes in the British Isles. *J. Hattori Bot. Lab.* **52**:275–281.

Steere, W. C. 1958. Evolution and speciation in mosses. *Am. Nat.* **42**:5–20.

———— 1972. Chromosome numbers in bryophytes. *J. Hattori Bot. Lab.* **35**:100–125.

Szweykowski, J. 1983. Genetic differentiation of liverwort populations and its significance for bryotaxonomy and bryogeography. *J. Hattori Bot. Lab.* **52**:21–28.

Tatuno, S. 1959. Chromosomen von *Takakia lepidozioides* und eine Studie zur Chromosomenevolution der Bryophyten. *Cytologia* **24**:138–147.

Vaarama, A. 1964. Notes on certain details of the karyological technique with mosses. *Portug. Acta Biol. ser. A.* **8**:81–94.

———— 1968. Structurally and functionally deviating chromosome types in Bryophyta. *Nucleus* (Suppl.):285–294.

———— 1976. The cytotaxonomic approach to the study of bryophytes. *J. Hattori Bot. Lab.* **41**:7–12.

Vitt, D. H. 1968. Sex determination in mosses. *Mich. Bot.* **7**:195–203.

Wylie, A. P. 1957. The chromosome numbers of mosses. *Trans. Br. Bryol. Soc.* **3**:260–278.

22 Chemistry

Most of the detailed knowledge of bryophyte chemistry has appeared since 1960. This has accompanied improvements in technology which permit accurate identification of compounds derived from very small amounts of plant material. Identification of stable compounds has progressed rapidly since 1960, but the assessment of unstable compounds has lagged, and the interpretation of the function of the diversity of compounds remains in very early stages of understanding.

Most chemical analyses are based on air-dried pulverized material from which an ether or a methanol extract is taken. Thin layer (TLC) and gas chromatography, combined with mass spectra, have been utilized for analysis, and identification of the substances has relied, in part, on comparison with chromatograms and mass spectra of previously identified compounds. Considerable purification of compounds is needed for satisfactory determination.

The greatest concentration of detailed effort has been applied to the liverworts, and the taxonomic implications of this information have been evaluated. Some research has concentrated on the presence of certain substances as a protection against animal predation. Of particular interest is the identification of the substances in some species of *Frullania* that cause serious contact dermatitis.

TERPENOIDS

Terpenoids are ubiquitous in hepatics. Monoterpenoids have been studied in more than 60, and are important in giving many hepatics their distinctive odor. Sesquiterpenoids have been noted from numerous hepatics, and many of these are unique to particular species and can serve as useful taxonomic markers. It is possible that these substances protect the bryophytes from consumption by animals. Indeed, experimental evidence strongly supports this assumption;

283

pinguisone, found in *Aneura pinguis,* is reported to inhibit insects from feeding. The largest group of sesquiterpenoids is the eudesmanes. Sesquiterpene lactones of some *Frullania* species cause allergic contact dermatitis. These substances, named (+)-frullanolide and (−)-frullanolide (Fig. 22-1), produce skin rash on the contact area. Sensitization can occur in some individuals after frequent contact with the substance. Since sesquiterpene lactones recur also in flowering plants, especially members of the Asteraceae, cross-sensitization may occur. The diterpenoid ent-16-kauranol (Fig. 22-1) is thought to contribute to the bluish tint of the moss *Saelania glaucescens*; it also is the main component that gives a similar waxy coating to the hepatic genus *Anthelia*. The triterpenes, such as hopenes and fernenes, are known ir mosses. The liverwort *Takakia* also produces hopanoids. Phytosterol appear to be ubiquitous in bryophytes, as it is in lignified plants.

FIGURE 22-1
Structural representations of organic compounds found in bryophytes. 1: (+)-frullan-olide; 2: (−)-frullanolide; 3: ent-16-kauranol; 4: lunularic acid; 5: pellepi-phyllin. (Provided by Y. Asakawa.)

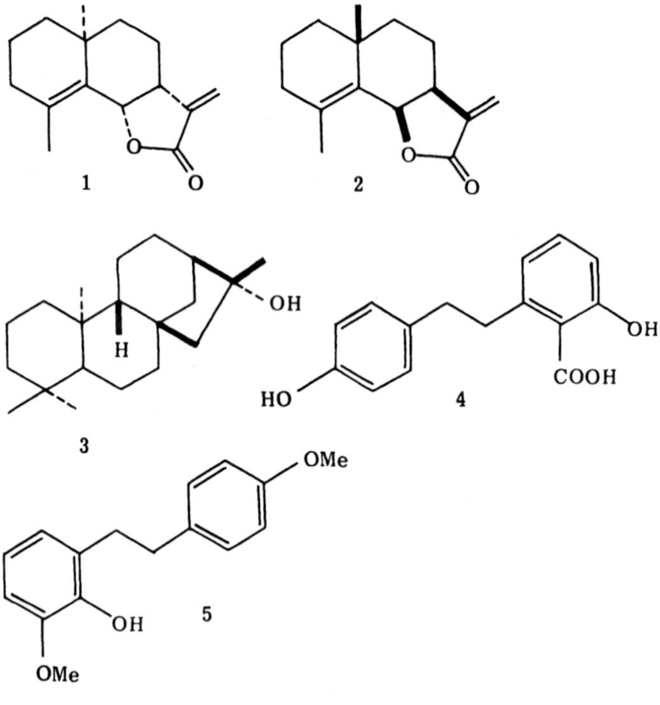

FLAVONOIDS

Flavonoids are widely distributed in mosses and hepatics, but at present there is no evidence that they are present in hornworts. Most detailed studies have concentrated on the hepatics. Flavone glycosides are most frequently encountered and a wide range of structural types have been identified. Some bryophytes have a great variety. The moss *Hedwigia ciliata,* for example, contains at least 12 flavonoids, four of which occur in major amounts, and the hepatic order Marchantiales exhibits a great diversity of flavone glycosides. Although flavonols have been reported for a small number of hepatics, they appear to be absent in mosses. A few examples of dihydroflavonoids have been reported for the hepatics, and a single biflavonyl (5'8"-biluteolin) has been noted in the moss *Dicranum scoparium,* the sole example known from a nonlignified plant, although it is possible that it is present in the completely unrelated moss, *Hedwigia ciliata.* In spite of the wide distribution of red pigments in mosses and hepatics, anthocyanins appear to be infrequent and have been reported convincingly from only two moss genera, *Bryum* and *Splachnum.* In the moss genus *Sphagnum,* sphagnorubin, an anthocyanidin-derived compound, appears to be the main reddish pigment, and this appears to be bonded to the cell walls.

OTHER ORGANIC COMPOUNDS

Lignin appears to be absent in the bryophytes, although it has been reported from *Sphagnum* and the polytrichaceous moss genera *Dendroligotrichum* and *Dawsonia,* but the reports remain controversial. There is no evidence of lignin from either hepatics or hornworts.

The dihydrostilbene, lunularic acid (Fig. 22-1), appears to induce dormancy in some hepatics, but it is unknown in other bryophytes. Another dihydrostilbene, pellepiphyllin (Fig. 22-1), has been noted in a few thallose hepatics. Recently more than 25 dihydrostilbenes were isolated from hepatics. Many other substances have been identified in the bryophytes, including aromatic esters, alkanes, alkanoic acids, starch, carotenes, free sugars, and alkaloids.

TAXONOMIC IMPLICATIONS

In spite of the limited body of data, some researchers have endeavored to speculate on the chemotaxonomic significance of various compounds in bryophytes. Since the bryophytes are such a heterogeneous group, one expects this to be reflected in their chemistry.

Phenolic chemistry is likely to prove helpful in drawing conclusions concerning affinities among genera of Hepaticae. In the Marchantiales, for example, there is a predominance of flavone-O-glycosides, while in the Jungermanniales, the flavone-C-glycosides predominate. Within the Jungermanniales, a number of families share an essential identity in phenolic chemistry, and this information, supplemented by morphological similarities, has been used to support fusion of some families, for example, the Radulaceae with the Madothecaceae (= Porellaceae).

The nearly ubiquitous presence of flavonoids in bryophytes contrasts with the green algae, in which they are generally absent, except in the charophytes, but relates the bryophytes to the lignified plants, where flavonoids are universally present. Indeed the detailed nature of some of these flavonoids appears to be identical to that of the angiosperms. In the Marchantiales, for example, the antheridiophores possess an aurone glycoside; in the angiosperms this substance is confined mainly to the flowers. There are numerous parallels to equivalent evolution of chemical compounds in bryophytes performing equivalent functions in the lignified plants. These similarities suggest a common ancestry as well as similar selective pressures. It is of particular interest that the calobryalean hepatics, which are morphologically similar to mosses in many features, also have the richest representation of flavonoids; but in most hepatics flavonoids are absent or less abundant than in mosses.

The nature of flavonoids has proven to be useful as a supplementary series of characters in distinguishing species, especially of the hepatics. Within some genera, e.g., *Marchantia,* each of the species examined has a readily distinguishable flavonoid pattern. In *Reboulia hemisphaerica* and *Asterella australis,* taxa difficult to distinguish on vegetative features, the flavonoid patterns are readily distinguished. There are in some species, however, different biochemical races, as for example in *Conocephalum conicum,* where the flavonoid patterns differ in various geographic areas. Terpenoids and lipophilic aromatic compounds are also important endogenous chemical characters of Hepaticae, and these substances are also applied to chemosystematic studies of Hepaticae.

In terms of gross fatty acid production, the hepatics, in particular, are more closely allied to the algae than to the seed plants. The base composition of DNAs for the mosses is more similar to the lignified plants than to the green algae, in which a much wider range is exhibited.

ELEMENT CONCENTRATION

Some bryophytes can concentrate elements many times higher than in their substratum. Barium, copper, lead, strontium, and zinc may be concentrated up to 200 times greater than in the substratum. Many bryophytes are distributed in relation to the chemistry of the substratum, and a few mosses appear to be positive indicators for some minerals, as is the case of the moss *Mielichhoferia,* which occurs most frequently on copper-rich substrata. Many other mosses are restricted to calcium-rich substrata, while others are never present on such substrata.

GENERALIZATIONS

The available biochemical information is generally consistent with assumptions concerning relationships among the bryophytes, as well as the speculation concerning affinities with the lignified land plants. As further information becomes available, it is probable that biochemical data will provide further insight into detailed relationships as well as into the function of the various chemical substances in the survival and life history of particular bryophytes.

FURTHER READING

Anderson, W. H., J. M. Hopkins, J. L. Gellerman, and H. Schlenk. 1974. Fatty acid composition as criterion in taxonomy of mosses. *J. Hattori Bot. Lab.* **38**:99–103.

Asakawa, Y. 1982. Chemical constituents of Hepaticae, in Herz, W., H. Grisebach and G. W. Kirby (eds.), *Prog. Chem. Organ. Nat. Prod.* **42**:1–285.

——— and E. O. Campbell. 1982. Chemosystematics of bryophytes XIV. Terpenoids and bibenzyls from some New Zealand liverworts. *Phytochemistry* **21**:2663–2667.

———, J.-C. Muller, G. Ourisson, J. Foussereau, and G. Ducombs. 1976. Nouvelles lactones sesquiterpéniques de *Frullania* (Hepaticae). Isolement, structures, propriétés allergisantes. *Bull. Soc. Chim. France (Chim. Mol.)*, 1465–1466.

——— and T. Takemoto. 1977. Sesquiterpene lactones and heterocyclic compounds of Bryophyta. *Heterocycles* **8**:563–582.

———, N. Tokunaga, T. Takemoto, S. Hattori, M. Mizutani, and C. Suire. 1980. Chemosystematics of bryophytes IV. The distribution of terpenoids and aromatic compounds in Hepaticae and Anthocerotae. *J. Hattori Bot. Lab.* **47**:135–164.

Benesova, V., and V. Herout. 1977. Components of liverworts. Their chemical strucutres and biological activity. *Bryoph. Bibl.* **13**:355–364.

Campbell, E. O., K. R. Markham, N. A. Moore, L. J. Porter, and J. W. Wallace. 1979. Taxonomic and phylogenetic implications of comparative flavonoid chemistry of species in the Family Marchantiaceae. *J. Hattori Bot. Lab.* **45**:185–199.

Douin, R. 1956. Pigments chlorophylliens des Bryophytes. Caroténoids des Bryales. *C.R. Acad Sci. Paris* **243**:1051–1054.

——— 1958. Pigments chlorophylliens des Bryophytes. Caroténoids des Andreaeales, des Sphagnales et des Hépatiques. *C.R. Acad Sci. Paris* **246**:1248–1251.

Erickson, M., and G. E. Miksche. 1974. On the occurrence of lignin or polyphenols in some mosses and liverworts. *Phytochemistry* **13**: 2295–2299.

Green, B. R. 1972. Isolation and base composition of DNAs of primitive land plants. *Biochim. Biophys. Acta* **277**:29–34.

Hébant, C. 1977. *The Conducting Tissues of Bryophytes.* Vaduz: J. Cramer.

Hegenaur, R. 1962. *Chemotaxonomie der Pflanzen,* Vol. 1, pp. 172–191. Basel: Birkhauser.

Huneck, S. 1969. Moosinhaltstoffe, eine Übersicht. *J. Hattori Bot. Lab* **32**:1–16.

———— 1981. The chemistry of some European liverworts, in Szweykowski, J. (ed.), *New Perspectives in Bryotaxonomy and Bryogeography,* pp. 73–76. Poznan: Adam Mickiewicz Univ.

———— 1983. Chemistry and biochemistry of bryophytes, in Schuster, R. M. (ed.), *New Manual of Bryology,* Vol. I, pp. 3–16. Nichinan: Hattori Botanical Laboratory.

Knoche, H., G. Ourisson, G. W. Perold, J. Foussereau, and J. Maleville. 1969. Allergic component of a liverwort: a sesquiterpene lactone. *Science* **166**:239–240.

Krzakowa, M. 1978. Isozymes as markers of inter- and intraspecific differentiation in hepatics. *Bryophyt. Bibl.* **13**:427–434.

———— 1980. Thin-layer chromatographic study of the phenolics of the *Pleurocladula* species (Hepaticae). *Acta Soc. Bot. Polon.* **49**:77–83.

———— and J. Szweykowski. 1979. Isoenzyme polymorphism in natural populations of a liverwort, *Plagiochila asplenioides. Genetics* **93**:711–719.

Lindberg, B., and O. Theander. 1952. Studies on *Sphagnum* peat II. Lignin in *Sphagnum. Acta Chem. Scand.* **6**:311–312.

Lowry, B., D. Lee, and C. Hébant. 1980. The origin of land plants: a new look at an old problem. *Taxon* **29**:183–197.

Markham, K. R., and L. J. Porter. 1978. Chemical constituents of the bryophytes. *Prog. Phytochem.* **5**:181–272.

————, L. J. Porter, and E. O. Campbell. 1978. The usefulness of flavonoid characters in studies of the taxonomy and phylogeny of liverworts. *Bryophyt. Bibl.* **13**:387–398.

Martensson, O., and E. Nilsson. 1974. On the morphological colour of the bryophytes. *Lindbergia* **2**:145–159.

McCleary, J. A., and D. L. Walkington. 1966. Mosses and antibiotics. *Rev. Bryol. Lichénol.* **34**:309–314.

McClure, J. W., and H. A. Miller. 1967. Moss chemotaxonomy. A survey for flavonoids and the taxonomic implications. *Nova Hedwigia* **14**:111–125.

Miksche, G. E., and Y. Yasuda. 1978. Lignin of "giant" mosses and some related species. *Phytochemistry* **17**:503–504.

Nagano, I. 1972. On the relation of the chemical composition of some mosses to their substrate rocks. *J. Hattori Bot. Lab.* **35**:391–398.

Nilsson, E., and G. Bendz. 1974. Flavonoids in bryophytes, in Bendz, G., and J. Santesson (eds.), *Chemistry in Botanical Classification,* pp. 112–120. New York: Academic Press.

Odryzykoski, I., M. A. Bobowicz, and M. Krzakowa. 1981. Variation in *Conocephalum conicum*—the existence of two genetically different forms in Europe, in Szweykowski, J. (ed.), *New Perspectives in Bryotaxonomy and Bryogeography,* pp. 29–32. Poznan: Adam Mickiewicz Univ.

Osterdahl, B-G. 1979. Isolation and identification of flavones and flavone glycosides. *Acta Univ. Upsaliensis* **516**:1–55.

Pokorny, M. 1974. D-methionine metabolic pathways in Bryophyta: a chemotaxonomic evaluation. *Phytochemistry* **13**:965–971.

Porter, L. J. 1981. Geographic races of *Conocephalum conicum* (Marchantiales) as defined by flavonoid chemistry. *Taxon* **30**:739–748.

Shacklette, H. T. 1965. Element content of bryophytes. *U. S. Geol. Survey Bull.* **1198D**:1–21.

———— 1967. Copper mosses are indicators of metal concentration. *U.S. Geol. Survey Bull.* **119G**:1–18.

Spencer, K. C. 1979. Chemical constituents of the Hepaticae. *Phytochem. Bull.* **12**:1–19.

———— 1980. Chemical constituents of the Musci. *Phytochem. Bull.* **13**:46–63.

Suire, C. 1975. Les données actuelles sur la chimie des bryophytes. *Rev. Bryol. Lichénol.* **41**:105–256.

———— and Y. Asakawa. 1981. Chimie et chemiotaxonomie des bryophytes: résultats essentiels et perspectives, in *Advances in Bryology,* Vol. 1, pp. 167–231.

Taylor, I. E. P., W. B. Schofield, and A. M. Elliot. 1970. Analysis of moss dehydrogenases by polyacrylamide disc electrophoresis. *Can. J. Bot.* **48**:367–369.

Wolters, B. 1964. Die Verbreitung antifungaler Eigenschaften bei Moosen. *Planta* **62**:88–96.

Zielinski, R., W. Prus-Glowacki, and M. Mendelak. 1981. Chemical variation in the central European *Pellia* taxa, in Szweykowski, J. (ed.), *New Perspectives in Bryotaxonomy and Bryogeography,* pp. 25–27. Poznan: Adam. Mickiewicz Univ.

23 Physiology

Bryophytes, as terrestrial photosynthetic plants, undergo the same physiological processes as do the seed plants. Most physiological information for the bryophytes has been derived from a very limited number of species of mosses. Details for hepatics are very limited, and for the hornworts information is extremely restricted. Since the gametophyte (particularly the gametophore) is the dominant photosynthetic phase of the life cycle, most data are based on this phase. This is in direct contrast to the information for seed plants, where the sporophyte is the dominant phase, and from which most plant physiological information has been derived. The dependence of the sporophyte upon the gametophore for its water and mineral supply makes an understanding of gametophore physiology fundamental in bryology.

WATER UPTAKE AND MOVEMENT

Bryophytes obtain most of their water and minerals from atmospheric moisture, while a few obtain these mainly from the substratum. Even during earlier developmental stages following spore germination, water and nutrients are mainly from atmospheric moisture rather than from the substratum. The physical and chemical nature of the substratum, however, are important in the establishment of the gametophyte, and markedly influence its survival. Water-holding capacity of the substratum is especially important.

Since bryophytes are relatively small plants, they live within a microenvironment that often differs conspicuously from the macroenvironment. Indeed, in some climates, bryophytes are restricted to microenvironments that favor their growth and reproduction.

Water content in bryophytes can vary from lower than 50% of their dry weight to as high as nearly 2000%. Water associated with bryophyte gametophores is apoplast water (held within cell walls), symplast water (held within the cell protoplast or lumen), and external capillary water. Each bryophyte, dependent on both its gross morphology and internal anatomy, has its own water potential. Water potential is controlled by environmental conditions in combination with the structure and physiology of the gametophore.

Among the bryophytes there is a diversity of means of water and mineral uptake. Most bryophytes are ectohydric; the gametophores of ectohydric bryophytes can absorb water and dissolved minerals over much of their surface, and these substances are immediately available in the sites where photosynthesis occurs. Since thalli of ectohydric hepatics are relatively thin, and leaves of ectohydric mosses and leafy hepatics are generally unistratose, the plants are well adapted to ectohydric water uptake. All of the Jungermanniales, Sphaerocarpales, Monocleales, and Anthocerotae are essentially ectohydric, and most of the mosses are also ectohydric, with the exception of the Polytrichidae, Tetraphidae, and some of the Bryidae. In ectohydric bryophytes water is conducted externally on the gametophore. It is possible that plications assist in longitudinal conduction in a leaf, and the higher the ratio of thickness of cell wall to the protoplast, the more rapid is the longitudinal conduction of water. In some ectohydric bryophytes (e.g., *Tortula*), the papillae on the surface of the cells are cutinized. Such papillae repel water, which is then directed to the capillary network of grooves among the papillae on the leaf surface and is taken up rapidly by the leaf cells. The leaves of some mosses are fully hydrated within a few seconds of immersion. Some leafy hepatics (including several species of *Scapania*) have silica deposited on the surfaces of the cells. The precise function of this is unknown, but it may be analagous to the cuticle on some moss leaves.

Movement of water up and down leafy shoots is affected by capillary networks in the felt of rhizoids, overlapping leaves, leaf bases or paraphyllia, or various combinations of these features. The growth habit of erect shoots, tufts of shoots, or loosely isolated shoots, affects movement of water among shoots. All of these features also are important in the uptake and retention of water by leafy bryophytes.

In many thallose bryophytes, including the hornworts, Monocleales, and many Metzgeriales, the rhizoids, besides forming an attachment to the substratum, also form a wick of capillary spaces that aids in external conduction of water along the thallus and improves retention of water that is ultimately absorbed by the thallus. The nature of ruffling of the thallus surface also serves to trap water, which is then absorbed through the upper surface of the thallus.

A number of mosses, especially Bryidae, and undoubtedly some Marchantiales and several Metzgeriales, absorb some moisture through the surface, but also conduct water internally. These are termed mixohydric. In some mosses hydroids form a central strand in the stem. Stereids of the stem also conduct water. Some Metzgeriales, including *Hymenophyton* and *Symphyogyna*, have a central conducting strand in the thallus; this strand is composed of elongate pitted cells and differs in this way from the hydroids of mosses. Many genera of the Marchantiales, several genera of the Metzgeriales, and *Monoclea* possess scattered pitted cells in the thallus. These cells allow free passage of water within the thallus, but generally do not form a continuous conduction strand.

In the Marchantiales the upper surface of the thallus is often cutinized, and cuticle is especially apparent around the mouths of the pores. This cuticle repels water and prevents clogging of the pores by water, and thus allows for free gas exchange. The undersurfaces of such thalli have rhizoids and scales that combine to form an effective capillary network for external conduction.

Finally, in the Polytrichidae a highly effective internal (endohydric) conducting system allows for upward movement of water in the stems. Since this system, composed of both hydroids and leptoids, is sometimes in continuous interconnection with the hydroids and leptoids of the leaves, it forms a system analagous to the vascular system of seed plants (Figs. 6-2, 6-3). In these mosses water is usually passed to the stem by the wick of capillary network of the rhizoids that are often abundant at the stem base, through the cortex to the hydrome, and thence to the leaves. In a saturated atmosphere, water is absorbed through the upper surfaces of the leaves as well. Only the uppermost cells of the lamellae are cutinized, and the rest of the abaxial surface of the leaf lacks cutin; it is here that absorption takes place. The adaxial surface of the leaf is often heavily cutinized. In some species of *Polytrichum* (e.g., *P. juniperinum*), the leaf lamina curves over the lamellae, enclosing them in a cutinized sheath (Fig. 6-4A, 6-17A). In such mosses water absorption is mainly through the stem.

In the Calobryales the leaves are relatively impermeable to water. The subterranean rhizomatous system, however, is an effective absorptive system, comparable to roots in lignified plants. The rhizomatous system has a central conducting strand of cells comparable to hydroids, and this strand is continuous with that of the leafy stem (Fig. 12-3G). In *Haplomitrium* the mucilage-rich cortical cells of the rhizome assist in water uptake. It is possible that endogenous fungi are also important in enhancing water and mineral uptake. In *Takakia*, too, fungi are frequent in the rhizomatous system; they are especially abundant near the mucilage cells.

In summary, water uptake in bryophytes is either ectohydric, mixohydric, or endohydric. Water is circulated through the gametophore either via stereids, thickened cell walls of the leaves, through plasmodesmata, thin cell walls, or pits from one cell to another, via hydroids or in a central conducting strand. In bryophytes, therefore, water can pass upward in the gametophore from a wet substratum, but in most bryophytes, water uptake is in the upper part of the gametophore, and passes downward.

WATER LOSS, DESICCATION, AND FREEZING TOLERANCE

Since water uptake is relatively rapid, water loss is also rapid. Many endohydric mosses have a cuticle that impedes water loss. In many leafy bryophytes the change in orientation of leaves and even in curling inward of branches can slow down water loss. Some thallose Marchantiales also curl inward when dry, and can resist desiccation to a certain degree. Many mosses and leafy hepatics show an extraordinarily high tolerance to extreme desiccation and resume normal metabolism very rapidly after rehydration. The moss *Tortula ruralis,* for example, can be air dried for 10 months, and recover normal photosynthesis within a few hours of rehydration. The process appears to be as follows: when a desiccated moss is rehydrated, respiration rapidly builds up to a rate much higher than dark respiration; at this time there is a net loss of assimilate ("leakage"). As the moss regains normal respiration, some of this leaked assimilate is reabsorbed.

Most research concerning desiccation tolerance has concerned mosses. Desiccation tolerance, although obviously most apparent in bryophytes in sites subject to recurrent dry periods, has been noted also in some bryophytes in other sites. Tolerance to drying is higher if the bryophyte dries out slowly; furthermore there is a seasonal acclimation to greater desiccation resistance. In temperate regions, resistance is greatest during periods of highest water stress, and lowest at times when water stress is low. Acclimation appears to be influenced by both photoperiod and temperature, but the exact means are not clearly understood. On the other hand, many bryophytes show little tolerance to desiccation during any season.

The morphology of some mosses allows them to prevent rapid water loss. Dry, colorless hair points on leaves of some mosses (e.g., some species of *Grimmia*) inhibit water loss by forming a boundary layer on the surface of the gametophore where evaporation does not take place. Tight turfs and cushions also retain moisture more effectively than loose turfs or wefts. These growth forms characterize the water regimes of sites and are useful as indicators of the microenvironment.

When dry, some bryophytes can tolerate both extremely high and extremely low temperatures. *Tortula ruralis,* for example, recovered after slow cooling to −196°C in liquid nitrogen. The same species, when dry, can tolerate temperatures above 100°C. The youngest part of the shoot tends to show greatest tolerance to desiccation or freezing. Tolerance to freezing tends to characterize bryophytes that are desiccation tolerant. Slow freezing is less damaging than rapid freezing. In freezing, dehydration is involved, and similar changes occur in the cell organelles as occur during desiccation at higher temperatures.

A number of hepatics are also tolerant to desiccation, but experimental work in these is scarce. Many species of *Frullania,* for example, are dehydrated much of the time, and some thallose hepatics, including *Targionia hypophylla,* are dry for extended periods.

Even in semiarid climates dew may be an important water source to some bryophytes. Condensation on pebbles makes water available near the pebble base, and a number of bryophytes exploit this site. The moss *Aschisma kansanum* is an apt example. This moss forms protonemata around the base of quartz pebbles of open gravelly sites in prairies of North America.

PHOTOSYNTHESIS AND RESPIRATION

Bryophytes have the same chlorophyll-protein complex comparable to other green land plants, but differences have been noted in the polypeptide patterns in isolated thylakoid membranes in mosses. Bryophytes seem to have a large amount of light-harvesting chlorophyll a/b-protein complex, an amount comparable to shade plants among the lignified plants. The action spectrum of light for both growth and photosynthesis is parallel to that of other green plants.

Photosynthesis can occur in bryophytes when moisture and available light permit. Maximum temperature for photosynthesis is typically 15–20°C, and net photosynthesis can be maintained to a temperature as low as −8 to −9°C. Lower temperature optima are known for some mosses of higher latitudes (5–10°C), and higher temperature optima have been reported (25–35°C). Light compensation values of mosses for net photosynthesis at optimum temperatures are often around 400 lux. In forest mosses, however, it is 300–700 lux, while in some mosses that grow in highly illuminated sites, it can be as high as 1000 lux. Many bryophytes photosynthesize effectively at far lower light intensities than do most seed plants. Very low CO_2 fixation rates characterize most bryophytes. During the year, with seasonal changes, optimal photosynthetic rates change, but the result is still lower carbon fixation than in seed plants. The precise mechanisms for this acclimation are unknown. Under continuous illumination, mosses in tundra habitats can maintain positive net assimilation continuously over 24 h, but sometimes chlorophyll destruction occurs and photosynthesis is reduced.

In some bryophytes, photosynthesis increases progressively to water contents of 500 to 1000% of dry weight. Measurable pyhotosynthesis occurs also at extremely low water levels (at −60 to −100 bar, but not as low as −150 bar).

After photosynthetic losses during nighttime respiration, it usually takes most mosses at least 2 h to reach compensation during the following day. In habitats where evaporation is rapid, bryophytes are scarcely able to compensate for nighttime losses. There tends to be an inherent selection for photosynthetic capacity at low moisture levels in bryophytes subject to such conditions. There is within some species a decidedly different photosynthetic rate in populations from different geographic localities or even from different habitats within a single geographic locality. This is generally a difference in response to temperature. The lowest temperatures at which photosynthesis can occur have been reported as −8 to −9°C in some mountain mosses; respiration could proceed as low as −14°C. Protein synthesis can proceed in *Tortula ruralis* as low as −2.5°C, and CO_2 fixation can be high at a few degrees above 0°C. This explains, in part, why some mosses grow during the winter in cold temperate regions. The main soluble photosynthetic product that accumulates is sucrose. Endohydric movement of photosynthate can be relatively rapid in some mosses that possess a leptome system. It has been estimated to be up to 32 cm/hr in *Polytrichum commune*. Ectohydric movement is much slower, and it may take several days to advance a single millimeter. Movement of photosynthate upward in shoots is more rapid than in the reverse direction.

SYMBIOSIS AND SAPROPHYTISM

In a few bryophytes, including the metzgerialean hepatic *Blasia pusilla,* a symbiotic relationship exists between the bryophyte and a nitrogen-fixing cyanobacterium, *Nostoc.* The *Nostoc* colonies are endogenous within a cavity from which filamentous extrusions extend among the algal cells. The *Nostoc* is thus enabled a more intimate contact with the hepatic and obtains metabolites from it. The hepatic growth is enhanced by nitrogen fixed by the cyanobacterium. In *Anthoceros* similar endogenous colonies exist and form a similar symbiosis. It is possible that algae and cyanobacteria that grow among bryophytes, through metabolite leakage from desiccation, also provide organic substances that are absorbed by the bryophyte, but these exogenous algae are apparently not intimately associated with the bryophyte.

Only the hepatic *Cryptothallus mirabilis* is an obviously saprophytic bryophyte. No chlorophyll is present in this hepatic, whose natural habitat is subterranean in organic substrata, often in partly decomposed peat. The fungi endogenous in the thallus provide the organic substances necessary for survival of the hepatic. In many other hepatics there are endophytic fungi, but their interaction with the bryophyte is not clearly understood. It is possible that they contribute to carbon nutrition. In the moss *Buxbaumia* it is possible that a similar relationship exists, and the Splachnaceae that occur on dung may obtain organic nutrients through the interaction with associated fungi. Soil bacteria, including *Agrobacterium,* can induce bud formation in protonemata of *Pylaisiella selwynii.* A species of *Agrobacterium* (*A. rhizogenes*), which induces root formation in seed plants, induces rhizoid formation in *Pylaisiella.* The bacteria may produce substances that affect the moss or they may alter the molecules released by the moss, and these, when reabsorbed, result in the induction of specific structures. This may be a mutualistic or commensal relationship.

MINERAL REQUIREMENTS AND UPTAKE

Mineral requirements for bryophytes are essentially the same as those in seed plants. Mineral uptake into the cells is controlled by a selectively permeable membrane. In leaves and thalli, mineral uptake is aided by the large surface area to volume ratio and to the low surface resistance to ion uptake in solution resulting from the general slight development of cuticle. Compared to lignified plants, the number of exposed cation exchange sites is high.

Elements associated with bryophytes are dissolved in external solution bathing the plant, and within the matrix of the cell walls. Mineral particles are also trapped by spaces among the overlapping plant organs. Ions are bound to exchange or chelating sites external to and including the plasma membrane and are soluble within the limits of the plasma membrane. Finally there is also insoluble material within the limits of the plasma membrane.

The very high cation exchange capacity in bryophytes is related to molecules in the cell wall. In *Sphagnum* it is correlated also to polygalacturonic acid or related molecules present within the cell wall, but in other bryophytes the specific nature of the wall materials in relation to cation exchange appears to be more complex, or at least variable among bryophytes on different substrata. There are some bryophytes that show anion exchange capacities, but this capacity is at least 100 times less than the cation exchange capacity. Dead plants have been observed to retain cations more effectively than living; in these, less proteinaceous material (for volume) is available than in living cells, hence possibly the enhanced exchange capacity.

Since divalent ions have a higher affinity for exchange sites compared to monovalent ions, they tend to displace them. In uptake, the soluble ions first establish an equilibrium with the binding sites of the cell exterior. This equilibrium is established very rapidly, and the remaining atoms are available for uptake by the cell. Any change in concentration or composition of the bathing medium alters the equilibrium.

It is possible that cell wall exchange sites buffer against excessive concentrations of ions that could be toxic; this may result from binding the incoming ions. Differences in this capacity restrict taxa to substrata and sites where toxic elements are absent or where their concentrations are low enough not to affect the bryophyte. There is no evidence in bryophytes that exchange sites can act as cation reservoirs for intracellular uptake of ions. Acid rain causes cations to be displaced from exchange sites and this could result in ion loss through persistent rainfall, and could ultimately prove toxic.

The minerals Na and K are mainly in solution within the cell, while Ca is mainly extracellular and in an exchangeable form. Intermediate patterns are shown by Mg and Zn. Element distribution within bryophyte gametophores differs dependent on the age of the plant part. Monovalent ions tend to be concentrated toward growing tips, while divalent ions increase in concentration with segment age. This is a general reflection of the fact that monovalent ions are mainly in solution within living cells. The increase in divalent ions makes the older shoots more effective cation exchange sites.

Calcium is essential to maintain the integrity of the cell membrane and to permit adhesion of cells, among other functions. It appears to be involved in differentiation of rhizoids in *Marchantia*, at least, which may also be true for *Funaria*, where it is necessary for bud formation. It is possible that calcicolous bryophytes have inherently leakier membranes than calcifuges maintained at the same Ca concentration, and consequently have greater requirements for available Ca to repair their membranes.

Nitrogen is as important to bryophytes as it is to all green plants. It has been observed that male gametophores in *Sphaerocarpos donnellii* may be smaller than the female gametophores because they more readily absorb ammonium ions; this results in a pH drop and may suppress growth. It is possible that nitrogen can be transferred from senescent or older parts of gametophores to the actively growing apices. Microbial growth among the older shoots may increase nitrogen availability in these shoots.

The significance of minerals made available through leachates from rainwash of surrounding forest vegetation is difficult to assess since it involves, besides the leachate, an interaction of the factors of illumination and evaporation. Unquestionably many minerals are available through both leachate and dust accumulated on the tree and washed down, during rainstorms, to the forest floor. From the leaves of trees K and P are readily leached, while Ca and Fe are often abundant in dust. The feces of insect larvae contribute P. Bryophytes nearest the tree base appear to be the most favorably located for nutrient availability and uptake, while more distant from the tree base there is a marked decline in element uptake in the terricolous bryophytes.

In grassland habitats, cation concentrations, especially K, were noted to be higher in bryophytes under shrubs. Leachates are considered the source, which make these protected sites more nutrient rich than the surrounding vegetation. Some bryophytes are confined mainly to such sites, where moisture persistence is also frequently more extended than in more open vegetation.

It is probable that bryophytes which occur in maritime localities receive mineral nutrients from salt spray. Some bryophytes, including the moss *Schistidium maritimum,* can tolerate salinity through their low permeability to cations as well as an active extrusion of sodium from the cells. Most other bryophytes that can live in salt marshes or rocks subject to erratic salt spray appear to be best represented in areas where annual precipitation is high and widely distributed through the year. It is possible that this leaches out any salts before they accumulate to become toxic. Since *Schistidium maritimum* is confined to maritime sites, it may receive some necessary specific minerals from salt spray.

SPORE GERMINATION AND GAMETOPHORE INITIATION

Spores of most bryophytes contain chlorophyll and oil spherules when mature. Sometimes this is masked by a heavily pigmented spore coat, as in many hepatics and hornworts. Occasionally chlorophyll is absent, as in the hepatic genus *Cryptothallus.* The oil spherules are especially conspicuous in spores of many mosses.

Spores vary in duration of their viability. Most can germinate within 24 h after they are shed. In the hepatics *Lophocolea cuspidata* and *Lunularia cruciata* germination occurs within 1 to 4 h after suitable conditions are provided. In other cases, germination occurs about a week after suitable conditions are provided (as in the moss *Andreaea rupestris*), or after 30 days (the mosses *Pseudoscleropodium purum* and *Polytrichum junipernium*).

It appears that jungermannialean hepatic spores have brief periods of viability. It is possible that those of mesic environments have briefer periods of viability than those of xeric environments. Some mosses, on the other hand, have been reported to have spores that are viable after many years. In *Sphagnum* 3 years has been noted, while *Oedipodium* has been noted to be viable after 20 years. Some moss spores may be viable after longer periods.

For germination, favorable moisture conditions coupled with available indirect light and suitable temperature are prerequisites. The latter varies among different species, but is generally above 10°C. A phytochrome system, under red action (660 nm), is apparently involved in initiating germination. Water uptake of the spore usually involves some increase in size of the spore, and increased respiration precedes active photosynthesis, which generally follows rupturing of the spore coat.

Germination patterns differ considerably between mosses and hepatics. In most hepatics (see pp. 147, 158, 182, 201, 214) there is no protonematal phase, or the protonema is reduced to a single cell, and an apical cell is rapidly differentiated, which initiates the vegetative gametophore. In hornworts, too, the protonematal phase is brief or absent (see p. 246). In mosses, on the other hand, a protonematal phase is general, and in a few cases this protonema is perennial (as in *Racelopus* and *Discelium*).

In most mosses, spore germination usually follows a distinctive pattern. After absorption of water there is usually asymmetric swelling of the spore, with a bulge generally appearing on the trilete face. Rupturing of the spore coat, usually at this face, extrudes a protonematal thread containing numerous chloroplasts. This thread may undergo several cell divisions as it elongates, and the cells have perpendicular cross-walls. This is the chloronema. From the chloronema emerge nonchlorophyllose rhizoids. These are also multicellular, but the cross-walls are oblique. While the chloronematal threads are positively phototropic, the rhizoids are negatively phototropic and positively geotropic (or gravitropic). The rhizoids function in attachment and are not important in absorption of water or minerals. Protonematal development is dependent on a carbon energy source (normally supplied by photosynthesis) and plant hormones, especially auxins and cytokinins.

After a time in many mosses (e.g., *Funaria*), the chloronema gives rise to cells in which the cross-walls are diagonal and in which chloroplasts are fewer and spindle-shaped, rather than nearly spherical, as in the chloronema. Auxin is necessary for this change, and its active transport from the protonematal apex toward its center is involved. The movement of most other substances, including nucleotides, sugars, and amino acids, is in the reverse direction. Inactive precursors appear to be carried from the center of the protonema to the tips of the branches where the auxin is activated and then transported backwards; here the change in filament morphology takes place. The walls of this filamentous stage, termed the caulonema, tend to be pigmented. Cells of the caulonema contain caulonema specific proteins that are absent from the chloronema, and only these cells can produce the buds that generate leafy gametophores. The switch to bud formation is assumed to be related to synthesis of these specific proteins, and extremely small amounts appear to be involved. In caulonemata, cytokinins are very loosely bound and are readily leached away. Buds are formed only if cytokinin is present in the substratum. Under natural conditions the protonema must reach a certain size to produce its own cytokinin in an amount sufficient for bud formation.

Normally protonema (at least in culture) releases phenolase into the medium; this can oxidize catechol into a colored product, and is responsible for browning of the wall in older caulonematal cells. The absence of O_2 prevents this reaction.

In some mosses, including *Sphagnum,* a filamentous protonema appears to be related to high potassium in the medium. Thus, since *Sphagnum* grows in extremely low potassium substrata, the filamentous stage is very reduced, and a unistratose thallus is initiated soon after spore germination. It is from this thallus that the buds are produced. When the protonemata of two species of mosses are growing on the same medium, auxin or related substances produced in one can leak into the medium and be taken up by the protonema of the other species, initiating bud production in that protonema before it is able to independently produce a concentration high enough.

The internal factors that control the differentiation of the morphology and anatomy of the gametophore have received little study. It is apparent that the events initiated by the formation of the apical cell serve as the initial organizer of the shoot apex. This is established early in the sporeling of hepatics, but in most mosses, changes leading to the appearance of the chloronematal phase already establish the oblique cross-walls, and an arresting of elongation of a chloronematal branch leads to the sequence of three cell divisions that set up the apical cell which generates the leafy gametophore. Succeeding cell divisons in the leafy gametophore lead to a very precisely organized sequence of events that produce stem segments and leaves from the three cells cut off from the apical cell. The same neat organization occurs in leafy gametophores of the Jungermanniales. Branching is also determined within a very clearly organized pattern.

In thallose hepatics and hornworts a very predictable sequence of cell divisions precedes the organization of layers within the thallus. In all cell formation and differentiation, the external environmental conditions can greatly alter the structure of the gametophore. Changes in illumination, moisture availability, and even associated organisms can result in these morphological changes.

Indole acetic acid (IAA) has been identified in *Marchantia polymorpha.* Its synthesis is localized at the thallus apex. The cells responsible for polar transport, at physiological concentrations, of endogenous auxin are localized within the ventral midrib region of the thallus, and are loosely organized into bundles. A polar basipetal flux is established by elongated cells as well as a presumed membrane permeability to auxin. If metabolic energy is depleted at either end of the thallus, the transport is blocked. The actual mechanism for active transport is currently unknown; possibly a carrier molecule is involved.

The presence of the auxin plays an important regulatory role in apical dominance, gemmae cup formation, and rhizoid formation. This is controlled through its influence on the balance between cell division and cell elongation. The extraordinary plasticity of the gametophore, especially in immature portions, is readily illustrated by the ease with which these cells can be induced to dedifferentiate to protonematal or sporeling-like stages. Injury of the gametophore sometimes results in dedifferentiation near the site of injury. Gametophores show strict apical dominance; when the apex of a shoot or thallus is destroyed, a lateral branch (or branches) appears near this point, and growth continues.

In hepatics it has been possible to alter the symmetry of leafy shoots by the application of hydroxyproline proteins. By this procedure, shoots were induced to produce underleaves of similar morphology to the lateral leaves. Indeed, some hepatics (e.g., *Scapania nemorosa*) were caused to produce underleaves, when in nature underleaves are unknown. Branching patterns of high specificity in some hepatics were also altered using the same methodology. It appears, therefore, that the biochemical factors influencing leaf shape and branching pattern can be readily influenced, which may provide a partial answer to the phenotypic plasticity of leafy hepatic gametophores, at least.

Growth of gametophores shows distinct periodicity in some bryophytes, independent of apparent suitable growth conditions. In others, however, gametophores appear to be able to grow at any time of the year when temperature and moisture conditions favor growth.

The production of rhizoids is enhanced by high moisture conditions in both mosses and hepatics. Rhizoids appear to be more sensitive to light than to gravitation. They grow toward higher moisture levels, but when bathed in water, a shoot produces rhizoids that grow in all directions.

Some pleurocarpous mosses, when growing on a hard substratum or on sand, produce abundant rhizoids, but when the same species grow interwoven with vascular plants, rhizoids are scarce. Production of rhizoids in acrocarpous mosses, on the other hand, appears to be independent of substratum. In most hepatics, when the apex of the rhizoid touches a hard surface, it forms a knobbed tip. This enhances its anchoring function. The mechanism for this tactile response is unknown. When young, rhizoids contain starch, but this is lost as they grow, and they grow toward moisture without it.

SEX ORGAN INITIATION

Most bryophytes have perennial gametophores and do not produce sex organs until after several years of vegetative growth. In some species of bryophytes, sex organs are unknown. The change from the vegetative phase to sex organ production is always related to temperature and moisture. There are also bryophytes in which photoperiod is one of the main controlling influences, but appropriate temperatures are also prerequisite. In most bryophytes the sex organs are initiated and mature during the same growing season.

In *Marchantia* and *Conocephalum*, long days (18 hr day, 6 hr night) are prerequisite for sex organ initiation, but the temperature must be near 21°C; if the temperature is as low as 10°C, sex organs do not appear. In *Sphagnum subnitens*, short days (8 hr day, 16 hr night) are necessary for sex organ formation. In *Pogonatum aloides*, a day neutral species, sex organs were initiated at 21°C, but not at 10°C, when vegetative growth of the gametophore is favored. On the other hand, in the moss *Physcomitrella patens*, also day neutral, vegetative growth is favored at 25°C, but when the temperature is lowered to 15°C, sex organs begin to appear. *Funaria hygrometrica* is also more or less day neutral. In monoicous bryophytes, the antheridia tend to appear first and the archegonia later.

In the hepatic genus *Riccia*, the presence of IAA induces female sex organ formation, while in some mosses (e.g., *Barbula* and *Bryum*), it enhances maleness. Since IAA is relatively common in the environment, its influence on bryophyte sex organ induction should be considered. Kinetin influences archegonial production in *Riccia crystallina* and in *Bryum argentum*. In the latter bryophyte, it inhibits antheridial production. In other bryophytes, however, kinetin has been demonstrated to have no effect on sex organ production.

In *Marchantia polymorpha*, which produces dioicous gametophores, the initiation of the sexual receptacles betrays sex organ initiation. When vegetative gametophores are subjected to continual daylight for 16 days under temperatures not exceeding 21°C, but not below 10°C, antheridiophores appear within approximately 16 days, while archegoniophores appear within approximately 21 days. The appearance of archegoniophores is characterized by a sharp rise in the C/N ratio and enhanced levels of the auxin IAA as well as protein and RNA. This is accompanied by a conspicuous increase in carbohydrate concentrated in the archegoniophore. With the appearance of antheridiophores, on the other hand, metabolic status is lowered; a decrease in IAA, RNA, and proteins is apparent; and the C/N ratio does not rise appreciably.

In the hornworts sex organs appear to be predominantly long-day induced. *Phaeoceros himalayensis*, for example, produces no sex organs under short day conditions, while they are produced under long day conditions, given favorable temperatures (around 18°C). This hornwort, when placed under short day conditions, is induced to increase tuber production, but upon a return to long days, tubers do not form. Some populations, however, lack the potential for tuber formation, presumably the result of a lack of the appropriate genetic system. *Phaeoceros laevis* has been reported to produce the sex organs only under a short day regime at 18°C.

Maturing sex organs are present in many bryophytes when the sporophytes from the preceding fertilization are maturing. This is especially apparent in many pleurocarpous mosses and jungermannialean hepatics. The maturation of sex organs is strongly dependent on favorable temperatures and moisture conditions. These factors vary among species and even with a single species, maturation is dependent on the geographic location of the population.

When antheridia are mature, their rupture is triggered by immersion in water. The archegonia, too, when mature, release the fluid residue from the disintegrated neck canal cells when they are immersed in water.

SPOROPHYTE PHYSIOLOGY

Fertilization establishes polarity in the zygote. In all bryophytes except the hornworts and *Monoclea,* the first division is transverse and horizontal; in the hornworts it is longitudinal, and in *Monoclea* it is reported to be free-nuclear up to the 26-nucleate stage.

If the developing embryo is removed from the gametophore, it tends to dedifferentiate, and a diploid gametophytic structure results. Thus in the mosses a protonematal phase generally appears, while in hepatics and hornworts, it is a sporeling. This diploid gametophyte behaves like a normal gametophyte.

The enlarging calyptra clearly influences the development of the sporophyte in both mosses and hepatics. The gametophore is vital in controlling sporophyte growth and differentiation. The interaction of these two controls is necessary for sporophyte growth and maturation.

In hepatics there is almost complete dependence of the sporophyte on photosynthate produced by the gametophore. Although there is chlorophyll in the developing sporophyte, gas exchange is highly inefficient, and it is unlikely that photosynthesis in the sporophyte is sufficient to balance any more than respiration losses. Thus for any growth, the photosynthate must pass from the gametophore through the transfer cells of the foot into the sporophyte. The usual presence of protective sleeves of tissue around the developing sporophyte, including perianths or perigynia, also decrease light availability to the sporophyte.

In the hornworts, on the other hand, the sporophyte usually possesses photosynthetic tissue well aerated by stomata. In *Phaeoceros laevis,* for example, the photosynthetic rate in the sporophyte is close to maintenance level. Continued growth of the sporophyte, however, is dependent on transfer of solutes from the gametophore. The haustorial base of the foot enhances more efficient transfer of these solutes to the sporophyte. There is some transfer in the reverse direction. The carbon absorbed by sporophytes accumulates in the spores as well as in the meristem and sporangium wall. Apparently the columella is more important in spore dispersal than in solute transport. Much of the solute transport appears to be symplastic, but uptake of gametophore-generated photosynthate is rapidly taken up in the intercalary meristem and passes on to the continually differentiating spores.

A well-aerated photosynthetic tissue is developed in many mosses. Stomata are of widespread occurrence in the sporangial wall of most mosses (excluding the Andreaeidae and Sphagnidae, and poorly developed in the Tetraphidae). Most sporangia have fewer than 20 stomata, and these tend to be more numerous near its base. Some, for example, *Polytrichum formosum,* have up to 200. Opening and closing of stomata is in response to light, but not all stomata close in the dark, at least in *Funaria hygrometrica.* Sporophytes are positively phototropic. In the growth of the seta, cytokinins are involved.

The moss sporophyte may be annual or perennial. In the latter case the sporophyte may take up to three years to reach maturity (e.g., in *Ptilium crista-castrensis*). The time of maturity of most moss sporophytes is directly related to the time of fertilization of the egg that initiated it. In temperate climates sporophytes frequently overwinter and do not shed their spores until the spring. In others, fertilization occurs in spring or summer, and spores are shed later in the same year. The ultimate maturation of the sporangium is strongly dependent on suitable temperature and available moisture.

Some mosses, for example, *Funaria hygrometrica,* are essentially self-supporting for the products of photosynthesis, at least during earlier stages of development; others, however, are heavily dependent on the gametophore. In *Mnium hornum* approximately 80% of the assimilate for the sporophyte is provided by the gametophore, while in *Pleuridium* approximately 90% is derived from the gametophore.

Even in *Funaria,* there is considerable variation in carbohydrate production in the sporophyte during its development. In early stages of sporangium expansion the sporophyte assimilates less than the gametophore, but when it is fully expanded, assimilation is approximately equal to that of the gametophore. As the green tissue of the sporophyte senesces, however, there is a rapid drop in photosynthesis until the sporangium dehisces. At this time the transfer cells in the haustrorium of the foot degenerate slowly, and transfer ceases completely during sporogenesis.

Respiration to photosynthesis ratio in moss sporophytes is commonly lower than in the gametophore. In sporophytes it tends to be less than 2:1, while in the gametophore 3–4:1 is common. Translocation of solutes from the gametophore to the sporophyte usually takes from 1 to 4 days.

The calyptra is important in the survival and morphogenesis of the sporophyte. It remains living until the sporangium enlarges. There is a coordination within the sporangium of gametophore action and that of the calyptra. Although the calyptra appears to provide no chemical influence, its physical presence is necessary for sporangial differentiation. Part of its function is protection of the apical meristem and the thin-walled dividing cells beneath, where desiccation could be lethal. Another function may be in physically controlling movement of materials within the differentiating sporangium. If the calyptra is removed very early in sporophyte development, the seta will continue to grow, but no sporangium differentiates. If the calyptra is removed just before sporangium expansion, however, the sporangium expands and differentiates normally. The seta elongates by intercalary cell division just below the apical zone, in contrast to elongation of the seta in hepatics, which is by rapid cell elongation after the sporangium is fully differentiated. The calyptra in the mosses, at least, appears to influence the establishment of endogenous reactions that inhibit the elongation of the seta and promote a series of cell divisions in the apical region, which result in the formation of the sporangium.

In most mosses the seta has a well-differentiated central conducting strand. This is frequently surrounded by a region of intercellular spaces with loosely arranged cells that have chlorophyll and starch when living. The outer cortex of the seta is of stereid cells and the surface is cutinized. The central strand sometimes has both hydroids (central) and leptoids (external to the hydroids), a common situation in the Polytrichidae, but also represented in some Buxbaumiidae and Bryidae. In many other mosses, however, only hydroids represent the internal conducting system of the seta. A number of mosses have no differentiated axial strand, but symplastic conduction has been demonstrated to occur in the stereids.

The foot of the sporophyte normally has well-developed transfer cells. These are often richly supplied with mitochondria and extensive systems of endoplasmic reticulum, and are thus implicated in active transport. The continued elongation of the seta would draw on solutes "pumped" through the transport cells and passing up the seta through the conducting system to the developing sporangium.

In summary, the physiology of bryophytes has opened up numerous questions concerning ion uptake, the factors leading to morphogenesis in both sporophytic and gametophytic generations, and has pointed out intriguing parallels of the reactions in the gametophyte with those that are essentially the same in sporophytes of lignified plants. These have been accorded great evolutionary significance by some researchers. It is reasonable to suggest that, in a finite biochemical and morphogenetic package, those features will be selected that best adapt the organism to its environment. It is unlikely to be important whether the sporophytic or gametophytic generation is involved.

FURTHER READING

Aro, E.-M. 1982. Polypeptide patterns of the thylakoid membranes of bryophytes. *Plant Sci. Lett.* **27**:335–345.

Ashton, N. W., N. H. Grimsley, and D. J. Cove. 1979. Analysis of gametophytic development in the moss, *Physcomitrella patens,* using auxin and cytokinin resistant mutants. *Planta* **144**:427–435.

Atanasiu, L. 1969. La photosynthèse et la respiration chez les mousses et lichens pendant l'hiver. *Rev. Bryol. Lichenol.* **36**:747–753.

Basiler, K. 1980. Fixation and uptake of nitrogen in *Sphagnum* blue-green algal associations. *Oikos* **34**:239–242.

Basile, D. V. 1980. A possible mode of action for morphoregulatory hydroxyproline-proteins. *Bull. Torrey Bot. Club* **107**:325–338.

——— and M. R. Basile. 1982. Evidence for a regulatory role of cell surface hydroxyproline-containing proteins in liverwort morphogenesis. *J. Hattori Bot. Lab.* **53**:221–227.

Bates, J. W. 1976. Cell permeability and regulation of intracellular sodium concentration in a halophytic and glycophytic moss. *New Phytol.* **77**:15–23.

——— 1982. The role of exchangeable calcium in calcicole and calcifuge mosses. *New Phytol.* **90**:239–252.

——— and D. H. Brown. 1974. The control of cation levels in seashore and inland mosses. *New Phytol.* **73**:483–495.

Bauer, L. 1963. On the stabilization of the male sexual tendency in Musci. *J. Linn. Soc. London (Botany)* **58**:337–342.

——— 1963. On the physiology of sporogonium differentiation in mosses. *J. Linn. Soc. London (Botany)* **58**:343–351.

——— 1967. Determination von Gametophyt und Sporophyt, in Ruhland, W. (ed.), *Encyclopedia of Plant Physiology XVIII*, pp. 235–256. Berlin: Springer.

Benson-Evans, K. 1964. Physiology of the reproduction of bryophytes. *Bryologist* **67**:431–445.

Berthier, J. 1978. Analyse des capacités morphogènes du filament des Eubryales, in Suire, C. (ed.), *Congr. Int. Bryol. Bryophyt. Bib.* **13**:222–241.

———, J. P. Larpent, and M. Larpent-Gorgaud. 1976. Light action on vegetative propagation in bryophytes. *J. Hattori Bot. Lab.* **41**:193–203.

Bewley, J. D. 1979. Physiological aspects of desiccation tolerance. *Ann. Rev. Plant Physiol.* **30**:199–238.

Boatman, D. J., and P. M. Lark. 1971. Inorganic nutrition of the protonemata of *Sphagnum papillosum* Lindb., *S. magellanicum* Brid., and *S. cuspidatum* Ehrh. *New Phytol.* **70**:1053–1059.

Bopp, M. 1961. Die Morphogenese der Laubmoose. *Biol. Rev.* **36**:237–280.

——— 1965. Entwicklungsphysiologie der Moose, in *Handbuch der Pflanzenphysiologie XV:* 802–843.

——— 1968. Control of differentiation in fern-allies and bryophytes. *Ann. Rev. Plant Physiol.* **19**:361–380.

——— 1976. External and internal regulation of the differentiation of the moss protonema. *J. Hattori Bot. Lab.* **41**:167–177.

——— 1981. Entwicklungsphysiologie der Moose, in Schultze-Motel, W. (ed.), *Advances in Bryology I*, pp. 11–77. Vaduz: J. Cramer.

——— 1982. How can external hormones regulate the morphogenesis of mosses? *J. Hattori Bot. Lab.* **53**:159–169.

——— 1983. Developmental physiology of bryophytes, in R. M. Schuster, (ed.), *New Manual of Bryology*, Vol. I, pp. 276–324. Nichinan: Hattori Botanical Laboratory.

——— and H. P. Weniger. 1971. Wassertransport vom Gametophyten zum Sporophyten bei Laubmoosen. *Z. Pflanzenphysiol.* **64**:190–198.

Brandes, H. 1973. Gametophyte development in ferns and bryophytes. *Ann. Rev. Plant Physiol.* **24**:115–128.

Bristol, B. M. 1916. On the remarkable retention of vitality of moss protonema. *New Phytol.* **15**:137–143.

Brown, D. H. 1982. Mineral nutrition, in Smith, A. J. E. (ed.), *Bryophyte Ecology*, pp. 383–444. London: Chapman & Hall.

Busby, J. R., and D. W. A. Whitfield. 1978. Water potential, water content, and net assimilation of some boreal forest mosses. *Can. J. Bot.* **56**:1151–1158.

Chopra, R. N., and P. K. Kumra. 1983. Hormonal regulation of growth and antheridial production in three mosses grown *in vitro*. *J. Bryol* **12**:491–502.

Clausen, E. 1964. The tolerance of hepatics to desiccation and temperature. *Bryologist* **67**:411–417.

Collins, N. J., and W. C. Oechel. 1974. The pattern of growth and translocation of photosynthate in a tundra moss, *Polytrichum alpinum*. *Can. J. Bot.* **52**:355–365.

Craigie, J. S., and W. S. G. Maass. 1966. The cation-exchanger in *Sphagnum* spp. *Ann. Bot.* **30**:153–154.

Dilks, T. J. K., and M. C. F. Proctor. 1979. Photosynthesis, respiration and water content in bryophytes. *New Phytol.* **82**:97–114.

Erichsen, J., B. Knoop, and M. Bopp. 1977. On the action mechanism of cytokinesis in mosses: caulonema specific proteins. *Planta* **135**:161–168.

———— 1978. Uptake, transport and metabolism of cytokinesis in moss protonema. *Plant Cell Physiol.* **19**:839–850.

French, J. C., and D. J. Paolillo. 1975. Intercalary meristematic activity in the sporophyte of *Funaria* (Musci). *Am. J. Bot.* **62**:86–96.

———— 1975. Effect of exogenously supplied growth regulators in intercalary meristematic activity and capsule expansion in *Funaria*. *Bryologist* **78**:431–437.

———— 1975. On the role of the calyptra in permitting expansion of capsules in the moss *Funaria*. *Bryologist* **78**:438–446.

Gorham, J. 1978. Effect of lunularic acid analogues on liverwort growth and IAA oxidation. *Phytochemistry* **17**:99–105.

Grubb, P. J. 1970. Observations on the structure and biology of *Haplomitrium* and *Takakia*, hepatics with roots. *New Phytol.* **69**:303–326.

————, O. P. Flint, and S. S. Gregory. 1969. Preliminary observations on the mineral nutrition of epiphytic mosses. *Trans. Br. Bryol. Soc.* **5**:802–817.

Gupta, R. K. 1978. Physiology of desiccation resistance in bryophytes: effect of pre-treatment on desiccation resistance. *Ind. J. Exp. Biol.* **16**:350–353.

Haberlandt, G. 1886. Beiträge zur Anatomie und Physiologie der Laubmoose. *Jb. Wiss. Bot.* **17**:359–498.

Hébant, C. 1977. *The Conducting Tissues of Bryophytes. Bryophyt. Bibl.,* Vol. 10. Lehre: Cramer.

Hosokawa, T., N. Odani, and H. Tagawa. 1964. Causality in the distribution of corticolous species in forests with special reference to the physioecological approach. *Bryologist* **67**:396–411.

Hughes, J. G. 1982. The effects of day-length on the development of sporophytes of *Polytrichium aloides* Hedw. and *P. piliferum* Hedw. *New Phytol.* **61**:266–273.

Kallio, P., and L. Kärenlampi. 1975. Photosynthesis in mosses and lichens, in Cooper, J. P. (ed.), *Photosynthesis and Productivity in Different Environments,* pp. 393–423. Cambridge: Cambridge University Press.

Kellomäki, S., and P. Hari. 1976. Rate of photosynthesis of some forest mosses as a function of temperature and light intensity and the effect of water content of the moss cushions on photosynthetic rate. *Silva Fenn.* **10**:288–295.

Koevenig, J. L. 1973. Effect of photoperiod, temperature and plant growth hormones on initiation or archegoniophore elongation in the thalloid liverwort *Reboulia hemisphaerica. Bryologist* **76**:501–504.

Larpent-Gourgaud, M., J. P. Larpent, and R. Jacques. 1974. Effet du phytochrome sur le développement des protonéma des Bryophytes. *Soc. Bot. France Coll. Les Problèmes Modernes de la Bryologie,* Bonnot, E.-S. (ed.), pp. 153–167.

Lockwood, L. G. 1975. The influence of photoperiod and exogenous nitrogen-containing compounds on the reproductive cycles of the liverwort *Cephalozia media. Am. J. Bot.* **62**:893–900.

Longton, R. E. 1980. Physiological ecology of mosses, in Taylor, R. J., and A. E. Leviton (eds.), *The Mosses of North America,* pp. 77–113. San Francisco: Pacific Division, AAAS.

———— and R. M. Schuster. 1983. Reproductive biology, in Schuster, R. M. (ed.), *New Manual of Bryology,* Vol. I, pp. 386–462. Nichinan: Hattori Botanical Laboratory.

Maeda, M. 1979. Isolation and culture of protoplasts from moss protonema. *Bot. Mag. Tokyo* **92**:105–110.

Maravolo, N. C. 1980. Control of development in hepatics. *Bull. Torrey Bot. Club* **107**:308–324.

————, D. J. Gaal, and S. J. Dufresne. 1982. On the control of endogenous auxin level in hepatics: transport of ¹⁴C-indoleacetic acid in *Marchantia polymorpha. J. Hattori Bot. Lab.* **53**:229–238.

McConaha, M. 1941. Ventral structures effecting capillarity in the Marchantiales. *Am. J. Bot.* **28**:301–306.

Paton, J. A., and J. W. Pearce. 1957. The occurrence, structure and functions of the stomata in British bryophytes. *Trans. Br. Bryol. Soc.* **3**:228–259.

Patterson, P. M., and J. S. Baker. 1961. Factors breaking vegetative dormancy in certain mosses. *Bryologist* **64**:336–338.

Proctor, M. C. F. 1977. Evidence on the carbon nutrition of moss sporophytes from ¹⁴CO₂ uptake and the subsequent movement of labelled assimilate. *J. Bryol.* **9**:375–386.

———— 1979. Surface wax on the leaves of some mosses. *J. Bryol.* **10**:531–538.

———— 1979. Structure and eco-physiological adaptation in bryophytes, in Clarke, G. C. S., and J. G. Duckett (eds.), *Bryophyte Systematics,* pp. 479–509. London: Academic Press.

———— 1981. Physiological ecology of bryophytes, in Schultze-Motel, W. (ed.), *Advances in Bryology I,* pp. 79–166. Vaduz: J. Cramer.

———— 1982. Physiological ecology: water relations, light and temperature responses, carbon balance, in Smith, A. J. E. (ed.), *Bryophyte Ecology,* pp. 333–381. London: Chapman & Hall.

———— 1982. ^{14}C experiments on the nutrition of liverwort sporophytes: *Pellia epiphylla, Cephalozia bicuspidata* and *Lophocolea heterophylla. J. Bryol.* **12**:279–285.

Proskauer, J. 1969. Studies on Anthocerotales. VIII. *Phytomorphology* **19**:52–66.

Rao, M. P., and V. S. R. Das. 1968. Metabolic changes during reproductive development in liverworts. *Z. Pflanzenphysiol.* **59**:87–99.

Rastorfer, J. R. 1962. Photosynthesis and respiration in moss sporophytes and gametophytes. *Phyton* (Argentina) **19**:169–177.

Rogers, G. A., and W. D. P. Stewart. 1977. The cyanophyte-hepatic symbiosis I. Morphology and physiology. *New Phytol.* **78**:441–458.

Scheirer, D. C. 1980. Differentiation of bryophyte conducting tissues: structure and histochemistry. *Bull. Torrey. Bot. Club* **107**:298–307.

———— and I. J. Goldklang. 1977. Pathway of water movement in hydroids of *Polytrichum commune* Hedw. (Bryopsida). *Am. J. Bot.* **64**:1046–1047.

Schönherr, J., and H. Ziegler. 1975. Hydrophobic cuticular ledges prevent water entering the air pores of liverwort thalli. *Planta* **125**:51–60.

Schwabe, W. W. 1975. Photoperiodism in liverworts, in Smith, H. (ed.),*Light and Plant Development,* pp. 371–382. London: Butterworths.

Sood, S., and K. D. Hackenberg. 1979. Interaction of auxin, antiauxin and cytokinin in relation to the formation of buds in moss protonema. *Z. Pflanzenphysiol.* **92**:385–397.

Spiess, L. D., B. B. Lippincott, and J. A. Lippincott. 1976. Comparative effects of growth substance and *Agrobacterium* on the moss protonema to gametophore phase change. *J. Hattori Bot. Lab.* **41**:185–192.

———— 1982. Bacteria-moss interaction in the regulation of protonemal growth and development. *J. Hattori Bot. Lab.* **53**:215–220.

Stewart, W. D. P., and G. A. Rogers. 1977. The Cyanophyceae-hepatic symbiosis. II. Nitrogen fixation and the interchange of nitrogen and carbon. *New Phytol.* **78**:459–471.

Tansley, A. G., and E. Chick. 1901. Notes on the conducting tissue system in Bryophyta. *Ann. Bot.* **15**:1–38.

Trachtenberg, S., and E. Zamski. 1979. The apoplastic conduction of water in *Polytrichum juniperinum* Willd. gametophytes. *New Phytol.* **83**:49–52.

Valanne, N. 1966. The germination of moss spores and their control by light. *Ann. Bot. Fenn.* **3**:1–60.

————, E.-M. Aro, and H. Niemi. 1982. Photosynthetic apparatus of *Ceratodon purpureus. J. Hattori Bot. Lab.* **53**:171–179.

Voth, P. D., and K. C. Hamner. 1940. Responses of *Marchantia polymorpha* to nutrient supply and photoperiod. *Bot. Gaz.* **102**:169–205.

Walker, R., and W. Pennington. 1939. The movements of the air pores of *Preissia quadrata* (Scop.). *New Phytol.* **38**:62–68.

24 Ecology

Bryophyte gametophores, rather than their dependent sporophytes, determine the ecology of each species. Survival is controlled by the conditions under which the spore germinates, the substrata to which it can become affixed and persist, and the nature of the site that influences survival. Many bryophytes of open or well-drained sites possess physiological and morphological features that allow them to tolerate desiccation.

SUBSTRATA COLONIZED BY BRYOPHYTES

Many bryophytes differ conspicuously from most seed plants in their ecological tolerance, since they can occupy hard surfaces, like bark or rock, that cannot be invaded by most seed plants. Bryophytes also colonize soil surfaces. Their persistence on soil, however, is sometimes brief, since the bryophyte cover often initiates changes in the substratum that make it suitable for colonization by seed plants, and these rooted plants generally alter the site so that the initial bryophyte colonists cannot survive. Bryophytes are important stabilizers of substrata that later become suitable for seed plant colonization.

Several bryophytes are restricted mainly to bare rock surfaces, including species of *Andreaea, Hedwigia, Schistidium,* and *Dicnemoloma* among the mosses, and some species of *Marsupella, Frullania,* and *Gymnomitrion* among the hepatics. Most of these bryophytes are highly tolerant of extended periods of desiccation. This tolerance is important in initial establishment as well as survival. Either spores or vegetative diaspores establish initial colonists. The anchoring of diaspores depends upon moisture availability long enough to permit production of attached protonemata and fixed gametophores in mosses, and rhizoids in hepatics and hornworts, after which an extended period of desiccation can be tolerated.

The particular bryophytes that colonize rock surfaces are determined by shade, available moisture, smoothness of the surface, and chemistry of the sub-

stratum. In moist crevices and mouths of caverns of siliceous cliffs, for example, the luminous moss *Schistostega pennata* is found in the Northern Hemisphere, while the same habitat in Australasia is occupied by another luminous
moss, *Mittenia plumula*. Some species of *Andreaea* and *Grimmia* are able to
occupy relatively smooth exposed surfaces; the same is true for the hepatic
genera *Gymnomitrion*, *Frullania*, and *Radula*.

The bryofloras of calcareous and siliceous rock surfaces differ conspicuously. Most calcicolous bryophytes appear to be less tolerant of desiccation
that those on siliceous substrata. Species of *Seligeria*, for example, are restricted mainly to calcareous rock surfaces that are shaded or where some
seepage exists. Another calcicole, *Andreaeobryum macrosporum*, occupies
calcareous cliffs where there is seepage or where water irrigates the surface
much of the year.

The environment that is at the opposite extreme to a rock surface is an aquatic site. It is surprising that the diversity of aquatic bryophytes is so low since
bryophyte morphology seems adapted in so many ways to an aquatic existence.
Several species of *Sphagnum* are aquatic. *S. macrophyllum*, for example, is
submerged, while *S. cuspidatum* frequently forms floating mats. The hepatic
Ricciocarpus natans often produces extensive colonies that float on the surface
of quiet waters, while *Riccia fluitans* is usually submerged just below the water
surface. The hepatic *Riella* is a submerged aquatic attached to the bottom muds
of shallow quiet water bodies.

Some bryophytes are confined to surfaces exposed to splashing or rapidly
moving water. The moss genera *Scouleria* and *Hygrohypnum* grow mainly in
such habitats, and a number of species of the hepatic genus *Scapania* form extensive colonies in these habitats. Several species of the moss genus *Fontinalis*
are attached aquatics submerged in moving water. The moss *Fissidens gran-
difrons* is usually confined to calcium-rich waters and thrives either submerged
or on continually irrigated surfaces.

Sphagnum is well-known as a mat former on pond and lake margins, especially in terrain where siliceous substrata predominate, and a number of
pleurocarpous Bryidae form floating mats around ponds that are in calcareous
terrain. Species of *Calliergon*, *Cratoneuron*, and *Drepanocladus*, for example,
often form thick floating mats, initiate the filling in of water bodies through accumulation of organic material, and serve as a surface upon which calcium car-
bonate precipitates. In hot springs, some mosses, including *Cratoneuron com-
mutatum* and *Eucladium verticillatum* often become encrusted by calcium car-
bonate and serve as "rock-formers."

Peat surfaces are often invaded by bryophytes. In peat cuttings, for example,
the moss *Dicranella cerviculata* sometimes forms extensive colonies, and the
hepatic genera *Mylia*, *Calypogeia*, and *Kurzia* produce interwoven mats. Some
species of *Campylopus* and *Dicranodontium* are also frequent on peaty substrata.

In peatland the species of *Sphagnum* are restricted to specific levels in the hummock and hollow topography of the site. They are important in determining the zonation of peatlands and controlling the pattern and dynamics of succession.

BRYOPHYTES AND SEED PLANT ASSOCIATIONS: FORESTED SITES

Many bryophytes intimately associated with vascular plants are confined to particular locations in vascular plant communities. Indeed, bryophytes are important in the dynamics of vascular plant associations. Extensive bryophyte mats, for example, can be significant in the water balance of the forest. At times when precipitation is low, all of the moisture may be taken up by the bryophyte mat, and none passes through to the roots of the seed plants. At other times, a loose deep bryophyte carpet enclosing many capillary spaces can prevent rapid water loss from the upper soil layers.

Bryophyte cover on the forest floor sometimes influences the dynamics of the vegetation in other ways as well. Rainwash from the tree canopy contributes nutrients that are rapidly absorbed by the bryophytes, and these nutrients may never be cycled into the rest of the ecosystem. Furthermore, a moist bryophyte carpet sometimes forms a suitable seedbed for tree seedlings. If the mat is very thick, however, it may prevent the roots of the seedlings from penetrating to the mineral layer; during a dry period, when the bryophyte mat dries out and contracts, death of the seedlings results. Physiognomic structure of the seed plant community controls the pattern of coverage by the bryophyte carpet which, in turn, influences regeneration of the seed plants.

Diversity of bryophytes tends to be low in forests where the bryophyte carpets are extensive; in boreal regions, for example, there may be considerable tracts of a single species of moss (Fig. 24-1). Bryophyte coverage on forest floors tends to be greater in gymnosperm than in angiosperm forests. This appears to be influenced by the leaf litter which, in a broadleafed forest, can inhibit formation of extensive bryophyte carpets.

Epiphytes, on the other hand, tend to be more abundant in angiosperm than in gymnosperm forests (Fig. 24-2). Much of the bryophyte growth in deciduous-leaf angiosperm forests is confined to periods when the canopy is leafless. Pattern and abundance of bryophytes vertically in a forest are determined by the balance between water and nutrient availability and tolerance to periods of desiccation. In needle-leafed trees there is greater uniformity in the epiphytic habitat than in broad-leafed deciduous angiosperm trees. This environmental difference is increased seasonally in deciduous-leafed angiosperms. In angiosperms, the shape of the tree crown may lead to greater environmental diversity in the different parts of the tree. This diversity is reflected in both floristic richness and biomass extent in the bryophytes of angiospermous trees. The bryophytes that occupy the canopy differ physiologically from those at the tree base. Respiratory efficiency of bryophytes in the canopy requires higher il-

FIGURE 24-1
A loose weft carpet of the
moss *Hylocomium splendens*.
(Provided by James Reid.)

lumination than is required by those of the base. Furthermore, the bryophytes of the canopy are more desiccation tolerant than those of the tree base. Bark textures also differ markedly in the crown compared to the tree base.

Among the diverse sites available for epiphytes are the nutrient rich tree base, the fissures and ridges of the bark, irregular surfaces on twigs, depressions at branch bases, and exposed surfaces on branches and trunks. These different sites vary in environmental conditions according to elevation in the tree and position in relation to the outside of the crown. Such differences affect nutrient availability, illumination, and humidity, and these interacting factors influence which bryophytes can colonize and persist on the tree. Epiphytic bryophytes often occur in company with lichens. Sometimes colonization by lichens prevents the establishment of bryophytes, or the lichens may overgrow the bryophytes and ultimately destroy them.

FIGURE 24-2
Epiphytic bryophytes. The mosses *Neckera douglasii* and *Isothecium stoloniferum* sheathing trunks of maples (*Acer circinatum*) in humid second-growth forest in coastal British Columbia, Canada. (Provided by James Reid.)

Fallen trees are occupied initially by the epiphytes of the living tree, but with change in illumination and moisture, these bryophytes tend to be replaced by others essentially restricted to moister decomposing bark and later to the decaying wood of the log. Within some forests a predictable succession by bryophytes is apparent, with the replacement of desiccation-tolerant bryophytes by those that are relatively intolerant of desiccation.

Forests at tropical latitudes, besides showing floristic differences in the bryophytes, show a number of features unique for the bryophyte vegetation. In tropical rain forests the bryophyte cover and floristic richness tend to be greatest on the tree bases, especially on trees with buttresses. The upper trunks have appressed hepatics, especially a number of Jungermanniales, and appressed mosses, particularly pleurocarpous Bryidae. Higher on the trunks and branches bryophytes attach to irregular surfaces, often at branch bases and near branch tips. The bryoflora is rich neither in diversity nor in abundance in the tropical rain forest. The forest floor tends to be very poor in bryophyte coverage.

At higher elevations in the tropics the foggy forests are often extraordinarily lush with bryophytes, which form thick carpets on the forest floor, sheathe tree trunks and branches, and form pendulous masses on the branches. Such extensive masses of bryophytes are very important in interception of precipitation, frequently more than 25%, which moisture often increases the weight of the bryophytes as much as four times. This increase in weight sometimes breaks the branches upon which the bryophytes grow. Undoubtedly such masses of bryophytes also profit from nutrients made available through rainwash of their tree substratum. Such foggy forests are especially well developed in the African mountains and the Malaysian archipelago.

Such "mossy" forests are found also in temperate oceanic areas and are especially well developed in New Zealand, Tasmania, and southern South America. On the northern Pacific coast of North America, especially in southeastern Alaska and adjacent Canada, the gymnosperm forests of wet climates exhibit a similar luxuriance in the bryophyte vegetation. The forest floor is often dominated by hepatics in such humid locales, in contrast to the extensive moss carpets in many gymnospermous forests in the boreal portion of the Northern Hemisphere.

Epiphyllous bryophytes are often frequent in humid tropical forests, particularly near water courses. They tend to be confined to evergreen leaves that have "wettable" surfaces. Although hepatics, especially Jungermanniales, tend to be best represented, a number of mosses, especially pleurocarpous Bryidae, are obligate epiphylls. The moss *Ephemeropsis* occurs predominantly on evergreen broad leaves in the Australasian and Malaysian area, while scattered throughout the tropics a number of moss and hepatic species are confined to living leaves of vascular plants.

BRYOPHYTE VEGETATION OF NONFORESTED SITES

In nonforested sites bryophyte coverage reaches is greatest extent and floristic diversity in alpine and polar climates. In such areas, however, lichens exhibit similar tolerances and are often intimately associated with bryophytes. Wetland habitats and areas of extended snow persistence often harbor the richest growth of bryophytes. In late-snow areas the bryophyte cover is markedly influenced by the duration of snow persistence; if a site is free of snow for a brief period, bryophyte cover is usually absent.

Turf-forming bryophytes often dominate the vegetation in alpine and polar climates, and floristic diversity is often very high within a limited space, with considerable mixture of shoots of many different mosses and hepatics. Crowding enhances survival, for which improved water retention is undoubtedly a prominent factor. The considerable floristic diversity within limited space (up to 20 species in 100 cm²) poses interesting problems concerning niche specificity.

Rock deserts of arctic areas often have richer bryophyte cover near the mouths of animal burrows and near nesting sites of birds. Undoubtedly manuring plays a role in this richness. In frost-boil areas, especially near the perimeter of boils, a number of bryophytes form well-defined local associations. Fine-textured soils are usually richer in bryophyte cover than coarser materials. Mosses usually predominate among the bryophytes while hepatics often occur as strands growing among mosses. In sites where organic material accumulates, the bryoflora is usually richer, and a more stabilized vegetation results, usually with vascular plants colonizing the same site.

Water is a limiting factor in much of the arctic bryoflora since precipitation is low, especially during the summer months. In consequence, turf-forming bryophytes often dominate the communities. Such a growth form retains moisture longer than more open and loose mats. Even pleurocarpous mosses often produce firm turfs in arctic regions, while the same taxa form loose, open mats in forested regions or in areas where precipitation is higher.

Arid and semiarid climates of the middle latitudes show a restricted bryoflora, and most bryophytes occur in rock crevices, in the shelter of rocks, or on somewhat shaded banks. Several, however, grow on fine textured soils that retain moisture longer than coarser soils. On soil crusts formed by cyanobacteria (blue-green algae), the stabilized surfaces permit invasion by bryophytes with an annual cycle of growth and spore production that is completed during a relatively brief period when moisture is available. Some bryophytes survive the unfavorable season as protonemata or portions of gametophores. A few bryophytes occupy exposed soils. Among these, especially in temperate climates, are species of the moss genera *Tortula, Barbula, Ceratodon,* and *Funaria.* Among the hepatics are the genera *Cephaloziella* and *Riccia.* Temporary water bodies within arid areas are often highly saline and have a number of bryophytes restricted to them, among which are *Riella helicophylla* and *R. numidica* of North Africa. In Australia and southern Africa the hepatic genus *Carrpos* is confined to gypsum rich "salt pans."

In dunes, whether in semiarid climates or maritime areas, a number of bryophytes help to stabilize the sand by extension of rhizoids among the sand grains and formation of a dense turf. Some mosses, through hygroscopic movements of the leaves can shed sand and prevent their burial. Most mosses of dune areas can produce innovations while buried, and these new branches penetrate the sand surface if they are not too deeply buried.

Other devices that appear to give bryophytes tolerance to high illumination are lamellae or filaments enclosed by incurved laminae or filaments with cutinized or reflective apical cells that appear to protect the cells beneath from high illumination. Many such bryophytes curl up when dehydrated and have a physiological tolerance to drought. The cushion habit also protects bryophytes from desiccation. The formation of a thick layer of still air over the cushion prevents the rapid dissipation of moisture. Hair points on leaves, frequent in mosses of sites subject to periods of aridity, also inhibit rapid water loss.

Grasslands also have a limited bryophyte vegetation. Mosses appear to be more successful than liverworts in grassland. The mosses form loose tangled threads among the flowering plants and sometimes occupy extensive areas, especially when the biological efficiency of the flowering plants is inhibited by overgrazing in pastureland or by excessive mowing in lawns. Continued annual repetition of the cycle of excessive mowing or overgazing results in deterioration of the grass and herb cover, and often in an increase in moss coverage. In temperate areas, because bryophytes are able to photosynthesize at somewhat lower temperatures than can the associated flowering plants, they are able to thrive in the cooler seasons when snow cover is not sufficiently deep to cut off adequate illumination.

BRYOPHYTES IN MAN-MADE HABITATS

Bryophytes also occupy man-made habitats. Rooftops, especially in regions with extended periods of wet weather, are often colonized by mosses (Fig. 24-3). Substrata that retain moisture particularly favor bryophyte colonization, with thatched rooftops sometimes harboring a rich moss mat and adding to the attractiveness of the roof. Slate, wooden, and pebble-impregnated shingles also favor moss colonization. Epiphytic and epilithic mosses usually predominate on rooftops, although several mosses of open soils are also represented, for example, *Ceratodon purpureus, Tortula ruralis,* and *Polytrichum juniperinum.* Among the epilithic mosses that invade rooftops are species of *Rhacomitrium, Grimmia,* and *Schistidium;* the epiphytes *Ulota* and *Orthotrichum* are frequent. On cement surfaces such as mortar among stones and bricks, and sheltered areas of city pavement, mosses often form extensive coverings. Shade, especially of vascular plants, enhances such coverings by improving moisture conditions and providing additional nutrients through rainwash of the tree canopy. Some bryophytes are more frequent in manmade habitats than in an undisturbed environment. The wide geographic range of the hepatics *Marchantia polymorpha* and *Lunularia cruciata* is enhanced by gardening activity, and disturbed banks, especially along roadways, are colonized by *Campylopus, Polytrichum,* and *Ceratodon.* The moss *Tortula muralis* is confined mainly to man-made substrata.

GROWTH FORMS

Growth form in bryophytes is the overall morphological structure of the gametophore. A number of growth forms have been recognized and show a positive correlation with specific environmental factors. Annual bryophytes are

not numerous and tend to characterize open sites. It is possible that some
bryophytes traditionally treated as annuals are perennial through long persistence of buried portions of the gametophyte, including rhizoidal gemmae and
fragments of protonemata in mosses, and portions of the thallus in hepatics.

Short turfs are composed of many shoots that form compact turfs less than 1
cm tall. Mosses of open sites frequently show this growth form, which usually
tolerates high illumination and a degree of desiccation. Tall turfs also consist of
upright shoots, but are taller than 1 cm. Although tall turfs are frequent in open
sites, they occur also in forested sites and often tolerate high illumination and
desiccation less than short turfs. Cushions are rounded tufts that tolerate some
desiccation, but the cushion habit tends to slow down rapid desiccation. Mats
formed of creeping shoots are usually firmly affixed to the substratum by
rhizoids. Mats are usually tolerant of a degree of desiccation or possess, as in
the marchantialean hepatics, structural devices that slow down desiccation.

Many bryophytes that grow on smooth surfaces possess a mat growth form; this includes epiphyllous and epilithic bryophytes, as well as epiphytes on tree bark.

Wefts are woven carpets with much branched shoots. They are usually loose on the substratum and are generally intolerant of high illumination and rapid desiccation (Fig. 24-1). Wefts abound on forest floors in temperate areas and in humid climates, where they also sheathe tree trunks and branches. Wefts retain moisture in capillary spaces of the carpet. Pendants are formed of shoots that hang down from branches and reach their greatest abundance in cloud forests of the tropics and highly humid forests in temperate coastal areas. Fans are bryophytes in which the branched gametophore fans out, usually in a single plane, from a suberect shoot. They are particularly characteristic of humid, shaded sites and are either terrestrial or epiphytic. Dendroids resemble miniature trees and possess an erect main stem terminated by a tuft of branches. Dendroids often have creeping stems from which the erect gametophores emerge. Dendroids are usually in damp sites.

Aquatic bryophytes have diverse growth forms. Some form loose tall turfs while others are essentially pendent or trailing in form, supported by the aquatic medium. Some are annuals, while others form perennial floating mats or entangled mats on the bottoms of water bodies.

It is possible to characterize microclimate of a site based on proportionate representation of different growth forms. Some bryophytes can change their growth form to suit different habitats while others lack such plasticity.

"Moss balls" are formed where the moss cushion is continually overturned because of freeze-thaw movements of the soil or ice beneath, creating "rolling stones" with a covering of moss on all surfaces. Another type of moss ball is formed on lake bottoms where aquatic mosses are rolled into spherical masses through water movement.

SAMPLING BRYOPHYTE VEGETATION AND RECOGNIZING COMMUNITIES

Sampling bryophyte vegetation has several inherent problems. First, since species sometimes cannot be recognized in the field, material must be removed from the sample plots for laboratory determination; often bryophyte vegetation is composed of a great diversity of taxa, and many of these taxa are indeterminable from vegetative material. Second, limits of the microenvironment are extremely difficult to determine, which can result in the collecting of data from a nonuniform environment. Third, it is often difficult to estimate either coverage or abundance, since bryophyte shoots are often intimately intertangled. Fourth, the appropriate size of sampling plots is difficult to establish, and plotless sampling in a nonuniform microenvironment can contribute a broad range of data derived from parts of adjacent microsites.

In spite of these problems, attempts have been made to describe bryophyte communities. The naming of communities has led to the comparison of similar communities from various parts of the world, but conclusions remain tentative, since the body of available data is small. Statistical techniques have been used in interpreting data derived from sampling, but the results are inconclusive largely because of the problems in sampling and interpretation of the limits of microenvironments.

PRODUCTIVITY

Nearly all information on bryophyte productivity is based on mosses. Net annual production is usually calculated on the basis of mean dry weight of the current season's growth. The limits of this growth are troublesome to determine, and consequently the results are rough estimates.

Estimates have shown unexpectedly high productivity for some mosses in Antarctica, while the same species in the Northern Hemisphere have shown notably lower productivity. Productivity in wet sites is considerably higher than in dry sites. The highest values have been given for *Sphagnum* peatland, where as much as 12 tons per hectare per annum have been calculated.

Decomposition appears to be slow in bryophytes, and much of the released nutrient appears to be transferred to the living shoots adjacent to or connected to the decomposing shoots. The slow decomposition results, in some sites, in the accumulation or organic bryophyte detritus. This is especially conspicuous in peatland, but also characterizes tundra habitats and, to a degree, forested sites.

ECOTYPES

Geographically differentiated ecotypes have been demonstrated for a few bryophytes. Ecotypes are defined as genetically differentiated and ecologically correlated populations of species that show physiological, biochemical, or slight morphological differentiation from other populations of the same species. While ecotypes have been demonstrated in sporophytes of many seed plants, they are represented by gametophores in the bryophytes. The evolution of ecotypes considerably enhances the survival in a varying environment since the species possesses overall heterogeneity even though each population may show uniformity.

The mosses *Polytrichum strictum, Bryum argenteum,* and *Pohlia nutans* possess ecotypes from different geographic areas that behave differently in a uniform environment. The Hepaticae *Conocephalum conicum* and *Plagiochila porelloides* exhibit biochemically different races, and some of these different populations show ecological segregation correlated with biochemical similarity.

POPULATION ECOLOGY

Population ecology is concerned with the changes of numbers of individuals within a population. It concerns itself with the factors that initiate those changes as well as the consequences of the changes. Several factors are of special importance in bryophyte population changes. These involve the sexuality of the gametophores, their potential for asexual reproduction, the fact that the sporophyte is a dependent generation, and the fact that the sperms can swim very short distances and therefore influence the rate of sexual reproduction between populations of the same species.

More than half of the species of bryophytes are dioicous. This increases the probability of outbreeding and possibility of evolutionary change within a population through sexual reproduction. Since the sperms must swim in a water film in order to reach the egg, both sexes of a dioicous species must be nearby (usually less than 1 m apart) in order for sexual reproduction to occur. Populations that consist of a single sex can expand by vegetative growth outward from the site of initial establishment and also through production of vegetative diaspores. Gemmae are usually small and are readily airborne or can be washed by surface water to sites distant from the population that produces them. This is another means of getting the two sexes close enough to each other to result in sexual reproduction.

In some bryophytes, especially leafy hepatics and bryidean mosses, asexual reproduction is often very highly developed during the period before the production of sex organs. Sex organs are unknown in a few bryophytes, and all reproduction is asexual. It is probable that such species have very high genotypic uniformity.

It is possible that sexuality has been important in the general structure of the bryoflora of a geographic region. Dioicy is predominant in some bryophytes, for example, in the Jungermanniales and the Polytrichidae. In environments where available water for sexual reproduction is limiting, it might be predicted that the monoicous conditon would predominate. To a degree, this is true, since in semiarid environments monoicy tends to predominate over dioicy. Here acrocarpous Bryidae and Marchantiales, in which monoicy is frequent, often characterize the flora. In the humid tropics, pleurocarpous Bryidae and Jungermannialean hepatics, both of which are mainly dioicous, are richly represented in the flora.

An ecological advantage of dioicy is that it permits each of the sexes to specialize in production of either sperms or eggs. The sperm-producing gametophore can be selected for the efficiency of morphologies that enhance more distant dispersal of living sperms, as in splash-cups which expose the antheridia. The egg-producing gametophores, on the other hand, can be selected for morphologies that protect the archegonia from exposure and keep them moist and the eggs viable long enough for the sperms to reach them. Furthermore, the protective perichaetia also protect the developing zygote. In monoicous mosses, at least, the perigonia and perichaetia are similar in gross structure, since both are usually bulbiform.

It appears that asexual reproduction is the main means of expansion of populations of bryophytes in many areas. In arctic and antarctic regions, for example, many bryophytes appear to produce sporophytes only rarely. The same is true for extensive carpets of mosses in temperate forests. This probably results in a great deal of genetic uniformity, which is possibly advantageous to plants in environmentally uniform landscapes. For long-term survival in a changing or nonuniform environment, however, following such an evolutionary pathway is hazardous.

Monoicy appears to be more conducive to sporophyte production than dioicy. Such sporophytes are undoubtedly often the product of inbreeding. This results in a simple duplication of genetic information in each gamete that gives rise to the zygote, and the spores from the sporophyte are genotypically identical to the gametophore on which the sporophyte was borne. There is a possibility of outbreeding in monoicous gametophores, but since many growth forms produce extensive clones, it is possibly low.

In dioicous bryophytes the ratio of male:female gametophores usually favors the female, and this, in turn, influences the frequency of sporophyte production. In some bryophytes antheridial plants are unknown, as in the hepatic *Takakia*; in others antheridial plants are rare, as in the moss *Rhytidium rugosum*. In the former, sporophytes have never been reported, while in the latter sporophytes are exceedingly rare. In spite of this apparent infrequency of sexual reproduction, this moss is widely distributed. The reasonable assumption is that the moss has great genotypic uniformity. It is sometimes difficult to determine sexuality of bryophytes since they may produce no sex organs for many years. In some species sex organs have not been noted.

Although the tiny spores of bryophytes would appear to be designed for easy dispersibility by passing air currents, it appears that most spores land within a few centimeters of the sporophyte that produces them, but the distances may exceed a meter or more for some spores. Since some spores have been found in the air more than a meter above the sporophyte, it is probable that they can be carried considerable distances. This makes it possible for dioicous species, in particular, to show considerable genotypic diversity.

Increased density in bryophyte populations enhances survival, whether of a single species or a diversity of species. The formation of dense turfs, mats, or cushions improves water-holding capacity. Although this may be of great advantage to the bryophytes, it may negatively or positively influence the seed plant communities with which the bryophytes are associated.

It is difficult to determine the age classes of bryophytes in a population or in a community. Since many bryophytes can perennate from gametophytic fragments, it is possible that resident gametophores are the persistent growing points of plants that became established on the site when the species first colonized it. The life span is often determined by the duration of time that the substratum or general environment has been available. Some bryophytes survive only in disturbed sites and disappear when the substratum becomes stabilized or persist for as long as the distrubance is maintained. Epiphytes persist until their host plant disappears from the community.

BRYOPHYTES AS SITE INDICATORS

European researchers, in particular, have utilized bryophytes to assist in characterizing seed plant communities. Some species are consistent inhabitants of very distinctive seed plant communities and can be used as character species. This positive relationship reflects specific interactions and, for the bryophytes, often denotes that the seed plant community provides a favorable microenvironment for them. Although few bryophytes are confined to a single type of seed plant vegetation, they are useful in characterizing the kind of site. A knowledge of bryophyte floristics as well as the bryophyte vegetation is often useful in characterizing a site. Indeed, in some areas it is possible to state fairly accurately the structure and productivity of the vascular plant community when only the bryophyte vegetation is documented.

Some bryophytes can persist in localized sites when the original vascular plant vegetation is destroyed. They become useful remnants to indicate the past existence of a forest or nonforest vegetation and can be used as indicators that this vegetation could be effectively regenerated on that site.

Some bryophytes serve as indicators of the substratum. The mosses *Mielichhoferia mielichhoferi* and *Scopelophila* spp. appear to be positively correlated with copper-rich substrata. Many mosses occur only on lime-rich substrata and are therefore indicators of this material. Although bryophytes are minor indicators of mineral deposits, they cannot be overlooked as useful tools in prospecting. Since some bryophytes concentrate elements in substantial quantities far greater than the substrata on which they grow, their chemical analysis serves as an inexpensive means of determining the presence of particular elements in the substratum.

RESPONSES OF BRYOPHYTES TO ENVIRONMENTAL POLLUTION

Cities and industrial areas are often characterized by a scarcity of bryophytes, although several mosses appear to be tolerant of some industrial pollution. In temperate areas, for example, the mosses *Ceratodon purpureus* and *Bryum argenteum* are frequent in urban areas. Epiphytic bryophytes, in particular, appear to be particularly susceptible to air pollution.

Many bryophytes that have greater tolerance of pollution have higher reproductive capacity in polluted areas than those that are intolerant. This tolerance is enhanced by a brief protonemal stage, a stage that is particularly sensitive to pollution, and very rapid production of gametophores, which are less sensitive. Growth rate of pollution tolerant species is usually much more rapid than that of intolerant species.

In the urban environment, bryophytes of stone walls and sheltered sites are often more tolerant of pollution than those of more open sites. Those of tree bases and calcareous substrata are usually also more tolerant. It has been suggested that increased pH, and increased ionization and oxidation of acidic pollutants, characterizes these sites. In more humid climates the tolerance to pollution is greatly increased, while in drier climates lower concentrations are toxic.

Some bryophytes are especially sensitive to particular pollutants, and their absence in areas where they are naturally distributed can serve as valuable pollution indicators. Such sensitive bryophytes can be exploited in the construction of a "bryo-meter" that can be used to determine atmospheric pollution in a particular site. Such a "bryo-meter" contains the living bryophytes within a chamber through which air is drawn. The recorded changes in the mortality of the bryophytes serves as a gauge to the amount of toxic material in the air.

Bryophytes are also passive concentrators of radioactive isotopes and heavy metals. Mosses concentrate nearly twice as much of these substances as do seed plants. This is interpreted to be the result of the greater surface area per unit of dry weight of tissue and also the growth form, which enhances trapping of particulate fallout. Furthermore, bryophytes are green throughout the year and pass readily in and out of dormancy, whereas tree leaves fall during some season and even evergreens are frequently dormant or physically inactive in some seasons. Chemical analysis of the bryophyte cover serves as a rapid and inexpensive method to survey heavy metal and radioactive isotope concentration in an ecosystem. Aquatic mosses are similarly useful as pollution indicators both because of their floristic diversity and that they concentrate heavy metals.

INITIAL COLONIZATION AND SUCCESSION

Open and often nutrient-poor sites are frequently colonized by bryophytes. On recently deposited volcanic soils, for example, some hepatics survive on substrata that are so nutrient poor that no other plant is able to persist. In time, however, the bryophytes build up an organic layer that is invaded by microorganisms, resulting in changes in the mineral substratum beneath. This increases nutrient availability and makes the site suitable for invastion by vascular plants. These, in turn, alter the microenvironment, and the initial colonists cannot survive.

The same process occurs on bare rock, but the speed of increasing nutrient availability is extremely slow. Some bryophytes can attach to relatively smooth surfaces and sustain extended periods of desiccation. Their ability to expand vegetatively, to extend their population through vegetative diaspores, and to obtain nutrients from rainwater, rainwash, and dust makes them ideal long-term colonists of such sites (Fig. 24-4). On steep cliffs colonies of bryophytes often occupy extensive areas. Sometimes these bryophyte mats are invaded by vascular plants, but these, as they increase biomass on perpendicular sufaces, often cause the whole mass to slide off the surface, and the site is reopened for colonization.

Other bare sites for colonization include trunks and branches of trees. Succession on living trees has been studied very little. The bryophyte colonists are determined by bark smoothness, texture, water retentiveness, and chemistry. Exposure of the trunk surface affects water availablity and exposure to the sun. In some epiphytic bryophytes, it appears that no succession occurs in some sites while the host tree is living; although there may be changes in size and shape of the colonies, the site appears to be continuously occupied by the same

FIGURE 24-4
Epilithic bryophytes: boulders covered mainly with the moss *Rhacomitrium canescens,* coastal British Columbia, Canada. (Provided by James Reid.)

species. Such is a common phenomenon in taxa that are tightly affixed to the bark. In those that form looser tufts, tangled masses, or turflike sleeves, other bryophyte opportunists frequently invade the colony, and buildup of organic material makes the site suitable for invasion by vascular plants. Such colonies sometimes become sodden during times of high precipitation; they become so heavy that the whole mass falls from the branch or so weights the branch that it is broken from the tree, carrying its epiphyte load with it. When such fallen bryophytes reach the forest floor, or when trees fall with their bryophytic colonists, a new succession is initiated with the change in moisture and illumination regimes and with decomposition of the wood.

There is some fluctuation in detailed pattern even in bryophyte vegetation of forest floors. The overall floristic pattern, however, is little altered. Since it is difficult to calculate ages of most bryophyte turfs and mats, it is difficult to determine succession in apparently stable bryophyte vegetation. It is possible that the identification of biochemical races of bryophytes may help to label populations and calculate their age but more research is needed before this is possible.

Bryophyte succession on horticultural sites after abandonment has shown a predictable pattern within limited geographic areas where uniform agricultural practices are followed, especially in timing of the tilling of the soil and harvesting of the crop. Initial invaders are opportunists that tolerate shade of the agricultural crop. They are frequently open turfs or annuals, and often are characterized by efficient vegetative propagation. Since available vascular plants in the flora successfully invade these sites, the bryoflora changes in response to the way these plants occupy the soil.

The bryophyte colonization of roadside banks is important in stabilizing these sites. Indeed, this could be exploited by transplantation of turf species or by scattering vegetative diaspores of appropriate species. Intelligent understanding of the hydrology of the slopes is necessary for successful stabilization of the site. Later invasion by available vascular plants gives the banks further stability.

BRYOPHYTES AND OTHER ORGANISMS

The growth form and nature of the chemical composition of bryophytes are the main factors that influence interaction with other organisms. Considered here are animals (other than man), algae, and fungi.

Water retention is a direct outcome of the morphology of the gametophore, and it is this feature that makes bryophyte vegetation an attractive habitat for many invertebrates. The bryophyte-invertebrate interactions are discussed by U. Gerson (1982), and his treatment should be consulted for more detail. A number of invertebrates seem to occur exclusively with bryophytes. *Sphagnum* seems to harbor the richest invertebrate fauna. One site revealed 145 species in a *Sphagnum* dominated site, while a comparable forest site with a non-*Sphag-*

num dominated bryophyte vegetation revealed 65. In the wetter habitat that usually characterizes *Sphagnum,* generally the number of individuals is also higher. In such habitats rhizopods, ciliates, and flagellates are well represented. Also present, sometimes within the hyaline cells of the *Sphagnum,* are rotifers, nematodes, various algae (including diatoms and desmids), and cyanobacteria.

Among the invertebrates represented, many can tolerate the periods of desiccation that affect their bryophyte substratum. Rotifers appear to be better represented in polar and alpine regions than in milder climates. They seem to feed on particles from the bryophytes. Sometimes rotifers can be seen in the lobules of the hepatic *Frullania.* Snails and slugs are frequent among both terrestrial and aquatic mosses and deposit egg masses upon them. Water bears (tardigrades) appear to live preferentially among mosses, especially turfs and cushions; these animals (and rotifers) can go in and out of dormancy very quickly. These tiny animals pierce the cells of their host plants and suck out the contents.

Many arthropods are found in bryophyte colonies: spiders and mites, millipedes, centipedes, and various crustaceans, the latter mainly in the aquatic environment. Insects are the most richly represented arthropods in bryophytes and include 21 of the 29 orders. Some insects feed on bryophytes by browsing on the gametophores, sucking the juices from the cells, and harvesting the sporangia. Many insects associated with bryophytes deposit their eggs there, and the larval stages often browse on the gametophores. Ants make extensive colonies in *Sphagnum* and *Dicranum* hummocks in peatland. In such sites elevated humidity would be favorable and saturation would be infrequent, in contrast to most of the peatland.

Within the bryophyte vegetation are bryophyte-feeding insects that are preyed upon by other insects. Bryophyte turfs that accumulate heavy metals or radioactive fallout initiate the passing of increased concentrations through the food chain. The bryophyte mat as a site for survival of juvenile stages of insects that later cause serious damage to forest trees is also important. Some of the pesticide "solutions" to this problem have led to accumulation of these substances in the bryophyte mat, and the results to the entire ecosystem have been serious.

The importance of bryophytes to invertebrates as shelters from adverse conditions in extreme environments, as in Antarctica and the arctic "deserts," is considerable.

Some insects have morphologies, surface patterns, or appendages that permit them to blend in with their bryophytic habitat. Many of them gain further protection by being sluggish or by spending long periods without moving. A few insects, including some weevils in New Guinea, paste a bryophytic "garden" on their wing cases and thus camouflage themselves against predation. In doing this, the bryophytes are transported to new areas when they become detached from the beetle.

The implication of invertebrates in carrying sperms from the antheridia to archegonia has been observed for a few mosses. In the moss family Splachnaceae, especially in the genus *Splachnum,* dipteran insects appear to carry the spores from the mature sporangium to the nitrogen-rich substratum that the moss inhabits. This is directed through the production of aromatic compounds in an expanded, and sometimes brightly colored, apophysis of the moss sporangium. This apophysis sometimes forms a convenient landing platform. The insect, when it lands, walks among spores that have fallen upon the platform. Then, when it moves to another aromatic substratum, especially dung or decaying animal remains, the spores are lost while the insect explores the new site for food.

It is highly probable that invertebrate activity within bryophyte communities enhances mineral cycling that favors the further growth of the bryophytes.

Bryophytes are also utilized by vertebrates. Birds sometimes use gametophores and even moss sporophytes for nest-building material. Any bryophyte that is easily removed from its substratum and can be woven into a nest seems appropriate.

Some fungi appear to be restricted to bryophytes. Discomycetes are especially frequent, but all major fungal groups are represented. In mosses the sporangia are more frequently infected than the gametophore. Fungi are often endophytic in hepatics. In bryophyte vegetation of cold climates fungal destruction of bryophytes is more apparent than in other environments, although in temperate areas, in particular, epiphytes are often infected. Some lichens overgrow bryophytes, while others are more frequent on bryophytes than on other substrata. In some sites colonization by *Cladina* and *Cladonia* is preceded by bryophyte vegetation.

A number of algae are frequent among aquatic bryophytes. In the terrestrial habitat cyanobacteria, especially *Nostoc,* are common, and green algae are frequent. Indeed, *Nostoc* is endophytic in some hepatics, for example *Blasia* (Metzgeriales), and in Anthocerotales, with which there is an obligate relationship. Since *Nostoc* is a nitrogen fixer, its advantage to the bryophyte is apparent.

FURTHER READING

Ando, H. 1979. Ecology of terrestrial plants in the Antarctic with particular reference to bryophytes. *Mem. Nat. Inst. Polar Res.* **11**:81–103.

Barkman, J. J. 1958. *Phytosociology and Ecology of Cryptogamic Epiphytes.* Assen: Van Gorcum.

Brooks, R. R. 1971. Bryophytes as a guide to mineralisation. *N. Z. J. Bot.* **9**:674–677.

Busby, J. R., L. C. Bliss, and C. D. Hamilton. 1978. Microclimate control of growth rates and habitats of the boreal forest mosses, *Tomenthypnum nitens* and *Hylocomium splendens. Ecol. Monogr.* **48**:95–110.

Clarke, G. C. S., S. W. Greene, and D. M. Green. 1971. Productivity of bryophytes in polar regions. *Ann. Bot.* **35**:99–108.

Collins, N. J. 1976. Growth and population dynamics of the moss *Polytrichum alpestre* in the maritime Antarctic. *Oikos* **27**:389–401.

During, H. J. 1979. Life strategies of bryophytes: a preliminary review. *Lindbergia* **5**:2–18.

Forman, R. T. T. 1964. Growth under controlled conditions to explain the hierarchial distribution of a moss, *Tetraphis pellucida. Ecol. Monogr.* **34**:1–35.

Gerson, U. 1982. Bryophytes and invertebrates, in Smith, A. J. E. (ed.), *Bryophyte Ecology*, pp. 291–332. London: Chapman & Hall.

Gimingham, C. Y., and E. M. Birse. 1957. Ecological studies on growth form in bryophytes. I. Correlations between growth form and habitat. *J. Ecol.* **45**:533–545.

—— and E. T. Robertson. 1950. Preliminary investigations on the strucutre of bryophyte communities. *Trans. Br. Bryol. Soc.* **1**:330–344.

Hamilton, E. S. 1953. Bryophyte life forms on slopes of contrasting exposures in central New Jersey. *Bull. Torrey Bot. Club* **80**:264–272.

Iwatsuki, Z. 1960. The epiphytic bryophyte communities in Japan. *J. Hattori Bot. Lab.* **22**:159–354.

Keever, C. 1957. Establishment of *Grimmia laevigata* on granite and its method of establishment. *Ecology* **38**:422–429.

——, H. J. Oosting, and L. E. Anderson. 1951. Plant succession on exposed granite of Rocky Face Mountain, Alexander County, North Carolina. *Bull. Torrey Bot. Club* **78**:402–421.

Leblanc, F., and D. N. Rao. 1974. A review of the literature on bryophytes with respect to air pollution, in Bonnot, E. J. (ed.), *Soc. Bot. France, Colloque: Les Problèmes Modernes de la Bryologie*, pp. 237–255.

Longton, R. E. 1974. Genecological differentiation in bryophytes. *J. Hattori Bot. Lab.* **38**:49–65.

—— 1976. Reproductive biology and evolutionary potential in bryophytes. *J. Hattori Bot. Lab.* **41**:205–233.

—— 1979. Studies on growth, reproduction and population ecology in relation to microclimate in the bipolar moss *Polytrichum alpestre. Bryologist* **82**:325–367.

—— 1980. Physiological ecology of mosses, in Taylor, R. J., and A. E. Leviton (eds.), *The Mosses of North America*, pp. 77–113. San Francisco: Pacific Div. AAAS.

—— and C. J. Miles. 1982. Studies on the reproductive biology of mosses. *J. Hattori Bot. Lab.* **52**:219–240.

Muhle, H., and F. LeBlanc. 1975. Bryophyte and lichen succession on decaying logs. I. Analysis along an evaporational gradient in eastern Canada. *J. Hattori Bot. Lab.* **139**:1–33.

Nash, T. H., and E. H. Nash. 1974. Sensitivity of mosses to sulfur dioxide. *Oecologia* **17**:257–263.

Nicolas, G. 1932. Association des bryophytes avec d'autres organismes, in

Verdoorn, F. (ed.), *Manual of Bryology,* pp. 109–128. The Hague: Martinus Nijhoff.

Olarinmoye, O. L. 1974. Ecology of epiphyllous liverworts: growth in three natural habitats in western Nigeria. *J. Bryol.* **8**:275–289.

—— 1976. Studies on epiphyllous liverworts-phorophyte relationship. *Nova Hedwigia* **27**:647–654.

Pócs, T. 1978. Epiphyllous communities and their distribution in East Africa. *Bryophyt. Bibl.* **13**:681–713.

Proctor, M. C. F. 1979. Structure and eco-physiological adaptation in bryophytes, in Clarke, G. C. S., and J. G. Duckett (eds.), *Bryophyte Systematics,* pp. 479–509. London: Academic Press.

Rasmussen, L. 1975. The bryophyte epiphyte vegetation in the forest, Slotved Skov, Northern Jutland. *Lindbergia* **3**:15–38.

—— 1977. Epiphytic bryophytes as indicators of the changes in the background levels of airborne metals from 1951-75. *Environ. Pollut.* **14**:37–45.

—— and J. Hertig. 1977. Statistical investigation of interspecific phytosociological relations in epiphytic bryophyte communities. *Rev. Bryol. Lichén.* **43**:207–217.

Richardson, D. H. S. 1981. *The Biology of Mosses.* New York: Wiley.

Seidel, D. 1976. A quantitative and analytical investigation of the moss vegetation in the spruce *Picea abies* (L.) Karst. forests of the Schoenbuch and the Schwaebische Alp. (S. W. Germany). *Flora* (Jena)**165**:139–162.

—— 1976. Ecological experimental investigations on forest floor inhabiting mosses as a basis for a causal-analytical interpretation of synecological aspects. *Flora* (Jena) **165**:163–196.

Shacklette, H. T. 1965. Element content of bryophytes. U. S. Geol. Surv. Bull. 1198-D. Washington, D.C.

—— 1967. Copper mosses as indicators of metal concentrations. U. S. Geol. Surv. Bull. 1198-G. Washington, D.C.

Sjögren, E. 1974. Bryophytes as indicators of environmental factors in forest phytocoenoses, in Bonnot, E. J. (ed.), *Soc. Bot. France, Colloque: Les Problèmes Modernes de la Bryologie,* pp. 225–232.

Slack, N. G. 1976. Host specificity of bryophytic epiphytes in eastern North America. *J. Hattori Bot. Lab.* **41**:107–132.

——, D. H. Vitt, and D. G. Horton. 1980. Vegetation gradients of minerotrophically rich fens in western Alberta. *Can. J. Bot.* **58**:330–350.

Smith, A. J. E. (ed.). 1982. *Bryophyte Ecology.* London: Chapman & Hall.

Streeter, D. T. 1970. Bryophyte ecology. *Sci. Prog.* (Oxford) **58**:419–434.

Tamm, C. O. 1953. Growth, yield and nutrition in carpets of a forest moss (*Hylocomium splendens*). *Medd. Skogsforskn. Inst. Stockholm* **43**:1–140.

Taoda, H. 1973. Bryo-meter, an instrument for measuring the phytotoxic air pollution. *Hikobia* **6**:224–228.

Tobiessen, P. L., K. A. Mott, and N. G. Slack. 1978. A comparative study of phytosynthesis, respiration and water relations in four species of epiphytic mosses in relation to their vertical distribution. *Bryophyt. Bib.* **13**:253–277.

Vitt, D. H., and N. G. Slack. 1975. An analysis of the vegetation of *Sphagnum*-dominated kettlehole bogs in relation to environmental gradients. *Can. J. Bot.* **53**:332–359.

Watson, M. A. 1979. Age structure and mortality within a group of closely related mosses. *Ecology* **60**:988–997.

—— 1980. Patterns of habitat occupation in mosses—relevance to considerations of the niche. *Bull. Torrey Bot. Club* **107**:346–372.

—— 1981. Chemically mediated interactions among juvenile mosses as possible determinants of their community structure. *J. Chem. Ecol.* **7**:367–376.

Whitehead, N. E., and R. R. Brooks. 1969. Aquatic bryophytes as indicators of uranium mineralization. *Bryologist* **72**:501–506.

Wyatt, R. 1982. Population ecology of Bryophytes. *J. Hattori Bot. Lab.* **52**:179–198.

Yarranton, G. A. 1967. Principal components analysis of data from saxicolous bryophyte vegetation at Steps Bridge, Devon. I–III. *Can. J. Bot.* **45**:93–115, 229–258.

—— 1970. Towards a mathematical model of limestone pavement vegetation. III. Estimation of the determinants of species frequencies. *Can. J. Bot.* **48**:1387–1404.

—— and W. J. Beasleigh. 1968. Towards a mathemetical model of limestone pavement vegetation. I. Vegetation and microtopography. *Can. J. Bot.* **46**:1591–1599.

—— 1969. Towards a mathematical model of limestone pavement vegetation. II. Microclimate, surface pH, and microtopography. *Can. J. Bot.* **47**:959–974.

25 Geography

Bryophytes show much wider distribution patterns than the seed plants. Many species are found in all continents of the Northern Hemisphere. Indeed, in arctic and boreal areas, the basic bryophyte flora is the same in North America, Asia, and Europe. In the tropics the similarities are less strong, but many of the same families and genera show wide distributions on all continents. In spite of this floristic similarity within climatic regions, numerous bryophytes also exhibit the restricted patterns of distribution shown by the seed plants, and the same factors appear to have shaped these shared patterns.

Many bryophytes in temperate and tropical climates strongly depend on the presence of the seed plant vegetation that produces both the microclimatic conditions and the substratum for the bryophytes. A high proportion of bryophytes require highly humid conditions and relatively indirect illumination and are restricted to forested sites. In consequence, elimination of the forest leads to the destruction of the resident bryophytes.

The study of bryophyte distributions has been somewhat hampered by inadequate documentation. This reflects the limited number of experienced collectors and researchers. Although a general knowledge of the kinds of bryophytes is reasonably well established, details of distribution patterns are better understood for those of temperate and arctic climates than for those of the tropics.

It is obvious that bryophyte diaspores can be dispersed readily by air currents. Diaspores, whether spores or gemmae, are very small, and once airborne, have the potential to be carried considerable distances. Substratum specificity and microclimatic restrictions of the gametophores cause bryophytes to show distribution patterns that coincide with the occurrence of these factors.

Among historical factors that have strongly influenced bryophyte distributions are the continental positions through time, the variation of climate through time (especially the most recent glaciations), and fluctuation in sea level. In recent time the activities of man have become increasingly important in altering distribution patterns in the bryophytes.

The fossil record for the bryophytes is so incomplete that it offers very fragmentary evidence of past distributions. Indeed, the time of origin of the bryophytes is uncertain, although circumstantial evidence indicates that hepatics, at least, were present in Devonian time. If the spores of earlier fossil deposits belong to bryophytes (which is possible, based on their structure), the bryophytes could have been among the first of the land plants. The beautifully preserved hepatics in Baltic amber of Eocene age provide evidence that modern genera which now show subtropical distribution were once present in Europe. Other fossil material has been extremely instructive in interpreting the genesis of modern distribution patterns in the bryophytes and has supported interpretations based on modern patterns of distribution.

Since the absence of convincing fossil evidence does not permit any certainty concerning the time of origin of most bryophyte genera, it is difficult to entertain seriously any absolute statements concerning the splitting of once continuous ranges of bryophyte taxa through the events of continental drift. It is possible that these assumptions are correct, but there is usually little or no solid evidence to support them.

DIASPORES AND VECTORS OF DISPERSAL

More than half of the bryophytes are dioicous. Although this is clearly an advantage in enhancing variability through outbreeding, it poses serious problems in sexual reproduction when the gametophores of both sexes are not close enough for the sperms to reach the egg. It appears that the effective distance that a sperm can travel is only a few centimeters in most cases. For bryophytes in which the antheridia are exposed in a "splash-cup," as in some mosses, effective travelling distance is somewhat increased. The same is true for some hepatics, especially Marchantiales that possess a receptacle which raises the antheridia above the thallus surface.

In some dioicous bryophytes it has been demonstrated that two of the spores in a tetrad produce antheridium-bearing gametophores, while the other two produce archegonium-bearing gametophores. For successful sexual reproduction, it is necessary for the spores of both sexes to germinate in proximity. If this does not occur, vegetative expansion of clones of each sex may eventually overlap, and sexual reproduction occur in the overlapping areas.

Vegetative diaspores are produced by many bryophytes, and these can be carried by wind, water, or animals from the gametophore of the opposite sex to a unisexual clone and ultimately permit sexual reproduction. Such diaspores are usually small and are often produced in abundance.

Some bryophytes are known only as asexual populations, and sporophytes are extremely rare in others. *Takakia,* for example, occurs only as archegonium-producing populations, yet this genus shows a rather wide distribution in a wide arc from the Himalayas, North Borneo, Japan, and thence to the Pacific coast of North America (Fig. 25-1). The moss *Rhytidium rugosum* is

FIGURE 25-1
World distribution of the genus *Takakia* (Calobryales) disjunctively distributed mainly on the east and west coasts of the Pacific Ocean.

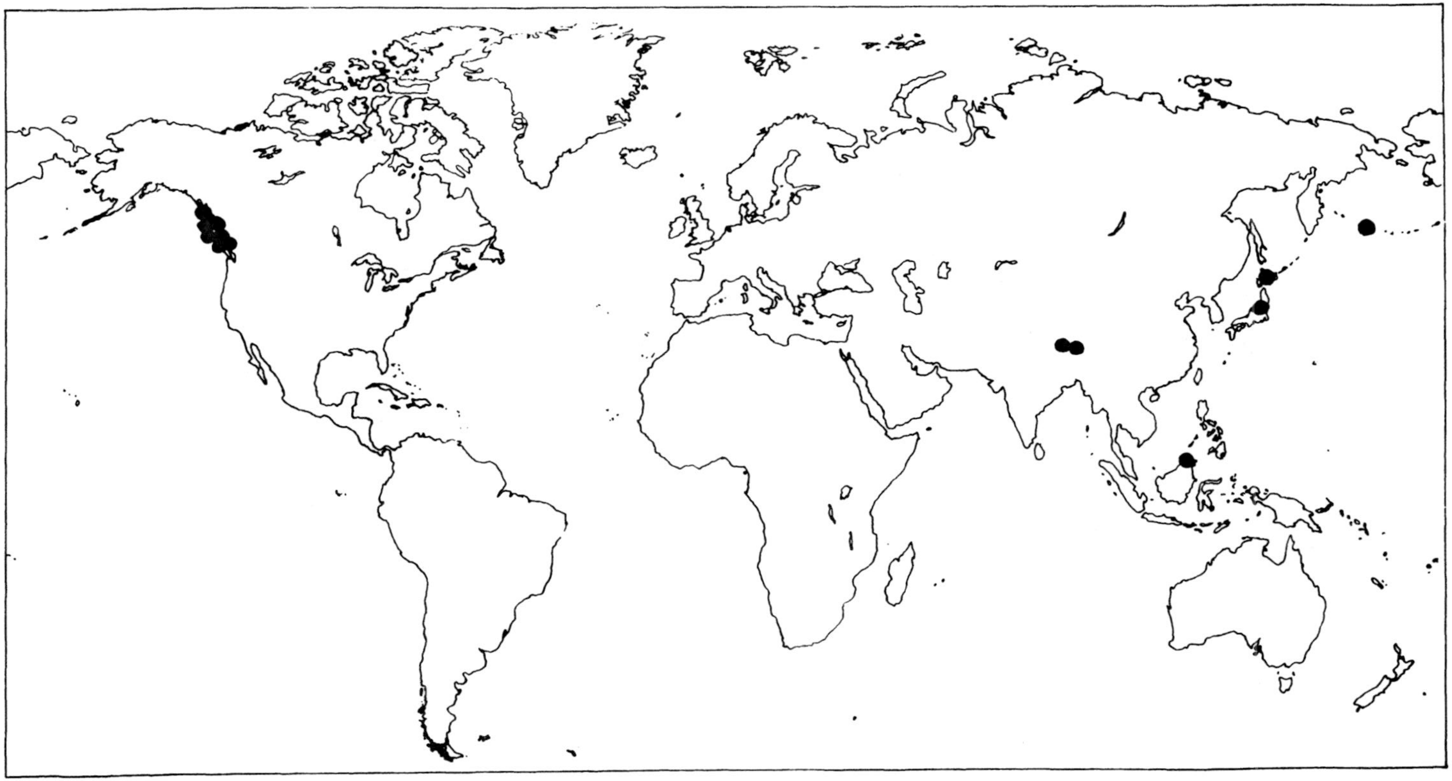

very widespread in the Northern Hemisphere (in North America from arctic regions, southward in the boreal forest, and down mountain chains into Mexico, and equally widespread in Europe and Asia), yet sporophytes are extremely rare, and the gametophores are able to produce no small vegetative diaspores. Gametophores are readily fragmented when dry, but regeneration from fragments would appear to be a very slow means of expanding populations. In spite of this, the species has achieved most of its present distribution since the glacial retreat some 10,000 years ago.

The problems faced by long-distance dispersal in bryophyte diaspores are numerous. Although structures in the bryophyte sporangium enhance ejection of the spores into air currents, and although gemmae are usually exposed on or near apices of shoots, the diaspores are raised only a few centimeters into the moving air. To be carried long distances, the spores must move into upper air masses. In very open flat terrain the opportunities for this are vastly increased, but in irregular terrain, or within closed vegetation of woody plants, the chances of diaspores getting into the upper air are reduced drastically. These changes in dispersibility can help to explain how bryophytes expanded their range so rapidly on deglaciated terrain following Pleistocene glaciations. It also helps to explain why bryophytes show such wide distributions in Northern Hemisphere areas that were affected by glaciation.

MAN'S INFLUENCE ON BRYOPHYTE DISTRIBUTION

The activities of man have destroyed vast areas of bryophyte habitat, and in the process it is highly probable that some taxa have been extinguished before their presence was recorded. Regrettably, floristic field work has a very low priority among research granting agencies, and many biologists treat such floristic research as decidedly old-fashioned. Fortunately, gifted amateurs have greatly enriched the floristic knowledge of areas of easy accessibility. In many countries, however, the free time available to residents, after subsistence, has not been sufficient for such leisure activities. The consequence has probably been a permanent loss of valuable floristic information.

The data base in North Temperate areas is reasonably sound. It is less complete for tropical and southern temperate areas, and a considerable amount of exploratory research is needed to determine whether the patterns revealed at present are real or are, at least in part, the result of collecting patterns.

A number of bryophytes have been introduced well beyond their natural range through man's activities; for example, the hepatics *Marchantia polymorpha* and *Lunularia cruciata* and the mosses *Funaria hygrometrica* and *Leptobryum pyriforme* are common greenhouse weeds in temperate climates. These bryophytes have a broad natural range which appears to have been considerably expanded through man's activities. Among the mosses, a number of species appear to have been inadvertently introduced to a wide range of widely separated geographic areas. *Pseudoscleropodium purum* is a convincing example (Fig. 25-2). This species, with a natural range in Europe and adjacent western Asia, has been transplanted, probably as packing material in nursery stock of horticultural plants, to many parts of the world.

Man has created new sites for colonization by bryophytes. Roadside plantings of trees, stone wall construction, and creation of open roadsides have extended the ranges of a number of bryophytes once restricted in distribution. *Dicranella pacifica,* for example, whose natural habitat is seepage sites of gravelly or sandy cliffs in a very restricted natural range, is distributed widely on road-cuts and in ditches in western North America. *Tortula muralis* seems to be rather uncommon in habitats that are away from the influence of man, yet on concrete walls within cities the species is often exceedingly common in both the Northern and Southern Hemispheres.

The moss *Orthodontium lineare* was introduced, possibly in lumber from the Southern Hemisphere, to Europe. It was noted in Great Britain as early as 1911 and has spread rapidly through Britain. It was noted in continental Europe in 1939, and has spread widely in Germany and the Netherlands since that time. Another Southern Hemisphere moss, *Campylopus introflexus,* also has expanded its range rapidly in Europe, especially in Great Britain.

BRYOPHYTES OF VERY WIDE DISTRIBUTION

Few bryophyte species are truly cosmopolitan (i.e., distributed throughout all climatic regions on all major land masses), although within a broad climatic region many species are widely distributed. Bryophytes of exposed habitats frequently show wide distributions, while those of closed sites are often more restricted. In arctic and in arid areas, in consequence, many of the same species occur in all parts of the world where such conditions are found. For arctic areas that share the same historical phenomena, especially the recent Pleistocene glaciations, interpretation of this pattern is relatively easy. The constant winds, the generally flat terrain over extended areas, the exposure of the land surface through most of the year, and the easy wind transportation of diaspores across sea ice from one island to another, are important factors. The relative uniformity of available suitable sites throughout the arctic is also of critical importance. Since much of the arctic terrain has been available for plant colonization only since the glaciers disappeared from the land surface, available time has been relatively brief for the establishment of these wide distributions, in most cases less than 10,000 years. The population from which these bryophytes are derived would be in areas that escaped glaciation. Such limited areas were present within arctic latitudes and existed south of the glacial boundaries (Fig. 25-3).

FIGURE 25-2
World distribution of the
genus *Pseudoscleropodium*
(Bryidae), with its indigenous
distribution in Europe and adjacent areas and its adventive
distribution in the rest of the
world.

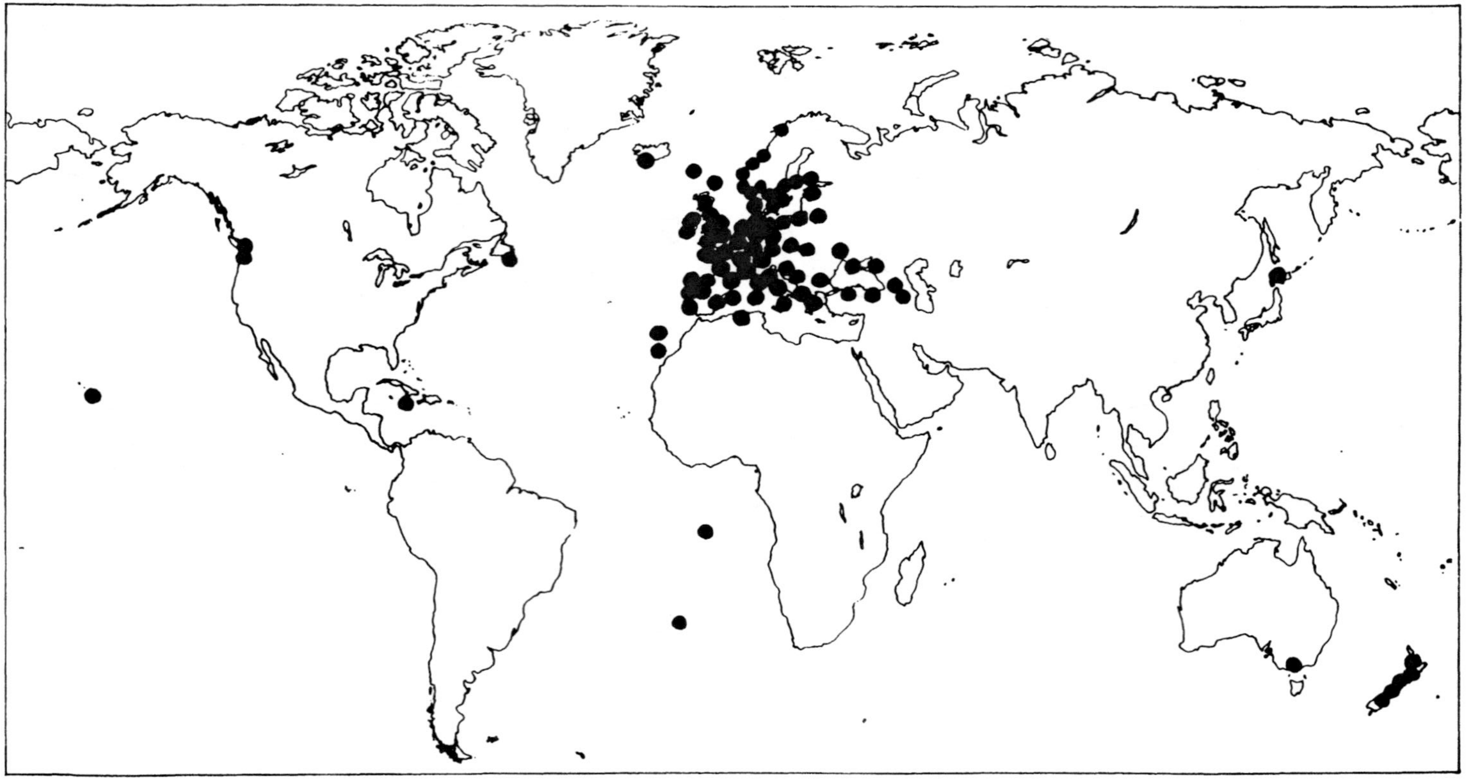

FIGURE 25-3
The approximate extent of the Pleistocene glaciations.

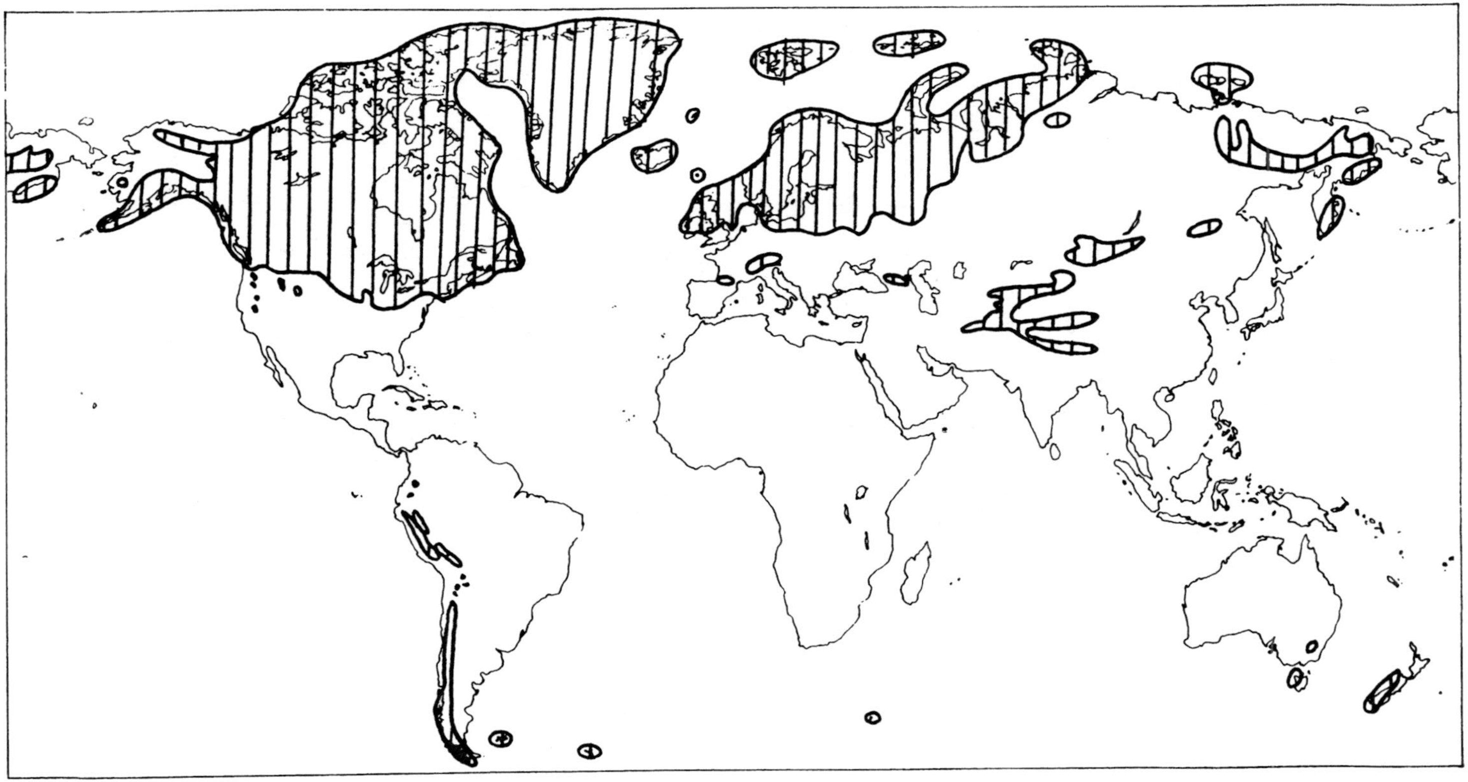

Deserts, on the other hand, show a very interrupted distribution on the earth's surface. Deserts also were affected by the conditions that produced the Pleistocene glaciations. Indeed, areas that are now arid were, in some cases, much more humid during glacial times. Even within generally humid conditions, sites of limited area that are well drained can be occupied by arid-adapted bryophytes and permit them to persist in any given region, although the general conditions are not arid. From such sites, when arid conditions become more widespread, the bryophytes can expand their range to exploit the more extensive areas. It seems probable that long-distance dispersal of diaspores might be important in giving arid region bryophytes such a wide geographical range. There is ample evidence that desert sands are carried extensive distances beyond the desert area. A number of bryophytes in deserts also show very local ranges; for these restricted patterns one must seek other explanations.

Compared to seed plants, many bryophyte families shown an extremely wide range throughout the world, independent of climate. Often the same genera, and rarely, the same species, are represented through all climatic regions. The following moss families are found throughout the world. These represent very large families in most cases, and this list is not comprehensive:

Sphagnaceae	Pottiaceae	Fontinalaceae
Andreaeaceae	Grimmiaceae	Hedwigiaceae
Polytrichaceae	Funariaceae	Neckeraceae
Buxbaumiaceae	Splachnaceae	Leskeaceae
Ditrichaceae	Bryaceae	Amblystegiaceae
Dicranaceae	Mniaceae	Plagiotheciaceae
Encalyptaceae	Aulacomniaceae	Sematophyllaceae
Fissidentaceae	Orthotrichaceae	Hypnaceae

For the hepatics the majority of the large families are also widely distributed and also include genera that are widely ranging, independent of climatic differences. The following are representative:

Blepharostomaceae	Scapaniaceae	Frullaniaceae
Lepidoziaceae	Lophocoleaceae	Fossombroniaceae
Calypogeiaceae	Plagiochilaceae	Dilaenaceae
Lophoziaceae	Cephaloziaceae	Aneuraceae
Jungermanniaceae	Radulaceae	Metzgeriaceae
Marsupellaceae	Porellaceae	Marchantiaceae

The very wide distribution of these families suggests that they are extremely ancient and were possibly in existence when Pangaea existed preceding Permian time (Fig. 25-4). A second explanation would emphasize the ready dispersibility of bryophyte diaspores over very wide distances. It is possible that both major factors are involved in determining such wide ranges of bryophyte families and genera.

Many other families show a wide range throughout the tropics, whether of

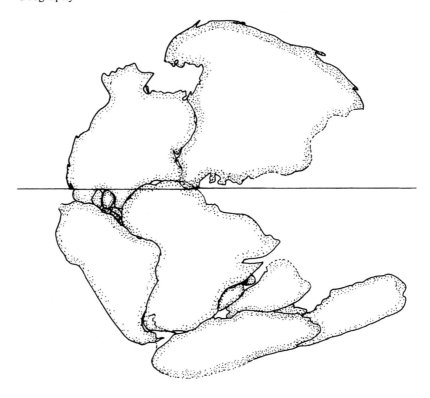

the Old or New World. These families also extend, in many cases, into the warmer temperate areas of both the Northern and Southern Hemispheres. Representative moss families are:

Calymperaceae	Sematophyllaceae	Phyllogoniaceae
Rhizogoniaceae	Trachypodaceae	Hookeriaceae
Racopilaceae	Pterobryaceae	Fabroniaceae
Cryphaeaceae	Meteoriaceae	

Representative hepatic familes are considerably fewer; the following are representative:

Isotachidaceae	Trichocoleaceae	Adelanthaceae
Lepicoleaceae	Acrobolbaceae	Lejeuneaceae

As in the families, numerous genera show a distribution throughout the world. Many of these genera are more richly represented in the Northern Hemisphere than elsewhere, suggesting that they originated in north temperate regions. The very wide distribution suggests that the genera are ancient, since they appear not to show any greater adaptation to rapid or long-distance dispersal than do bryophytes that exhibit a more restricted range. Among these widespread genera are the following mosses:

Amblystegium	*Drepanocladus*	*Plagiothecium*
Amphidium	*Encalypta*	*Pogonatum*
Andreaea	*Entosthodon*	*Pohlia*
Atrichum	*Eurhynchium*	*Polytrichum*
Barbula	*Fissidens*	*Pottia*
Bartramia	*Fontinalis*	*Rhacomitrium*
Brachythecium	*Funaria*	*Seligeria*
Bryum	*Grimmia*	*Sphagnum*
Calliergon	*Hedwigia*	*Tayloria*
Campylium	*Hypnum*	*Thuidium*
Campylopus	*Isopterygium*	*Tortella*
Ceratodon	*Mnium*	*Tortula*
Conostomum	*Neckera*	*Trematodon*
Dicranella	*Orthotrichum*	*Ulota*
Dicranum	*Philonotis*	*Zygodon*
Ditrichum	*Physcomitrium*	

Among the hepatics are the following:

Lepidozia	*Lophocolea*	*Fossombronia*
Bazzania	*Plagiochila*	*Aneura*
Calypogeia	*Cephaloziella*	*Riccardia*
Lophozia	*Cephalozia*	*Metzgeria*
Jungermannia	*Porella*	*Marchantia*
Marsupella	*Frullania*	*Riccia*
Scapania		

Among the hornworts, *Anthoceros* shows a remarkably wide distribution throughout the world (Fig. 25-5).

Within tropical and subtropical latitudes numerous genera are widely distributed. Among these are the following:

Barbella	*Erpodium*	*Papillaria*
Brachymenium	*Floribundaria*	*Pinnatella*
Breutelia	*Hookeria*	*Porotrichum*
Brotherella	*Hookeriopsis*	*Racopilum*
Calymperes	*Hyophila*	*Rhodobryum*
Ctenidium	*Leptodontium*	*Rhynchostegium*
Daltonia	*Leucobryum*	*Schlotheimia*
Ectropothecium	*Macromitrium*	*Sematophyllum*
Entodon	*Neckeropsis*	*Syrrhopodon*
Eriopus	*Octoblepharum*	*Vesicularia*

Among the hepatics, widespread tropical genera are predominantly those also

FIGURE 25-5
World distribution of
Phaeoceros laevis
(Anthocerotae).

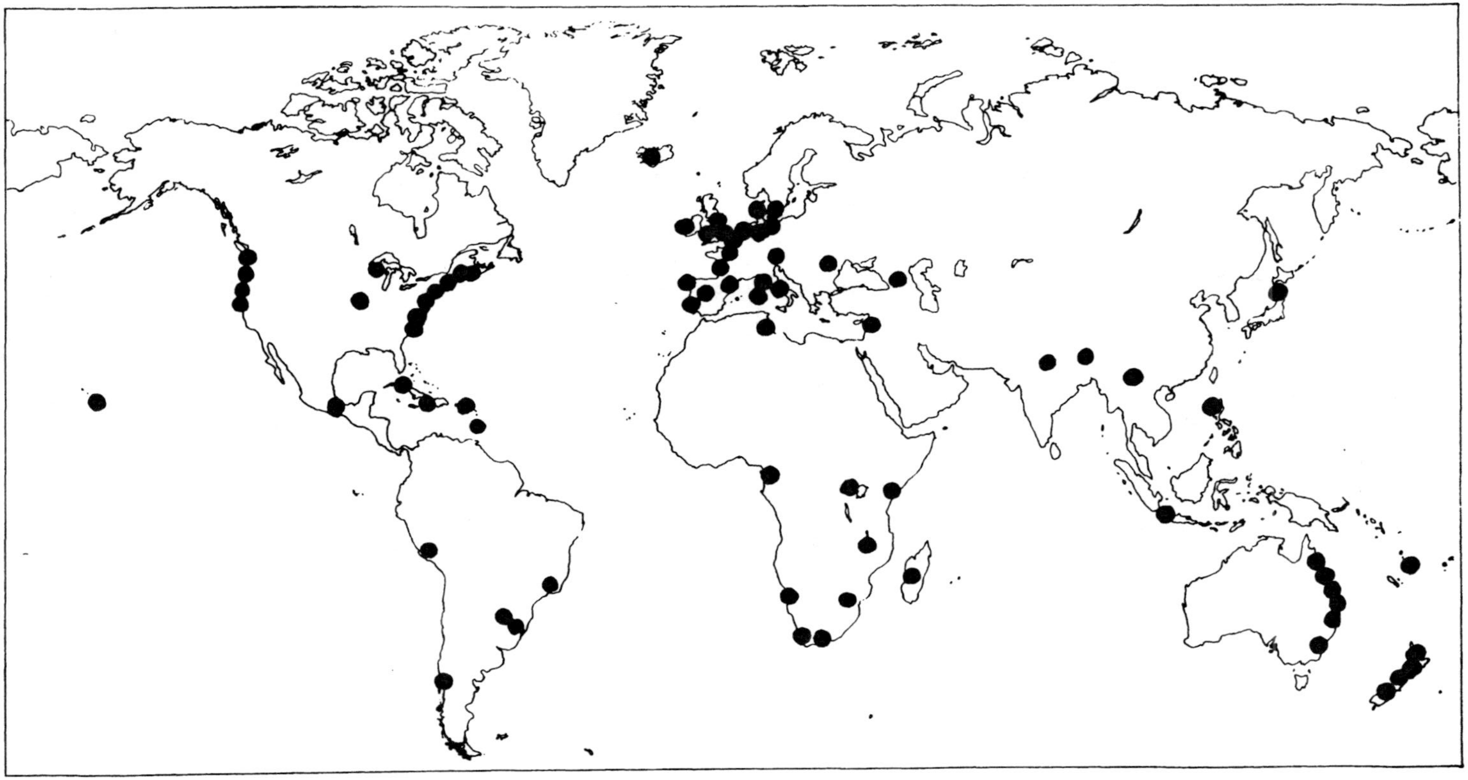

found in temperate areas. A number, however, are predominantly tropical and subtropical:

Herbertus	*Symphyogyna*	*Mastigolejeunea*
Anastrophyllum	*Asterella*	*Ptychanthus*
Trichocolea	*Dumortiera*	*Cheilolejeunea*
Acrobolbus	*Acrolejeunea*	*Colura*
Pallavicinia		*Brachiolejeunea*

A number of other hepatic genera, although widespread in the world, are most richly represented in the tropics. Among these are:

Riccardia	*Porella*	*Frullania*
Metzgeria	*Radula*	*Lejeunea*
Plagiochila		*Marchantia*

Among the hornworts, the genera *Megaceros* and *Dendroceros* are predominantly tropical and subtropical throughout the world.

Using the present distributions, it is possible to suggest an interpretation of the origin of these patterns. The fossil evidence supporting such speculation is absent, and therefore the arguments are weakened. The interpretations of continental positions just preceding continental drift in the Permian and the timing of separation of land masses does show a strong relationship with the broad distribution of bryophyte families and suggests that many families and genera had originated before Permian time and had achieved a wide distribution before the breakup of the supercontinent, Pangaea (Fig. 25-4).

On the basis of fossil evidence of the lignified plants, it is apparent that this immense land mass showed climatically differentiated areas, with more temperate conditions influencing the flora of the northern regions. Fortunately some fossil evidence exists for bryophytes from these regions. The fragmentary nature of fossil remains creates problems in accurate identification. It cannot be demonstrated that the bryophytes present were not of modern families, but they cannot be placed into any modern families with any confidence. On such a gigantic land mass there were clearly many environmental extremes, and these factors would have presented conditions that could have segregated out a diverse bryoflora.

More than 60% of the bryophyte families show a wide world distribution. A number of families, most of them containing a single genus, and often a single species, are restricted to the temperate or frigid climatic portions of the Northern Hemisphere. It is possible that these families originated in the northern portions of Pangaea or in the Laurasian subcontinental mass, and were unable to reach the southern subcontinental mass of Gondwanaland. Some of these families show a very interrupted distribution in the Northern Hemisphere. This interrupted distribution is related to the interrupted nature of suitable habitats in some cases, while in others historical events have fragmented a possibly more continuous past distribution.

Finally, many families are now restricted mainly to one or several parts of the land that was once Gondwanaland (Fig. 25-6). This involves somewhat fewer than 30% of the bryophyte families of the world. Representative families include the following:

MUSCI

Dicnemonaceae	Sorapillaceae	Ptychomniaceae
Pleurophascaceae	Calomniaceae	Hypopterygiaceae
Bryobartramiaceae	Hypnodendraceae	Lepyrodontaceae
Gigaspermaceae	Spiridentaceae	Rutenbergiaceae
Phyllodrepaniaceae	Wardiaceae	Phyllogoniaceae
Eustichiaceae	Hydropogonaceae	Lembophyllaceae
		Echinodiaceae

HEPATICAE

Haplomitriaceae	Lepidolaenaceae	Perssoniellaceae
Vetaformaceae	Balantiopsidaceae	Goebeliellaceae
Chaetocoleaceae	Schistochilaceae	Treubiaceae
Isotachidaceae	Chonocoleaceae	Hymenophytaceae
Chaetophyllopsidaceae		Monocleaceae

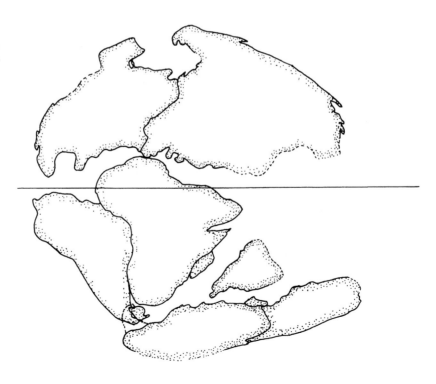

FIGURE 25-6
Reconstruction of continental positions during Triassic time, showing Gondwanaland breaking up in the Southern Hemisphere, and Laurasia beginning to break up in the Northern Hemisphere.

It appears that the basic distribution of most bryophyte families was established before the Permian, while subsequent events, particularly the separation of Laurasia from Gondwanaland, established the gross distributional pattern of most of the remaining families.

ENDEMISM AT THE FAMILY LEVEL

Among the bryophytes, endemism at all taxonomic levels tends to show restrictions similar to those of seed plants, but more closely parallels patterns exhibited by other nonseed bearing land plants. Restricted endemism at the family level is not frequent in the bryophytes, but the following can be given as representative of specific geographic areas.

Holarctic

The holarctic region encompasses much of the Northern Hemisphere. It includes most of North America, Europe, and Asia, but excludes the Arabian peninsula, the Indian subcontinent and adjacent areas south of the Himalayas. This immense area has a remarkable uniformity in the basic bryoflora, even to the species level. Among the moss families that are restricted essentially to this area and show a wide range are:

Bryoxiphiaceae (Fig. 25-7)	Tetraphidaceae (predominantly)
Disceliaceae	Hylocomiaceae (predominantly)
	Timmiaceae (predominantly)

Other moss families show a more restricted endemism; for example, Pseudoditrichaceae are known only from the type locality in subarctic Canada, and Pleuroziopsidaceae are found along the Pacific coast of North America and in Eastern Asia.

Among the hepatics, the situation is similar, and families that are restricted to, and widespread in, the Holarctic include only the Conocephalaceae and Southbyaceae. Families of more restricted endemism are Takakiaceae of Pacific North America, which extend to Japan, Borneo, and the Himalayas; and Gyrothyraceae, restricted to the Pacific coast of North America (Fig. 25-8).

Neotropics

This area incorporates much of South America and extends northward to include much of Mexico and the Caribbean islands. A few moss families are restricted to this area, including the Helicophyllaceae with only *Helicophyllum torquatum,* a species of wide neotropical distribution, and the Hydropogonaceae with two monotypic genera *Hydropogon* and *Hydropogonella,* restricted to tropical and subtropical South America. Among the hepatic families are the

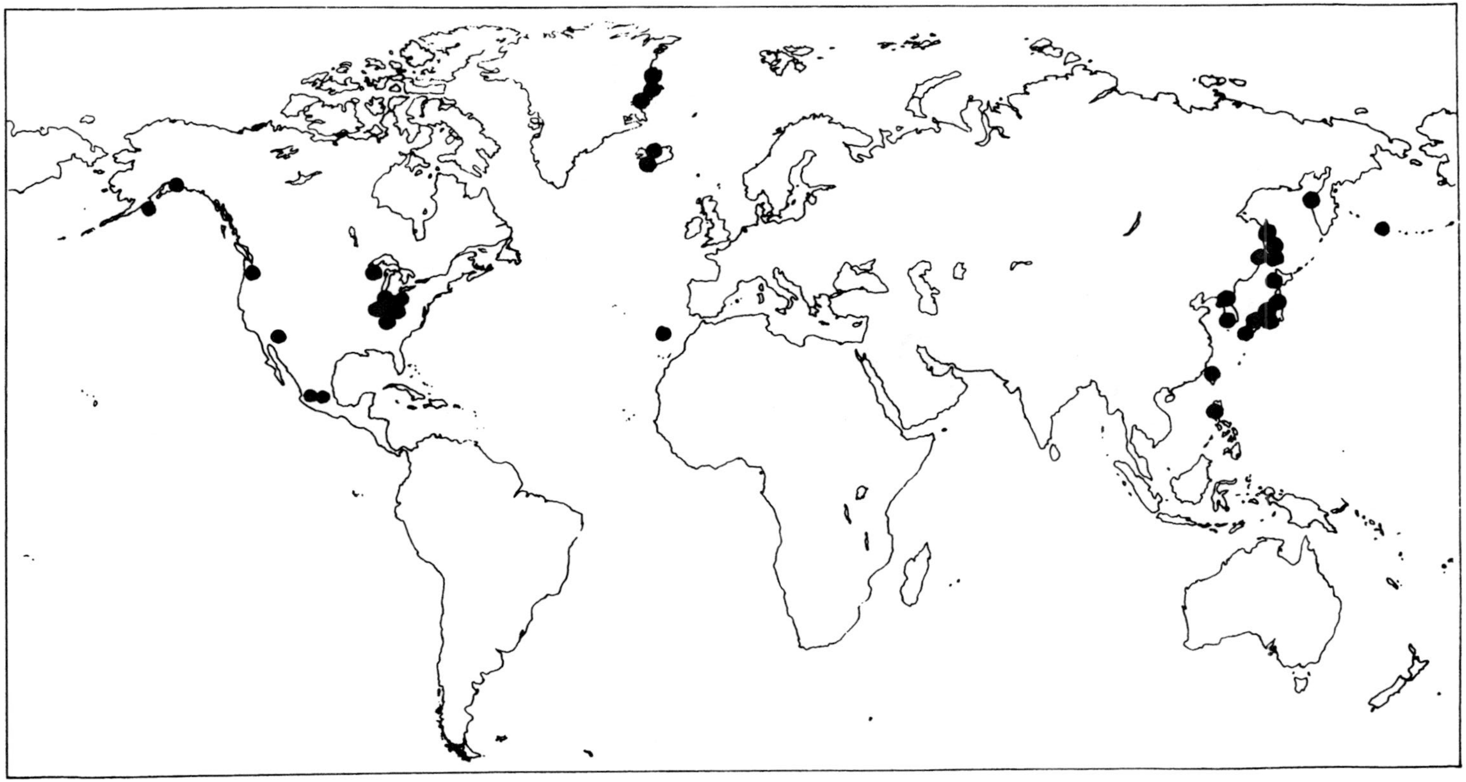

FIGURE 25-7
World distribution of the
moss family Bryoxiphiciaceae
(Bryidae).

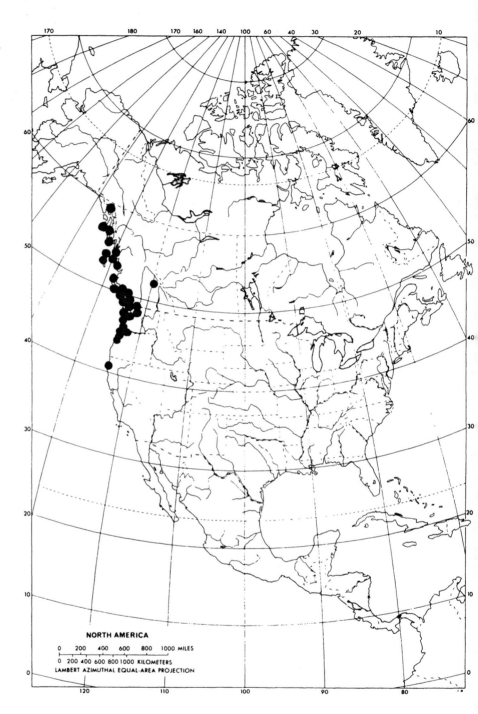

FIGURE 25-8
World distribution of the hepatic family Gyrothyraceae (Jungermanniales).

Chaetocoleaceae, Phycolepidoziaceae and Chonecoleaceae, each of which contains a single monotypic genus.

Paleotropics

The paleotropics include most of Africa, the Arabian peninsula, most of the Indian subcontinent, and the Malaysian area plus most of the islands of the Pacific Ocean. Few bryophyte families are restricted to this immense area. The moss family Nanobryaceae contains a single genus and is found in Africa and Australia. In the moss family Spiridentaceae, *Spiridens* (10 spp.) shows a wide distribution, but *Franciella* (1 species) is restricted to New Caledonia. The moss family Rutenbergiaceae has two genera; *Rutenbergia* (3 species) is restricted to Madagascar and adjacent islands, while *Neorutenbergia* is found locally on the adjacent African mainland (Fig. 25-9). In the hepatics, the monotypic family Perssoniellaceae is restricted to New Caledonia, and the family Metzgeriopsaceae is confined to New Caledonia, Java, North Borneo and Malaya. The remainder of the paleotropical hepatic flora appears to be made up of families of worldwide distribution.

Australasia

This area, including both Australia and New Zealand, has the monotypic moss families Pleurophascaceae and Bryobartramiaceae confined to it, and among the hepatics, Chaetophyllopsidaceae, with two genera, and Verdoorniaceae (monotypic) are also so restricted, while Carrpaceae (with only *Carrpos*) is found in both Australia and South Africa.

Antarctic

This area is generally interpreted to include New Zealand, the subantarctic islands, south-temperate and frigid South America, and the Antarctic continent. There appear to be no moss familes restricted to this area. Among the hepatics are the families Vetaformaceae (monotypic) and Phyllothalliaceae with a single genus and two species (Fig. 25-10).

Cape Area of South Africa

This area contains the endemic moss family Wardiaceae (with only *Wardia*), while no hepatic families are restricted to the area.

ENDEMISM AT THE GENUS LEVEL

Generic endemism shows patterns similar to the families, but there are more instances of marked restriction of natural range. Several genera are widely distributed in the holarctic region and are essentially restricted to it. Several of

FIGURE 25-9
World distribution of the moss family Rutenbergiaceae (Bryidae) with *Neorutenbergia* on the African mainland and *Rutenbergia* on Malagasy and adjacent islands.

348

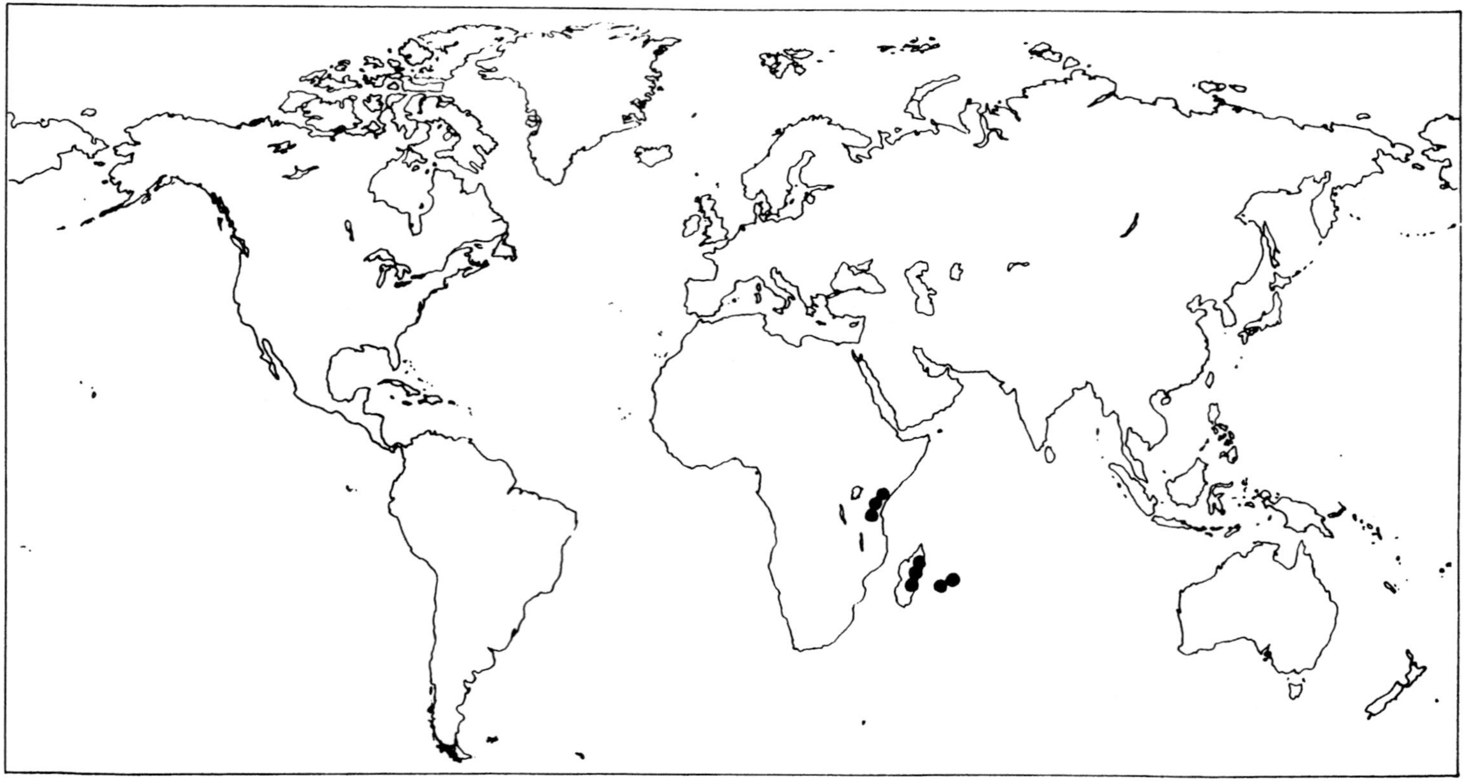

FIGURE 25-10
World distribution of the hepatic family Phyllothalliaceae (Metzgeriales) with *Phyllothallia nivicola* in New Zealand and the vicariant *P. fuegiana* in Southern South America.

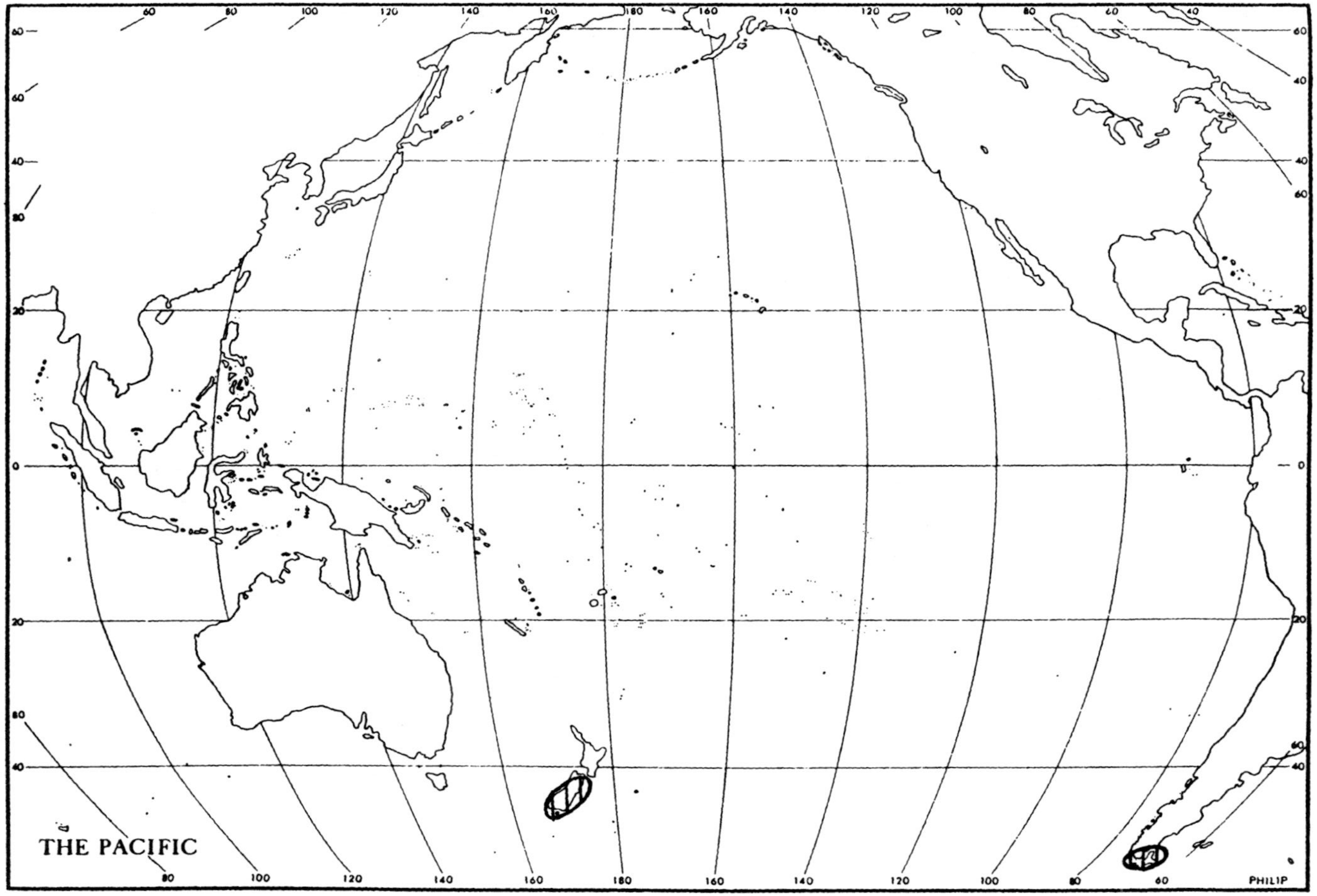

THE PACIFIC

PHILIP

these genera dominate the bryophyte vegetation of a large portion of that geographic area. Among the mosses are:

Abietinella	*Heterocladium*	*Pseudostereodon*
Bryobrittonia	*Loeskhypnum*	*Ptilium*
Catoscopium	*Myurella*	*Pylaisiella*
Cyrtomnium	*Paludella*	*Rhytidiadelphus*
Discelium	*Pleurozium*	*Rhytidium*
Helodium	*Pseudobryum*	*Schistostega*
		Tetraphis (Fig. 25-11)

Within the holarctic region a number of areas possess their own spectrum of endemic moss genera. Along the Pacific coast of North America, confined predominantly to the region west of the Rocky Mountains, are the monotypic genera:

Andreaeobryum	*Dendroalsia*	*Rhytidiopsis* (Fig. 25-12)
Alsia	*Leucolepis*	*Roellia*
Bestia	*Pseudobraunia*	*Trachybryum*

Eastern North America, principally in the Appalachians and eastward, has the following endemic moss genera:

Aphanorrhegma	*Bryoandersonia*	*Donrichardsia*
Brachelyma	*Bryocrumia*	

The number of endemic moss genera is extremely low in Europe, and includes the monotypic genera *Pleuroweisia*, *Ptychodium*, *Stylostegium*, and *Trochobryum*.

Southeast Asia, especially Japan and the adjacent Asiatic mainland, has a number of endemic moss genera, including:

Bissetia	*Fauriella*	*Orthoamblystegium*
Boulaya	*Giraldiella*	*Pseudatrichum*
Brachymeniopsis	*Hondaella*	*Pseudopterobryum*
Bryonoguchia	*Kurohimehypnum*	*Sakuraia*
Cratoneurella	*Microdendron*	*Sasaokea*
Dolichomitriopsis	*Miyabea*	*Scabridens*
Dozya	*Myuriopsis*	*Sciaromiopsis*
Eurohypnum	*Okamuraea*	*Tutigaea*

In the Himalayan region the following moss genera are endemic:

Ditrichopsis	*Ortholimnobium*	*Sphaerotheciella*
Dixonia	*Osterwaldiella*	*Stenotheciopsis*
Handeliobryum	*Penzigiella*	*Struckia*
Gammiella	*Pleuridiella*	*Symphyodon*
Leiodontium	*Prionodontium*	*Trolliella*
Orontobryum	*Pylaisiopsis*	

FIGURE 25-11
World distribution of the circumpolar Northern Hemisphere moss species *Tetraphis pellucida* (Tetraphidae).

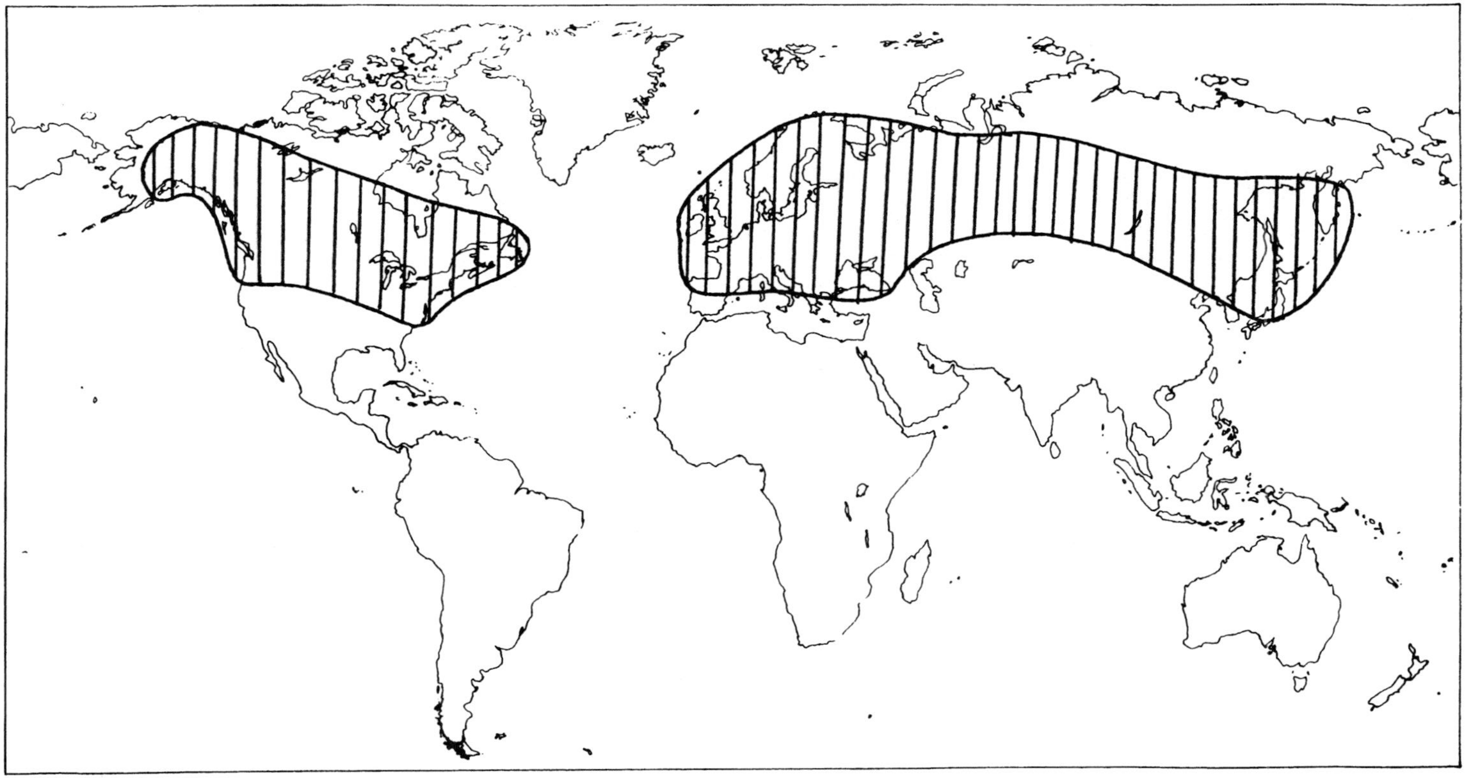

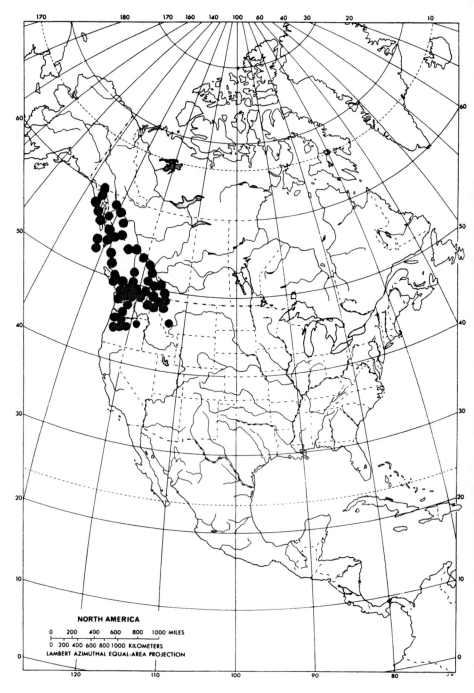

FIGURE 25-12
World distribution of the restricted endemic moss *Rhytidiopsis robusta* (Bryidae).

Details concerning generic endemism are less well documented for the hepatics, but in the holarctic area endemism is extremely low. Western North America has three endemic hepatic genera, all of which are monotypic (*Geothallus, Gyrothyra, Schofieldia*), but no endemic hepatic genera are known from eastern North America. In Europe, the endemic genus *Cryptothallus* appears to be the only endemic genus of hepatics, and it is known also from eastern Greenland. Eastern Asia, especially in Japan and adjacent areas, has a number of endemic hepatic genera:

Cavicularia	*Makinoa*	*Nipponolejeunea*
Cryptocoleopsis	*Neohattoria*	*Tuzibeanthus*
Hattoria	*Neotrichocolea*	

ENDEMISM AT THE SPECIES LEVEL

The bryofloras for most parts of the world are insufficiently documented to make reliable assessments of species endemism. Careful monographs need to be constructed for numerous genera before such estimates become reasonably accurate. Table 25-1 presents information concerning endemism for a number of geographic areas. The figures are calculated based on published floras of the areas and, in most cases, represent reasonable guesses. The figures do suggest trends that are unlikely to be vastly altered with more detailed information.

Under some circumstances, as in western North America west of the Cordilleran mountain chain, generic and species endemism appear to be high because the taxa were able to persist in unglaciated areas during the Pleistocene, and now reach their richest representation in regions that have been available for colonization for less than 12,000 years. In this area the mosses have 10 endemic genera that represent about 5% endemism, and of the species about 18% are endemic. Among the hepatics of the same area there are three endemic genera, which represents about 3% of the genera present, while about 16% of the species are endemic. For eastern North America, mainly east of the Appalachian Mountains, there are six endemic moss genera, representing about 2% of the genera present, while about 16% of the species are endemic. In the hepatics of the same area, no genera are endemic, while about 20% of the species appear to be endemic. In the bryoflora of eastern North America, the tropical and subtropical affinities with Central and South America are relatively strong, while in western North America, the desert areas appear to have presented a barrier through which this element could not pass. Most bryophyte taxa of tropical and subtropical affinity in western North America are related to the flora of Southeastern Asia.

Endemism in the bryoflora of nearly all areas is positively related to three features: (1) the length of time during which the region has been available for colonization; (2) considerable environmental diversity, especially in the availability of atmospheric moisture; and (3) the length of time in comparative isola-

TABLE 25-1.
Endemism in Bryophyte
Genera and Species:
Representative Areas for
Which Information is
Sufficiently Complete

Area	Genera (%)	Species (%)
Hawaiian Islands[a]		
Mosses	1.6	51
(Hoe, 1974)		
Hepatics	0	ca. 67
(Miller, 1957)		
New Zealand[b]		
Bryophyta	ca. 5	ca. 28
(Van Zanten and Pocs, 1981)		
British Isles[c]		
Mosses	<1	ca. 2.5
Japan[d]		
Mosses	ca. 2	ca. 25
Hepatics	ca. 5	ca. 33
W. Indian Islands[e]		
Mosses (Crosby, 1969)	?	ca. 30
North America		
(north of Mexico)[f]		
Mosses	ca. 5	ca. 23
(Schofield, 1980)		
Hepatics	ca. 3	ca. 25

[a]Strongest affinities with Malaysia (ca. 61% of spp.).
[b]Strongest affinities with Australia (81% genera, 61% spp.).
[c]Strongest affinities Holarctic > 50% spp.
[d]Strongest affinities Holarctic?
[e]Strongest affinities with Central and S. America
[f]Strongest affinities Holarctic > 50% spp.

tion. Islands, in particular, show high endemism; in Madagascar, for example, about 70% of the moss species are endemic, while in the Hawaiian Islands, about 51% of the species of mosses and about 67% of the hepatic species are endemic. Other islands that show high species endemism are Réunion, New Caledonia, and The Comores. The Macaronesian islands also show strong species endemism, but in any of these islands, it rarely exceeds 20% of the moss flora.

DISJUNCTIONS

A conspicuously interrupted pattern of world distribution is shown by a number of bryophytes, with populations disjunct from each other by considerable distances. Most of these disjunctive patterns also characterize other land plants. As in other patterns of distribution shown by bryophytes, it appears that distribution in all land plants has been shaped by the same basic factors.

A few representative disjunctions illustrate the dramatic patterns shown, and present possible explanations of these patterns. It could be argued that some disjunctions represent the inadequate understanding of total distributions of taxa. Those that are presented here appear to represent genuine disjunctions, and even if further populations are discovered, it is unlikely that the basic pattern will change.

Bipolar

A sizeable group of mosses and a few hepatics show a very wide distribution in the Northern Hemisphere, and reappear (sometimes very locally), in temperate portions of the Southern Hemisphere. Approximately 100 species of mosses and fewer than 10 species of hepatics show this pattern. A few of the mosses are found also in high mountains nearer equatorial latitudes, but the majority appear to be restricted entirely to temperate or near-polar latitudes (Figs. 25-13, 25-14).

Since these disjuncts are identical species and have achieved their extremely broad holarctic distribution within less than 10,000 years, it is reasonable to assume that the Southern Hemisphere populations may have reached there through long-distance dispersal from the Northern Hemisphere. It has been suggested that dispersal to the Southern Hemisphere could have occurred during the Pleistocene glaciations when there may have been extensive populations of these taxa further south in the Northern Hemisphere, and cooler conditions of more southern mountain ranges could have provided a possible stepwise dispersal route for southern expansion of range.

It is also of interest that at least 10 species of mosses have attained a bipolar distribution as inadvertent introductions by man. Most of these species were introduced from Europe to New Zealand and Australia, in particular, but at least two species were introduced from the Southern to the Northern Hemisphere (*Orthodontium lineare* and *Campylopus introflexus*).

High Latitude South American–High Latitude Australasian

At least 50 bryophytes, most of them mosses, are disjunctive between southern South America and New Zealand and southern Australia. These are either of humid forests or of alpine to subalpine sites and are mainly the same species, although sometimes, as in the hepatic genus *Phyllothallia*, one species, *P. nivicola*, is in New Zealand, while the other, *P. fuegiana*, is in southern South America (Fig. 25-10). Among the mosses, *Dendroligotrichum dendroides* occupies the same forest floor habitat in South America and New Zealand, and the epiphytic moss *Weymouthia cochlearifolia* also has an identical habitat in both areas of the disjunction. Since the forest habitat is a closed one, and the possibility of long distance dispersal from the forest of one continent to that of another is highly complicated by a remarkable sequence of coin-

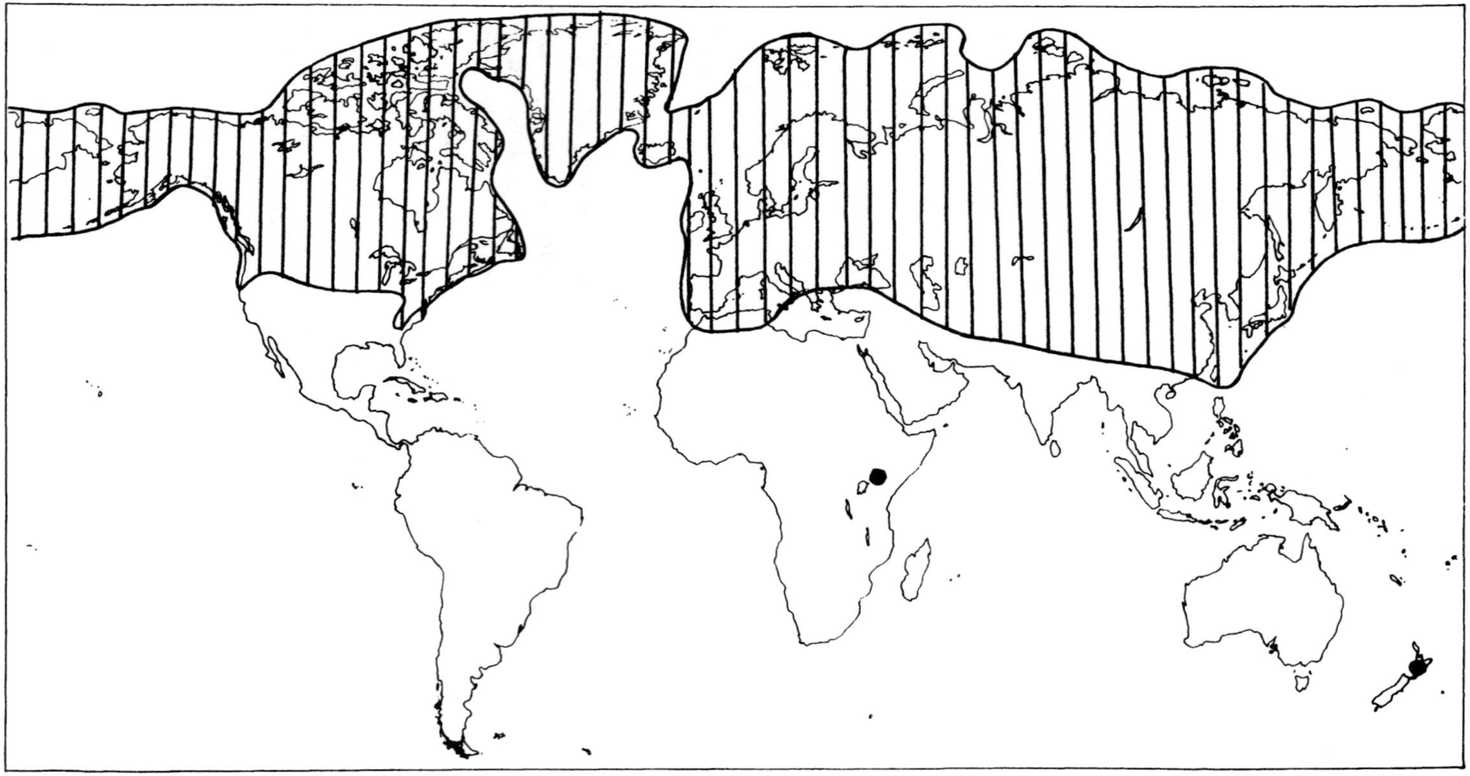

FIGURE 25-13
World distribution of the bi-
polar disjunctive moss
Hylocomium splendens
(Bryidae).

FIGURE 25-14
World distribution of the bi-polar disjunctive hepatic *Ptilidium ciliare* (Jungermanniales).

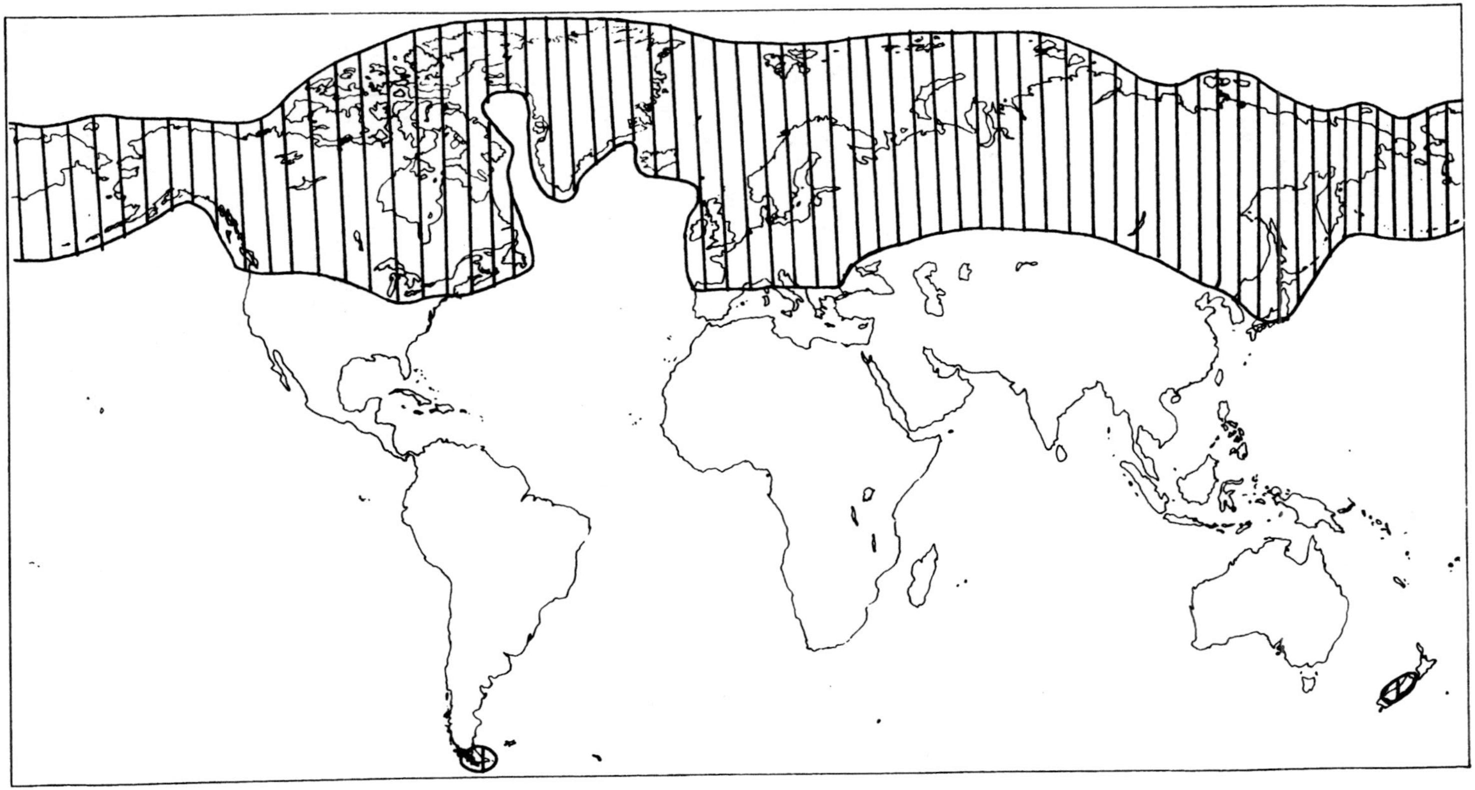

cidences of both events and timing of the events, it seems highly probable that these bryofloras shared a common origin in Gondwanaland before continental drift, or at a time during breakup when free exchange of floras was possible between the various portions of Gondwanaland. It is not impossible that some of the species extended their range from one continent to another in a stepwise manner via subantarctic islands. A number of bryophytes show a relatively continuous range between the continents. These are generally species of open exposed sites, but a number of them are epiphytic or grow on the floor of the scrubby woody vegetation that exists on a number of the islands.

Western North American–Western European

Several bryophyte species are confined either to mountains or to highly humid climates of western North America and western Europe. An unusual feature of these bryophytes is that many of the species are not known to reproduce sexually, and several produce no specialized asexual diaspores. These bryophytes occupy many types of habitat. The mosses *Plagiothecium undulatum* (Fig. 25-15) and *Dicranum tauricum* are common in both areas of disjunction, but in forests in which the species of seed plants that make up the forest are very different. The hepatic *Porella cordaeana* is also frequent in damper areas of both areas of disjunction, occurring either as an epiphyte or on humid shaded rock surfaces (Fig. 25-16).

Several species are disjunctive in Mediterranean climates of both areas. The mosses *Antitrichia californica* and *Metaneckera menziesii* are representative. In extremely oceanic climates of the two areas, the hepatics *Anastrophyllum donianum* and *Moerckia hibernica* are good examples, as are the mosses *Leptodontium recurvifolium*, *Campylopus schwarzii*, and *Ditrichum zonatum*.

Since many seed-plant vegetation types are represented, and since the species involved are elements of a well-integrated vegetation in both areas of disjunction, it appears that the bryophytes are surviving remnants of a very ancient flora. The bryoflora has persisted partially intact over a very long period of time, possibly protected through the microclimatic uniformity through time, while the drastic changes in macroclimate have led to the extinction of the seed plant species or have led to the evolution of different species in the disjunctive areas.

Eastern Asian–Eastern North American

This disjunction also appears to be extremely ancient. While it is represented among the seed plants mainly by vicariad species in genera that show this disjunction, the bryophytes tend to show the same species in each area of disjunction. Vicariad species have had a common origin from a single ancestral population, and have diverged, in their isolated disjunctive areas, to become entirely independent species. Such vicariads in the bryophytes are the following:

FIGURE 25-15
World distribution of the western European-western North American disjunctive moss *Plagiothecium undulatum* (Bryidae).

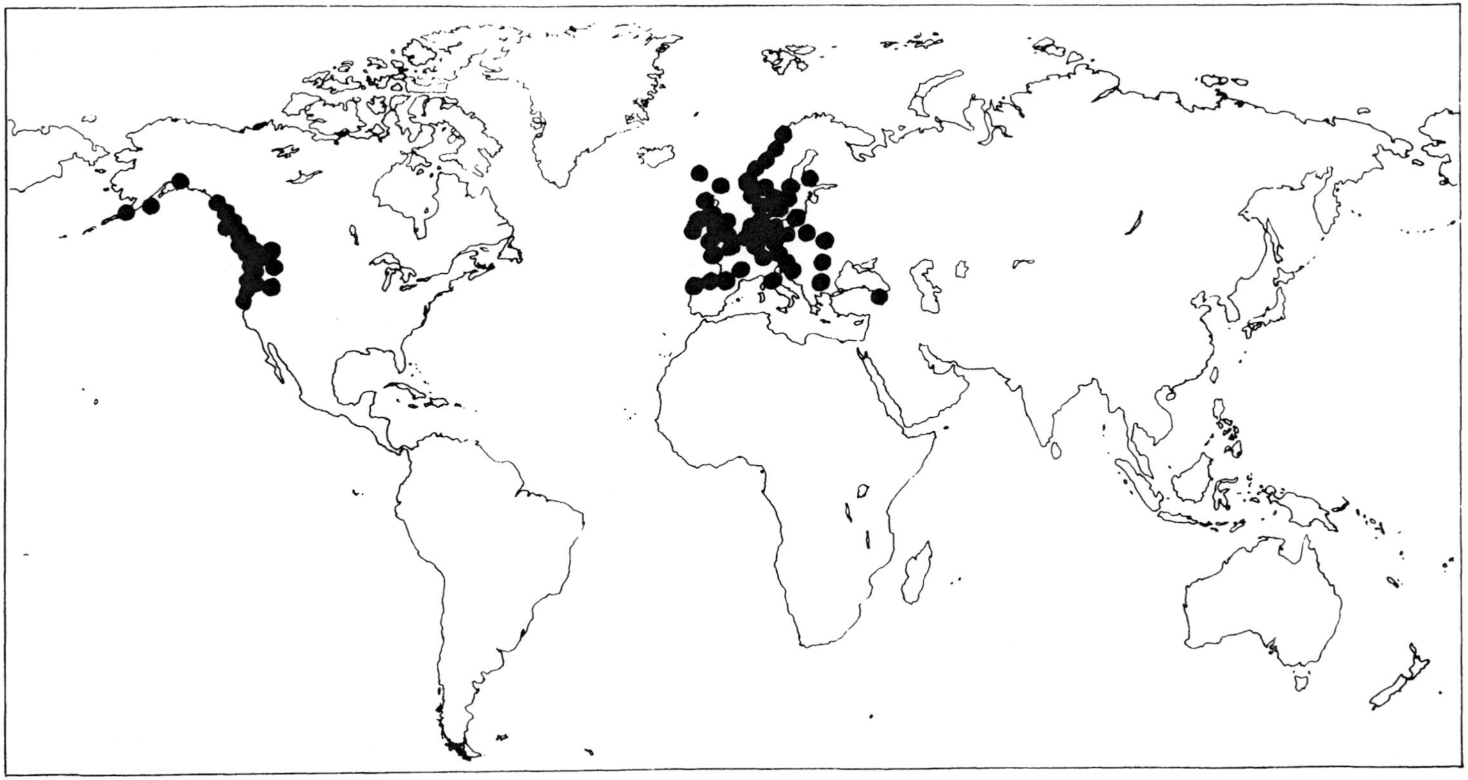

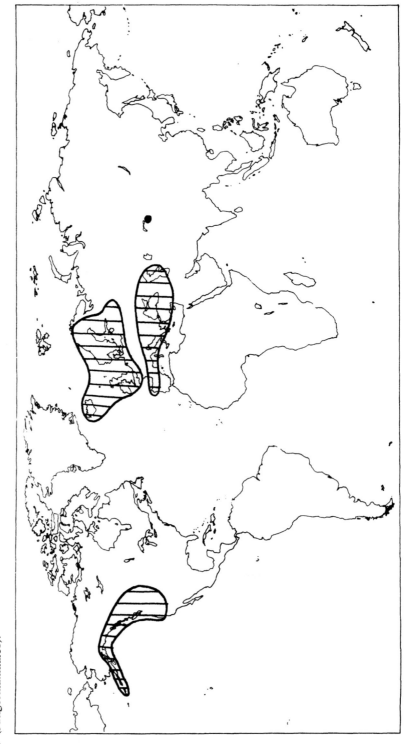

FIGURE 25-16
World distribution of the
western European–western
North American disjunctive
hepatic *Porella cordaeana*
(Jungermanniales).

Eastern North America	Eastern Asia
Bruchia sullivantii	*B. microspora*
Climacium americanum	*C. japonicum*
Plagiomnium carolinianum	*P. maximowiczii*
Diphyscium cumberlandianum	*D. involutum*

In most cases these vicariads are very weakly differentiated from each other. Numerous species show this disjunction, most of which belong to the very closed, mixed broadleafed deciduous angiosperm forests and occupy a broad range of habitats within such forests. Although the supportive fossil evidence is very scant for the bryophytes, a single example suggests that this bryoflora was more widespread during Tertiary time and may have been relatively continuous with the seed plant vegetation that shows evidence of a relatively continuous Tertiary distribution from eastern Asia across North America.

This evidence is in the moss genus *Aulacomnium*. *A. heterostichum* now shows the eastern Asia–eastern North America disjunction. During the Tertiary, *A. heterostichoides,* an extinct species extremely similar to *A. heterostichum,* was present in western North America (British Columbia) (Fig. 25-17). It grew with a vegetation similar to that found in eastern North America and eastern Asia at that time, and fragments of that vegetation now persist in these disjunctive areas, while it has been largely extinguished in western North America. Occasionally a species has persisted in the modern flora in the intervening geographic area (Fig. 25-18).

Amphi-Atlantic

The Amphi-Atlantic disjunction shows two distinctive parts, those species restricted to near-Atlantic coastal areas on the eastern and western coast bounding the Atlantic Ocean, and those found in tropical and subtropical portions of west Africa and eastern portions of Central and South America (Figs. 25-19, 25-20). Most of the North Temperate bryophytes of this disjunctive pattern occupy open sites and strongly suggest long-distance dispersed species. Those of the northern portion tend to be best represented in terrain that was thoroughly glaciated and achieved most of their present range since the melting of the Pleistocene glaciers. In those that exhibit a more southern distribution, however, the taxa are of more closed forest habitats and may represent floras that were disjointed through continental drift. The examples of this latter pattern are not numerous, represent mainly identical species, and tend not to be numerically abundant in either part of their disjunction.

FLORISTIC AFFINITIES

In determining the historical development of a flora, it is valuable to compare the geographic relationships of the flora of one area with those of other parts of

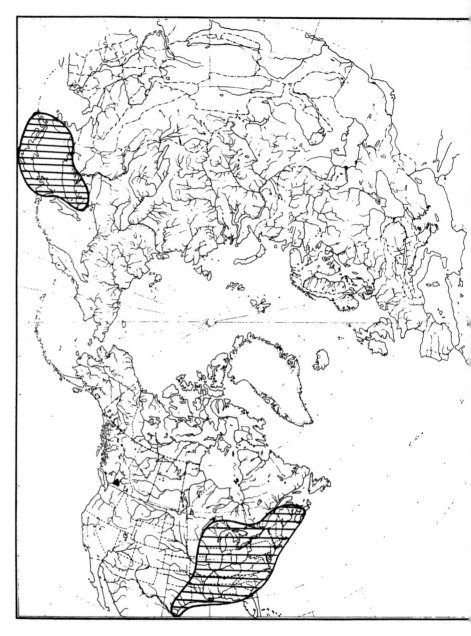

FIGURE 25-17
World distribution of the eastern Asian-eastern North American disjunctive moss *Aulacomnium heterostichum* (Bryidae), showing also the fossil species *A. heterostichoides* by a closed triangle.

FIGURE 25-18
World distribution of the eastern Asian-eastern North American disjunctive hepatic *Acrobolbus ciliatus* (Jungermanniales).

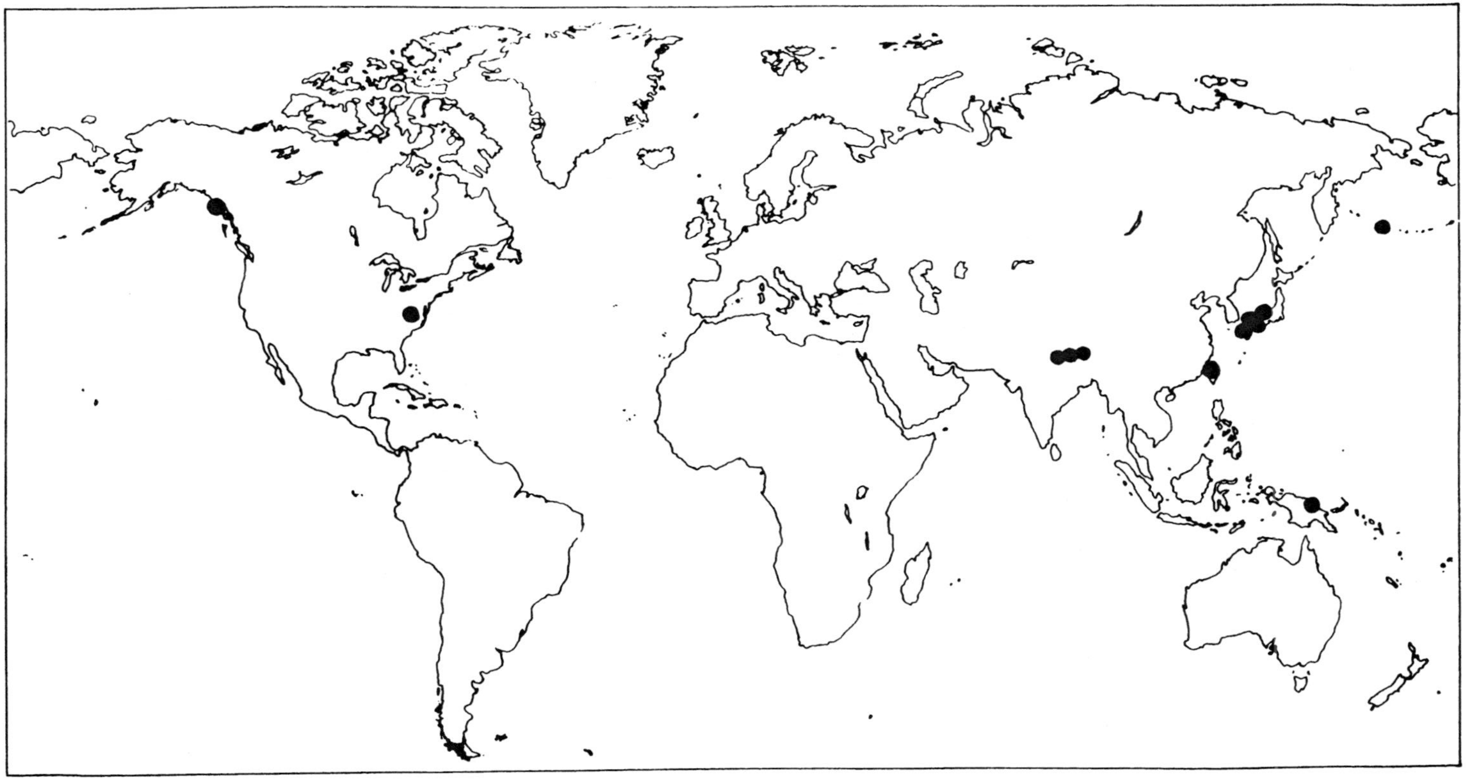

FIGURE 25-19
World distribution of the amphi-Atlantic disjunctive peat moss *Sphagnum angermanicum* (Sphagnidae).

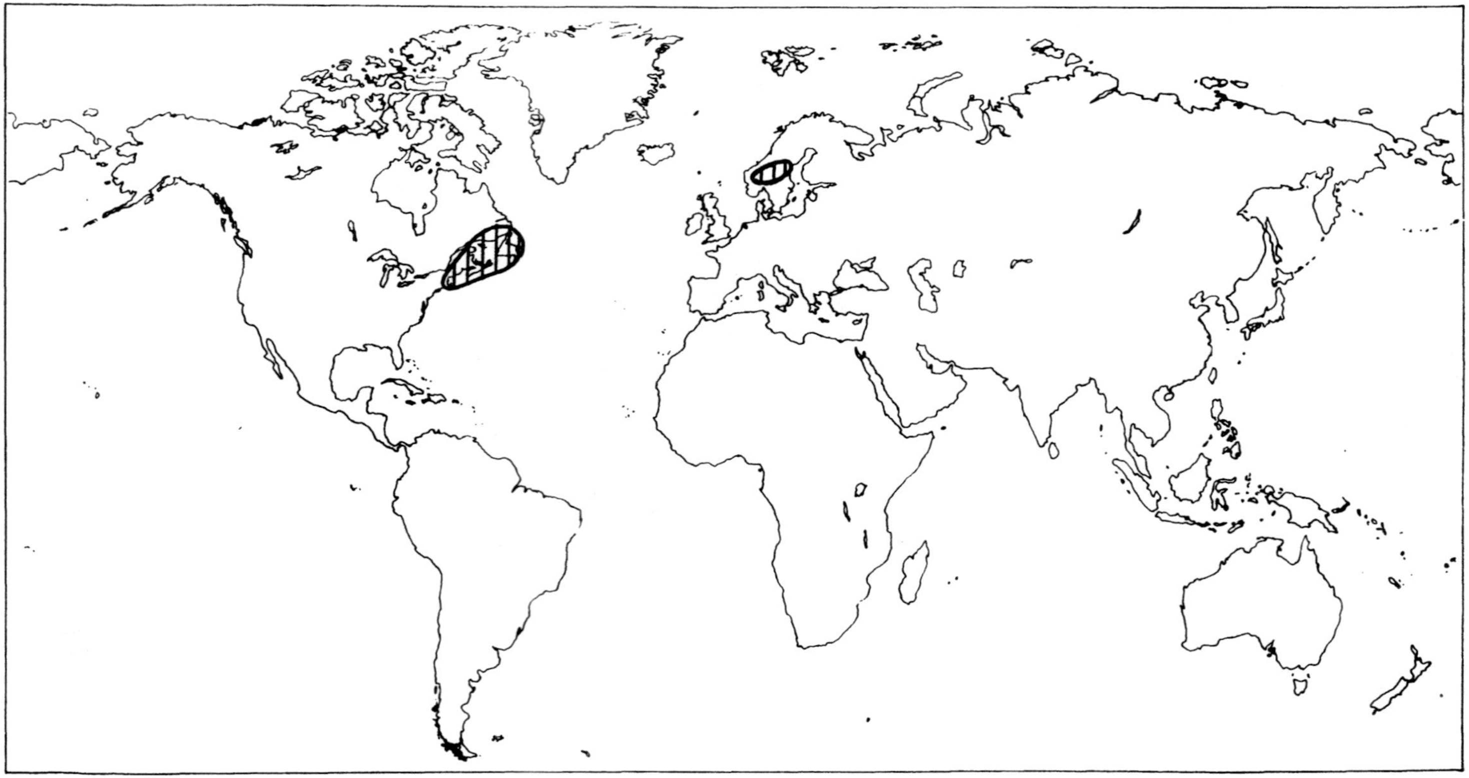

364

FIGURE 25-20
World distribution of the amphi-Atlantic disjunctive hepatic *Radula voluta* (Jungermanniales).

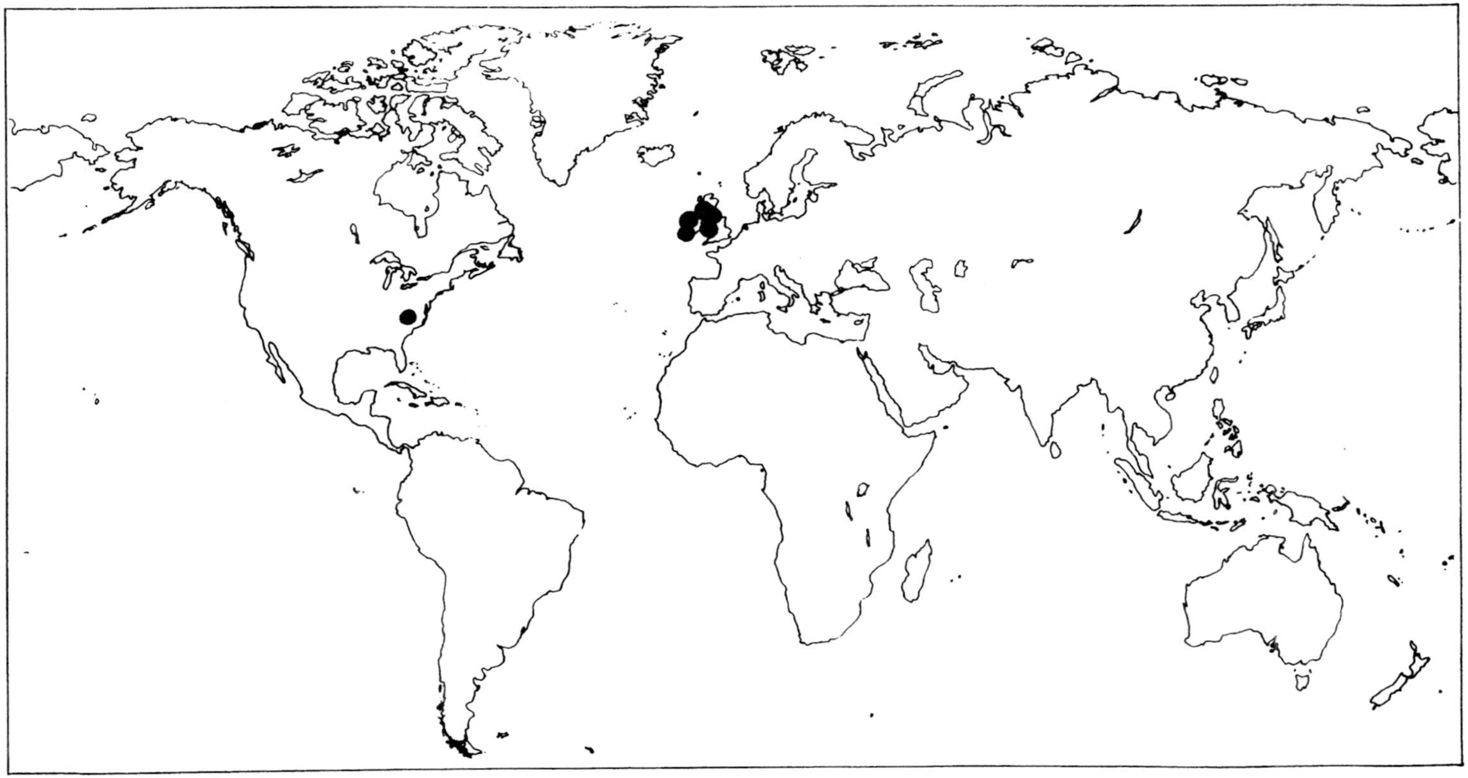

the world. Such information can suggest past continuity of the flora or can indicate dissemination routes of a flora to an area. Such affinities may be derived from the basic composition of the flora, plus the disjunctions and affinities of the endemics, and correlated with the geographic affinities of the rest of the biota. When these data are integrated with the past history of the areas, particularly in the timing of past events and their favorability to the survival and expansion or restriction of floras, it is possible to speculate on the factors that led to the building of a flora. Fortunately bryophytes tend to be relatively free from major introduction by man and can be used reliably to construct natural distribution patterns. Such information can be useful in interpreting distribution patterns of other organisms.

It is important to emphasize that bryofloras have developed by no single series of events. The flora of a single area has elements that reached there through long-distance dispersal, others that have reached the area through slow expansion from the area of origin of the taxa, and still others that have been isolated through geologic events, particularly continental drift, orogeny, and climatic change.

FURTHER READING

Abramova, A. L., and I. I. Abramov. 1969. Eastern Asiatic affinities of the Caucasion bryophyta. *J. Hattori Bot. Lab.* **32**:151–154.

Anderson, L. E. 1943. The distribution of *Tortula pagorum* (Milde) DeNot. in North America. *Bryologist* **46**:47–66.

——— 1971. Geographical relationships of the mosses of the Southern Appalachian Mountains, in Holt, P. C. (ed.), *The Distributional History of the Biota of the Southern Appalachians. Part II Flora.* Res. Div. Monogr., Vol. 2, pp. 101–115. Blacksburg, Va.: Virginia Polytechn. Inst. and St. Univ.

——— and R. H. Zander. 1973. The mosses of the Southern Blue Ridge Province and their phytogeographic relationship. *J. Elisha Mitchell Sci. Soc.* **89**:15–60.

Ando, H. 1972. Distribution and speciation in the genus *Hypnum* in the circum-Pacific region. *J. Hattori Bot. Lab.* **35**:68–98.

——— 1979. A phytogeographical account of *Hypnum cupressiforme* Hedw. in Japan. *Bull. Yokohama Phytosoc. Soc. Japan* **16**:339–348.

——— 1982. *Hypnum* in Australasia and the southern Pacific. *J. Hattori Bot. Lab.* **52**:93–106.

Billings, W. D., and L. E. Anderson. 1966. Some microclimatic characteristics of habitats of endemic and disjunct bryophytes in the southern Blue Ridge. *Bryologist* **69**:79–95.

Bizot, M., T. Pocs, and A. J. Sharp. 1979. Results of a bryogeographical expedition to East Africa in 1968 II. *J. Hattori Bot. Lab.* **45**:145–165.

Bowers, F. D. 1970. High elevation mosses of Costa Rica. *J. Hattori Bot. Lab.* **33**:7–35.

Brassard, G. R. 1971. The mosses of Ellesmere Island, Arctic Canada I. Ecology and phytogeography, with an analysis for the Queen Elizabeth Islands. *Bryologist* **74**:233–281.

Brassard, G. R. 1974. The evolution of Arctic bryophytes. *J. Hattori Bot. Lab.* **38**:39–48.

Campbell, D. H. 1907. On the distribution of the Hepaticae and its significance. *New Phytol.* **6**:203–212.

Catcheside, D. G. 1982. The geographical affinities of the mosses of south Australia. *J. Hattori Bot. Lab.* **52**:57–64.

Croizat, L. 1962. Les hépatiques par devers la biogéographie mondiale. *Rev. Bryol. Lichénol.* **31**:5–22.

Crosby, M. R. 1969. Distribution patterns of West Indian mosses. *Ann. Missouri Bot. Garden* **56**:409–416.

Crum, H. A. 1966. Evolutionary and phytogeographic patterns in the Canadian moss flora, in Taylor, R. L., and R. A. Ludwig (eds.), *The Evolution of Canada's Flora,* pp. 28–42. Toronto: University of Toronto Press.

——— 1972. The geographic origins of the mosses of North America's eastern deciduous forest. *J. Hattori Bot. Lab.* **35**:269–298.

Delgadillo, M. C. 1979. Mosses and phytogeography of the *Liquidambar* forest of Mexico. *Bryologist* **82**:432–449.

Dickson, J. H. 1967. *Pseudoscleropodium purum* (Limpr.) Fleisch. on St. Helena and its arrival on Tristan da Cunha. *Bryologist* **70**:267–268.

Düll, R. 1980. Bryoflora und Bryogeographie der Insel la Palma, Canaren. *Cryptog., Bryol. Lichénol.* **1**:151–188.

Egunyomi, A. 1979. The viability of spores of some tropical moss species after long-time storage and their survival chances in nature. *J. Hattori Bot. Lab.* **45**:167–171.

Engel, J. J., and R. M. Schuster. 1973. On some tidal zone Hepaticae from South Chile, with comments on marine dispersal. *Bull. Torrey Bot. Club* **100**:29–35.

Florschütz, P. A., S. R. Gradstein, and W. V. Rubers. 1972. The spreading of *Fissidens crassipes* Wils. (Musci) in the Netherlands. *Acta Bot. Néerl.* **21**:174–179.

Frahm, J. P. 1972. Die Ausbreitung von *Campylopus introflexus* (Hedw.) Brid. in Mitteleuropa. *Herzogia* **2**:317–330.

Fulford, M. 1951. Distribution patterns of the genera of leafy Hepaticae in South America. *Evolution* **5**:243–264.

——— 1963. Continental drift and distribution patterns in the leafy Hepaticae. *Soc. Econ. Paleontol. Mineral. Special Paper* **1D**:140–145.

Gams, H. 1955. Zur Arealgeschichte der arktischen und arktisch-oreophytischen Moose. *Feddes Repertorium* **58**:80–92.

Gaume, R. 1952–1954. Les éléments de la flore bryologique de Bretagne. *Rev. Bryol. Lichénol.* **21**:229–234. Ibid. **22**:20–21, 141–147. Ibid. **23**:291–294.

Gemmell, A. R. 1950. Studies in the Bryophyta I. The influence of sexual mechanism on varietal production and distribution of British Musci. *New Phytol.* **49**:64–71.

———1952. Studies in the Bryophyta II. The distribution of the sexual groups of British Mosses. *New Phytol.* **51**:77–89.

Gradstein, S. R., and H. J. M. Sipman. 1978. Taxonomy and world distribution of *Campylopus introflexus* and *C. pilifer* (= *C. polytrichoides*: a new synthesis. *Bryologist* **81**:114–121.

———, and W. A. Weber 1982. Bryogeography of the Galapagos Islands. *J. Hattori Bot. Lab.* **52**:127–152.

Greene, S. W. 1967. Bryophyte distribution, in Bushnell, V. (ed.), *Terrestrial Life in Antarctica*, pp. 11–13. New York: Am. Geogr. Soc.

Greig-Smith, P. 1950. Evidence from hepatics on the history of British flora. *J. Ecol.* **38**:320–344.

Grolle, R. 1969. Grossdisjunctionen in Artenarealen Lateinamerikanischer Lebermoose, in Fittkau, E. J., et al (eds.), *Biogeography and Ecology in South America,* pp. 562–582. The Hague: W. Junk.

——— 1981. On hepatics in Baltic amber. Present knowledge and promisings, in Szweykowski, J. (ed.) *New Perspectives in Bryotaxonomy and Bryogeography,* pp. 83–88. Poznan: Adam Mickiewicz Univ.

Harvill, A. M. 1950. The western American-European element in the Alaskan moss flora. *Rev. Bryol. Lichénol.* **19**:32–34.

Hattori, S. 1951. On the distribution of the Hepaticae of Shikoku and Kyushu (Southern Japan). *Bryologist* **54**:103–118.

——— 1970. Geography and ecology of bryophytes—miscellaneous thoughts on bryology (1). *Miscell. Bryol. Lich.* **5**:70–71.

Herzog, T. 1926. *Geographie der Moose.* Jena: G. Fisher.

Hoe, W. J. 1974. Annotated checklist of Hawaiian mosses. *Lyonia* **1**:1–45.

Hong, W. S. 1966. The leafy Hepaticae of South Korea and their phytogeographic relationships, especially to the flora of North America. *Bryologist* **69**:393–426.

Horikawa, Y. 1955. Distributional studies of bryophytes in Japan and the adjacent regions. Hiroshima: Contrib. Phytotax. Geobot. Lab., Hiroshima Univ., N. Ser. No. 27.

Inoue, H. 1972. Distribution and speciation of certain sections of *Plagiochila.* *J. Hattori Bot. Lab.* **40**:24–30.

——— 1982. Speciation and distribution of *Plagiochila* in Australasia and the Pacific. *J. Hattori Bot. Lab.* **52**:45–56.

Irmscher, E. 1929. Pflanzenverbreitung und Entwicklung der Kontinente. II Teil. Weitere Beiträge zur genetischen Pflanzengeographie unter besonderer Berucksichtigung der Laubmoose. *Mitt. Inst. Allg. Bot. Hamburg* **8**:171–374, 16 plates.

Iwatsuki, Z. 1958. Correlations between the moss floras of Japan and of the Southern Appalachians. *J. Hattori Bot. Lab.* **20**:304–352.

——— 1972. Distribution of bryophytes common to Japan and the United

States, in Graham, A. (ed.), pp. 107–137. *Floristics and Paleofloristics of Asia and Eastern North America.* Amsterdam: Elsevier.

———— and A. J. Sharp. 1967. The bryogeographical relationships between Eastern Asia and North America I. *J. Hattori Bot. Lab.* **30**:152–170.

———— 1968. The bryogeographical relationships between Eastern Asia and North America II. *J. Hattori Bot. Lab.* **31**:55–58.

Koch, L. F. 1954. Distribution of California mosses. *Am. Midl. Nat.* **51**:515–538.

Koponen, A. 1982. The Family Splachnaceae in Australasia and the Pacific. *J. Hattori Bot. Lab.* **52**:87–91.

Koponen, T. 1979. On the taxonomy and phytogeography of *Mnium* Hedw. s.str. (Musci, Mniaceae). *Abstr. Bot.* **5**(Suppl. 3):63–73.

———— 1982. The Family Mniaceae in Australasia and the Pacific. *J. Hattori Bot. Lab.* **52**:75–86.

Lazarenko, A. S. 1957. Versuch einer Analyse der Laubmoosflora vom nordöstlichen Asien. *Rev. Bryol. Lichénol.* **26**:146–157.

———— 1958. Remote transportation of spores and its significance for the formation of moss ranges. *Urainkij Botanicij Zl.* **15**:71–77.

Martin, W. 1946. Geographic range and internal distribution of the mosses indigenous to New Zealand. *Trans. Roy. Soc. N. Z.* **76**:162–184.

———— 1949. Distribution of mosses indigenous to New Zealand. Suppl. I. *Trans. Roy. Soc. N. Z.* **77**:257–277.

———— 1952. Distribution of the mosses indigenous to New Zealand Supplement II. *Trans. Roy. Soc. N. Z.* **80**:197–205.

———— 1958. Survey of moss distribution in New Zealand. *Bryologist* **61**:105–115.

Muhle, H. 1970. Zur Ausbreitung von *Orthodontium lineare* Schwaegr.: *Orthodontium* in Schwarzwald. *Herzogia* **2**:107–112.

Müller, K. 1916. Zur geographischen Verbreitung der europäischen Lebermoose und ihrer Verwertung für die allgemeine Pflanzengeographie. *Ber. Deutsch. Bot. Gesell.* **34**:212–221.

———— 1954. Die pflanzengeographischen Elemente in der Lebermoosflora Deutschlands. *Rev. Bryol. Lichénol.* **23**:107–122.

Ochi, H. 1972. Some problems of distributional patterns and speciation in the subfamily Bryoideae in the regions including Eurasia, Africa and Oceania. *J. Hattori Bot. Lab.* **35**:50–67.

———— 1974. Some Bryaceous "Old World" mosses, also distributed in the New World. *J. Fac. Educ. Tottori Univ., Nat. Sci.* **24**:35–41.

————1982. A phytogeographical consideration of Australasian Bryoideae in relation to those of other continents. *J. Hattori Bot. Lab.* **52**:65–73.

Petet, E., and P. Szmajda. 1981. Remarques sur la distribution des mousses "eu-Atlantiques," in Szweykowski, J. (ed.), *New Perspectives in Bryotaxonomy and Bryogeography,* pp. 89–104. Poznan: Adam Mickiewicz Univ.

Pocs, T. 1975. Affinities between the bryoflora of East Africa and Madagascar. *Boissiera* **24**:125–128.

———— 1976. Correlations between the tropical African and Asian bryofloras. *J. Hattori Bot. Lab.* **41**:95–106.

———— 1982. Examples of the significance of historical factors in the compositions of bryofloras and in the speciation of liverworts. *Nova Hedw. Beiheft* **71**:305–311.

Pursell, R. A., and W. D. Reese. 1970. Phytogeographic affinities of the mosses of the Gulf coastal plain of the United States and Mexico. *J. Hattori Bot. Lab.* **33**:115–152.

Ratcliffe, D. A. 1968. An ecological account of Atlantic bryophytes in the British Isles. *New Phytol.* **67**:365–439.

Robinson, H. E. 1972. Observations on the origin and taxonomy of the Antarctic moss flora. *Antarctic Res. Ser.* **20**:163–177.

Rose, F., and E. C. Wallace. 1974. Changes in the bryophyte flora of Great Britain, in Hawksworth, D. L. (ed.), *The Changing Flora and Fauna of Britain*. Systematics Assoc. Special Vol. 6, pp. 27–46.

Sainsbury, G. O. K. 1965. Introduced mosses in New Zealand. *Bryologist* **68**:91–92.

Schofield, W. B. 1965. Correlations between the moss flora of Japan and British Columbia, Canada. *J. Hattori Bot. Lab.* **28**:17–42.

———— 1969. Phytogeography of Northwestern North America: Bryophytes and vascular plants. *Madroño* **20**:155–201.

———— 1972. Bryology in arctic and boreal North America and Greenland. *Can. J. Bot.* **50**:1111–1133.

———— 1974. Bipolar disjunctive mosses in the Southern Hemisphere, with particular reference to New Zealand. *J. Hattori Bot. Lab.* **38**:13–32.

———— 1980. Phytogeography of the mosses of North America (North of Mexico), in Taylor, R. J., and A. E. Leviton (eds.), *The Mosses of North America*, pp. 131–170. San Francisco: Pacific Div. AAAS.

———— and H. A. Crum 1972. Disjunctions in bryophytes. *Ann. Missouri Bot. Garden* **59**:174–202.

Schornherst, R. D. 1943. Phytogeographic studies of the mosses of northern Florida. *Am. Midl. Nat.* **29**:509–532.

Schulze-Motel, W. 1975. Die bryogeographische Stellung der Samoa-Inseln. *Bull. Soc. Bot. France. Colloque Bryol.* **1974**:295–298.

Schuster, R. M. 1958. Boreal Hepaticae, a manual of liverworts of Minnesota and adjacent regions III. Phytogeography. *Am. Midl. Nat.* **59**:257–332.

———— 1969. Problems of antipodal distribution in lower land plants. *Taxon* **18**:46–91.

———— 1972. Continental movements, "Wallace's line" and Indo-Malaysan-Australasian dispersal of land plants: some eclectic concepts. *Bot. Rev.* **38**:3–86.

———— 1979. On the persistence and dispersal of transantarctic Hepaticae. *Can. J. Bot.* **57**:2179–2225.

———— 1981. Paleoecology, origin, distribution through time, and evolution of

Hepaticae and Anthocerotes. *Palaeobot., Ecol. Evol.* 2:129–191.

—— 1982. Generic and familial endemism in the hepatic flora of Gondwana-land: origins and causes. *J. Hattori Bot. Lab.* **52**:3–35.

—— 1983. Phytogeography of the Bryophyta, in Schuster, R. M. (ed.), *New Manual of Bryology,* Vol. I, pp. 462–626. Nichinan: Hattori Botanical Laboratory.

Seppelt, R. D. 1982. *Ditrichum* and other genera of Ditrichaceae in Australasia and the Pacific. *J. Hattori Bot. Lab.* **52**:107–112.

Sharp, A. J. 1941. Some historical factors and the distribution of southern Appalachian bryophytes. *Bryologist* **44**:16–18.

—— 1958. Tropical bryophytes in the Southern Appalachians. *Ann. Bryol.* **11**:141–144.

—— 1972. Phytogeographical correlations between the bryophytes of eastern Asia and North America. *J. Hattori Bot. Lab.* **35**:263–268.

—— 1974. Some geographic relations in the Himalayan bryoflora. *J. Hattori Bot. Lab.* **38**:33–37.

—— and Z. Iwatsuki. 1965. A preliminary statement concerning mosses common to Japan and Mexico. *Ann. Missouri Bot. Garden* **52**:452–456.

Smith, G. L. 1972. Continental drift and distribution of Polytrichacae. *J. Hattori Bot. Lab.* **35**:41–49.

Steere, W. C. 1953. On the geographical distribution of arctic bryophytes. *Stanford Univ. Publ. Biol. Sci.* **II**:30–47.

—— 1965. The boreal bryophyte flora as affected by Quaternary glaciation, in Wright, H. E., and D. G. Frey (eds.), *The Quaternary of The United States.* Princeton: Princeton Univ. Press, pp. 485–495.

—— 1965. Asiatic elements in the bryophyte flora of western North America. *Bryologist* **72**:507–512.

—— 1976. Ecology, phytogeography and floristics of arctic Alaskan bryophytes. *J. Hattori Bot. Lab.* **41**:47–72.

Størmer, P. 1969. *Mosses with a Western and Southern Distribution in Norway.* Oslo: Universitetsfortlaget.

—— 1983. Characteristic features of the moss flora of the various parts of Europe, 91 pp. Norway: Erling Sem Offsettrykkeri A.S.

Takaki, N. 1972. Geographical distribution of Japanese *Dicranum* species in the Northern Hemisphere. *J. Hattori Bot. Lab.* **35**:31–40.

Tixier, P. 1969. Biogéographie: Ecologie ou Paléogéographie? Les cas des bryophytes. *C. R. Soc. Biogéog.* **406**:205–212.

Vitt, D. H. 1982. Populational variation and speciation in austral mosses. *J. Hattori Bot. Lab.* **52**:153–159.

—— and D. G. Horton. 1979. Mosses of the Nahanni and Liard ranges area, southwestern Northwest Territories. *Can. J. Bot.* **57**:269–283.

Whitehouse, H. L. K., and J. A. Paton. 1963. The distribution of *Tortula stanfordensis* in Cornwall. *Trans. Br. Bryol. Soc.* **4**:462–463.

Zanten, B. O. van. 1976. Preliminary report on germination experiments designed to estimate survival chances of moss spores during aerial trans-oceanic long-range dispersal in the Southern Hemisphere, with particular reference to New Zealand. *J. Hattori Bot. Lab.* **41**:133–146.

———— 1978. Experimental studies on trans-oceanic long range dispersal of moss-spores in the Southern Hemisphere. *J. Hattori Bot. Lab.* **44**:455–482.

———— and T. Pocs. 1981. Distribution and dispersal of bryophytes. *Ad. Bryol.* **1**:479–562.

Glossary

A **abaxial** facing away from the axis of the plant.

absorption the ability of one substance to swallow up another.

acclimation having the ability to change to accomodate changing conditions.

acidic describes substrata that neutralize and are neutralized by alkalis, and are dominated by a compound that is made up of hydrogen and another element or elements.

acrocarpous in mosses, bearing the sporophyte at the apex of the main shoot, with further growth of the shoot by lateral innovations.

acrogyny the condition in hepatics in which the female sex organs terminate the main shoot; the apical cell produces these organs (contrasted with anacrogyny).

acuminate gradually tapering to a narrow point.

adaxial facing toward the axis of the plant.

adsorption the ability to attract molecules of a liquid to a solid surface.

affinity relationship.

alar an area at the basal corner of a leaf where it is attached to a stem; alar cells are often different in shape or color from other cells of the leaf.

alkane a chemical compound consisting of hydrogen and carbon only (waxes are alkanes).

allopolyploidy polyploidy that results after hybridization of two different chromosome sets in which there was no effective pairing preceding spontaneous doubling of the chromosome number.

amber a yellowish, brittle fossil resin of vegetable origin.

amphigastria in leafy hepatics (Jungermanniales), the underleaves, i.e., leaves on the ventral surface of the stem; these leaves often differ in shape and size from the lateral leaves.

amphithecium the embryonic tissue of the sporangium enclosing the endothecium, forming the outer layers of cells that produce the sporangial jacket in all bryophytes, plus other layers in some.

anacrogyny   the condition in hepatics in which the female sex organs are produced by a lateral cell; thus the growth of the main shoot of the gametophore is indeterminate, and sporophytes are lateral.

anisospory   producing spores of two overlapping size classes in the same sporangium.

annual   completing life cycle within one year.

annular   forming a ring; descriptive of thickenings in some cells.

annulus   a ring of differentiated cells forming a somewhat elastic band of cells around the mouth of a sporangium, aiding in the shedding of the operculum.

antheridiophore   a specialized branch that bears the archegonia in the Marchantiales (Hepaticae).

antheridium   the multicellular male sex organ of bryophytes that consists of a stalked sac containing many sperms enclosed by a unistratose sterile jacket of cells.

anthocyanin   a water-soluble blue, purple, or red pigment found in the cell sap.

anticlinal   at right angles to the circumference of the surface.

apogamy   the formation of a sporophyte from a gametophyte without the intervention of fusion of gametes.

apolar   lacking a pole; descriptive of the manner in which a germ tube ruptures the spore coat at unpredictable sites.

apophysis   a swelling at the base of the sporangium where it joins the seta (= hypophysis).

apoplast   held within cell walls.

apospory   the formation of a gametophyte from a sporophyte without the intervention of spores; thus the new gametophyte is 2n.

archegoniophore   archegonium bearer; a specialized branch in the Marchantiales (Hepaticae) that bears the archegonia.

archegonium   the flask-shaped multicellular female sex organ that contains a single egg.

archesporium   the layer of cells in an embryonic sporangium that produces the spores (= sporogenous layer).

areolation   the cellular network of a leaf or thallus.

articulated   jointed; descriptive of peristome teeth which are segmented.

asexual   reproducing without the union of eggs and sperms.

assimilate   organic products of photosynthesis available for utilization by the plant.

asymmetric   lacking symmetry; describing objects in which the parts are arranged unevenly around a central axis.

autopolyploidy   polyploidy that arises from chromosomes of the same source through spontaneous doubling of chromosome number.

autotrophic   able to manufacture organic material from inorganic, utilizing an inorganic energy source.

auxin   growth hormone.

awn a bristle or hair point, describing moss leaves in which the costa (midrib) extends beyond the main blade of the leaf.

axil a point in the angle where the leaf is attached to the stem.

B **bar** a unit of atmospheric pressure.

basic descriptive of a substance capable of combining with an acid to form a salt (e.g., calcium-rich substrata).

bifid two-parted; dividing into two parts, as in a leaf with two elongate lobes.

biflagellate having two flagella, as in sperms of bryophytes.

bilateral with one side essentially a mirror image of the other.

bilobed two-lobed.

biomass weight of living material per unit area.

bipinnate branching pattern in one plane in which the main axis bears evenly spaced branches on two sides, each branch of which may also be evenly branched in the same manner.

bipolar describes a distribution pattern in which an organism is found in the temperate or polar portions of the Northern and Southern Hemispheres, but absent in middle latitudes.

bisbifid a two-lobed leaf in which each lobe is also divided into two equal parts.

bistratose two cell layers in thickness.

bivalent the two paired chromosomes at pairing.

bud a cell mass in mosses that arises from the protonema or stem containing the apical cell that produces the leafy gametophore or branch.

bulbiform in the shape of a bulb, rather like a miniture onion bulb.

C **calcareous** a substratum rich in calcium

calcicole an organism confined mainly to calcium-rich substrata.

calcifuge an organism intolerant of calcium in high concentrations.

callus a thickened mass of undifferentiated cells.

calyptra a thin hood that surrounds the developing sporangium or entire sporophyte, developed in large part from cells of the lower part of the archegonium.

campanulate shaped like a bell.

Carboniferous a period of geologic time beginning approximately 345 million years ago, and lasting for approximately 65 million years.

capillary hairlike, describing the thin hairlike spaces formed among leaves, branches, or scales.

capitulum a head, describing the crowded apical mass of branches in *Sphagnum*.

carpocephalum the specialized, essentially radially symmetrical receptacle that surmounts the perpendicular stalk; it bears the sporangia in many marchantialean hepatics.

caulid (ia) a term sometimes used in preference to the term "stem" in bryophytes.

caulonema   the portion of the protonema capable of producing buds that initiate leafy gametophores in mosses.

cellulose   a complex insoluble carbohydrate formed of glucose molecules attached end-to-end, which forms the chief component of cell walls in most plants.

central strand   a central strand of smaller cells in the center of stems of many mosses.

charophytes   a group of specialized green algae of fresh-water habitats in which branches are borne in whorls at nodes.

chelating agent   a chemical agent capable of forming loose bonds with a metal ion.

chlorenchyma   a specialized layer of chlorophyll-containing cells forming a tissue or tissue-like region.

chlorocyst   a leaf cell containing chlorophyll, used to distinguish these from cells in the same leaf that are dead and contain no chlorophyll (leucocysts)

chloronema   the chlorophyllose part of the protonema in which cell walls are usually not oblique between the cells; this part of the protonema often cannot produce "buds" that initiate gametophores in mosses.

chromatogram   the recorded result of a chromotographic analysis.

chromatography   a technique used to separate compounds based upon their differential solubilities in the solvents used to develop the chromatogram.

chromosome   threadlike form of DNA and associated nucleoproteins which replicates itself during each complete meiotic or mitotic cycle.

cilia   threadlike structures.

clone   a group of plants produced vegetatively from a single original plant.

coevolution   development of genetically determined traits in two species to facilitate some interaction, usually mutually beneficial.

columella   a central cylinder of sterile tissue in the sporangium.

compensation   a point at which respiration and photosynthesis are in balance.

complanate   flattened parallel to the substratum.

complex pore   a pore in the thallus of marchantialean hepatics which is bounded by a barrel-shaped circle several cells high.

complicately bilobed   two-lobed in which one lobe may be structurally complex.

cortex   the outer layers of cells in a stem

costa   the thickened midrib of a moss leaf or hepatic thallus.

Cretaceous   a period of geologic time beginning approximately 136 million years ago and lasting approximately 71 millions years.

cryptopolar   describing a spore in which the germ tube emerges from either of the poles or at the equator.

cryptopore   describing stomata that are immersed so that the pore is embedded within a chamber of overarching cells.

cucullate   hooded or hood-shaped, often used to describe the usual shape of the calyptra in mosses.

cushion of somewhat rounded or tufted growth form, where the stems diverge outward from a central point.

cuticle the noncellular waxy external wall coating of some cells.

cutin the material forming the cuticle.

cytokinins a group of chemically related plant hormones that promote cell divisions, among other effects.

D **dehiscence** means of rupturing or opening of sporangia.

dendroid shaped like a tree; describing gametophores in which a leafy unbranched stem is terminated by an apical cluster of branches.

derived a term used to imply that an organism has arisen from a more generalized ancestor.

dermatitis inflammation of skin by localized irritation.

desiccation drying-up; i.e., water loss until the object is nearly water-free.

Devonian a period of geologic time beginning approximately 395 million years ago and lasting for 50 million years.

disapore any structure that can reproduce a plant, whether vegetatively produced or a spore.

dichotomous equally forked, divided into two equal branches.

differentiated describing cells that are distinctly different in size, shape and/or color from adjacent cells.

dihydrostilbene a plant compound based on a C_6–C_2–C_6 skeleton.

dimorphic of two different forms.

dioicous a gametophore that bears the male sex organs only; the female sex organs are borne on a separate plant; opposite of monoicous.

diploid having two sets of chromosomes (2n), characteristic of the sporophyte generation.

diplolepidous describing peristome teeth in which each segment of the outer face of the articulated tooth is formed of parts of the walls of two cells; thus each segment is composed of two "scales."

diplospory polyploidy arising from unreduced spores (i.e., spores that are 2n).

disjunct referring to a distribution pattern that is conspicuously interrupted by areas where the taxon is absent.

distal away from the center of body or point of attachment.

divergent arranged to point at a broad angle from the point of attachment.

dorsal on the upper surface.

dwarf male miniature antheridial gametophores that are attached to the normal-size archegonium-bearing gametophores.

E **ecotype** a physiologically differentiated race of a species.

ectohydric conducting water externally.

egg the haploid gamete produced by the female sex organ.

elater elongate sterile cells, usually hygroscopic, admixed among the spores of most hepatics and hornworts.

elaterophore   a cylindrical mass of sterile cells to which elaters are attached in some metzgerialean hepatic sporangia (= elater-bearer).

elliptic   oblong in shape, in which both the sides are convex.

embryo   the early developmental stages of a sporophyte.

endemism   refers to taxa restricted to a specific geographic area.

endogenous   originating within, as in branches that arise from cells within the stem, rather than from the stem surface.

endohydric   conducting water internally.

endosporic   descriptive of a spore that undergoes several cell divisions before the spore coat is ruptured.

endostome   the inner ring of peristome teeth in mosses.

endothecium   in an embryonic sporangium, the cells that give rise to layers other than the jacket of the sporangium.

entire   without teeth, describing the margin of leaves or thalli.

epidermis   the outermost layer of cells in a structure several cells in thickness.

epiphragm   the membranelike expansion of the columella covering most of the mouth of the sporangium in Polytrichidae (hair-cap mosses).

epiphyte   a plant that grows perched upon another plant, and is normally not parasitic.

epilithic   growing on rock.

epiphyllous   growing on living leaves.

euchromatin   region of a chromosome that is not condensed during interphase.

eudesmane   a specific sesquiterpene.

exogenous   originating from the surface cells.

exosporic   a spore that undergoes cell division after the spore coat is ruptured.

exostome   the outer row (or rows) of peristome teeth in the mosses.

exothecium   the outermost layer of cells in the sporangium.

F   **fascicle**   the cluster of lateral branches arranged in spirals on the main stem of *Sphagnum*.

fatty acid   a long chain carbon-based acid common in all organisms

fibril   thin transverse thickened bands in the swollen nonchlorophyllose leaf cells of leaves, and sometimes stems, in *Sphagnum*.

flagelliform   a slender branch or stem, leafless or with leaves much smaller than those of the rest of the gametophore.

flagellum (-a)   the hairlike process that controls the movement of the sperm.

flavonoid   a group of plant pigments composed of C_6-C_3-C_6 compounds.

flavonol   a specific structural type of flavonoid.

floristic   referring to the flora, or kinds of plants.

flux   the rate of flow.

foot   the base of the sporophyte that attaches it to the gametophore.

foveolate   spore surface ornamented with regular circular depressions, giving it the appearance of a golf ball.

free-nuclear   an early stage in zygote development in which nuclear divisions are not accompanied by cell wall formation.

G **gas chromatography**   separation of mixtures in the gas phase.

germinate   to begin to grow.

gametophore   the stage in the gametophyte that bears the sex organs.

gametophyte   the haploid stage of the life cycle that ultimately produces the sex organs and gametes.

gemma (-ae)   a small vegetative reproductive structure that ultimately differentiates an apical cell, producing a gametophore.

generalized   of very general structure, used to describe organisms that most resemble in structure the primitive progenitor (opposite of derived or specialized).

geotropic   sensitive to gravity.

germling   the mass of undifferentiated cells that gives rise to the apical cell in an hepatic gametophore.

glycolate oxidase   an oxidizing enzyme in green leaves and stems.

Gondwanaland   a supercontinent that consisted of the fused land masses of what are now the major Southern Hemisphere continents plus the Indian subcontinent.

gravimetric   sensitive to gravity.

guard cells   pairs of specialized epidermal cells in the sporangium that surround a pore.

guide cells   a row of enlarged cells that extends across a costa as seen in transverse section, continuous with the cells of the unistratose lamina of the leaf.

H **haploid**   a state in which each chromosome is represented only once (n), characterisitc of the gametophytic generation.

haplolepidous   describing jointed peristome teeth in which the outer face of each joint is composed of a single face of one cell.

haustorium (-a)   the absorptive structure of a sporophyte that penetrates the gametophore and transfers nutrient and water from it to the sporophyte.

heterochromatin   region of a chromosome that becomes condensed during interphase.

heterotrichous   of diverse kinds of hairs, used to describe protonemata with branches that show differing form.

holarctic   a distributional pattern in which a taxon is widely distributed in the Northern Hemisphere.

humic   a humus, a dark soil substratum made up of decayed vegetable matter.

humidity   moisture content in air.

hyaline   clear or translucent.

hybrid   the offspring of two organisms of different species.

hydrated   combined with water; indicates that a dried plant has returned to a moisture condition suitable for metabolism.

hydroid   a specialized cell that conducts water internally in mosses.

hydrome   the system of water conducting cells (hydroids) that characterizes some mosses.

hygroscopic   readily taking up moisture.

hyphae   the tubular filaments that make up a fungal mycelium.

hypophysis   a swelling at the base of the sporangium where it joins the seta (= apophysis).

I   **imbricate**   closely appressed and overlapping.

innovation   a new shoot that originates laterally after the sex organs are mature.

intercalary   among the cells rather than on the surface.

interphase   the nuclear stage that occurs between two nuclear divisions.

involucre   a tube of thallus tissue that protects the archegonia, not derived from fused leaves.

involute   curving inward, as in leaves in which the margins curve inward over the upper face of the costa.

isodiametric   of equal length and width.

J   **jacket**   the surface cells that enclose a specific organ, as in a sporangium.

julaceous   wormlike in general appearance.

K   **karyotype**   the mitotic metaphase morphology of any given chromosome complement.

L   **lamellae**   green ridges or plates of cells that make up flaps of tissue characterizing the costa or blade of some moss leaves.

lamina   the main body (usually one cell in thickness) of the leaf.

lanceolate   lance-shaped; narrow and tapered from the base.

Laurasia   a supercontinent that consisted of the fused continents that now form North America and most of Eurasia.

lenticular   shaped like a lens with convex upper and lower surfaces.

leptoid   specialized cell that conducts synthesized foods from one part of a moss to another.

leptome   the system of leptoids in the moss.

leucocysts   colorless dead cells that make up some moss leaves.

lignified   containing lignin, the common compound of wood in seed plants.

lignin   one of the most important constituents of the secondary wall of vascular plants.

lipophilic   defining materials soluble in fats.

lobule   the smaller of two lobes.

longitudinal   arranged lengthwise.

lumen   the space in a dead cell bounded by the cell walls.

lux   a unit of illumination.

M **macronema (-ta)** rhizoids that originate from large cells that surround buds on a stem.

marsupium a subterranean pouch of stem material that encloses the archegonia in some jungermannialean hepatics, termed a marsupidium when greatly elongated.

mass spectrum the recorded result of a controlled decomposition of a compound by electron bombardment.

meiosis chromosomal division in which the number of chromosomes is reduced from the diploid (2n) to haploid (n) state.

meristem an area of active cell division.

metaphase the phase of chromosomal arrangement nearly in a single plane at the equator of the cell.

microclimate climate within a very restricted area as, for example, the climate at the soil level.

micronema (-ta) rhizoids that originate randomly on the moss stem.

middle lamella the layer of intercellular material rich in pectic materials that cements the primary walls of adjacent cells.

mitosis cell division in which chromosome number in new cells is the same as in the parental cell.

mitrate a term used to describe a calyptra shaped like a bishop's miter; i.e., conic and undivided or equally lobed at base.

mixohydric conducting water both externally and internally.

monoicous a gametophore that possesses both male and female sex organs on the same plant; opposite of dioicous.

monoterpenoid a terpenoid built of two isoprene units (see terpenoid).

monotypic refers to a genus that contains only one species.

mucilage cells cells that exude gumlike matter; such cells are often hairlike (mucilage hairs) and are near the growing points and among the sex organs of many bryophytes.

multistratose composed of many cell layers.

mycorrhizal containing fungal cells within the cells of a plant.

N **neotropic** in the New World (i.e., American) tropics.

nurse cells sterile cells scattered among the spores in some hepatic sporangia, which usually disintegrate as the spores mature.

O **obovoid** shaped like an egg with the expanded part uppermost.

operculum (-a) the circular lid that closes the opening in a moss sporangium.

ovoid shaped like an egg with the enlarged part at the base.

P **palaeotropic** Old World (i.e., African and Asian) tropics.

palynology the study of spores.

Pangaea the geographical designation given to a huge ancient continental land mass that consisted of the fused parts of present continents.

papilla (-ae)   wartlike protuberance.

paraphyllium (-a)   small green, irregularly leaflike outgrowths scattered on the stems of some mosses and a few hepatics.

paraphysis (-es)   filamentous sterile structures intermixed among the sex organs of most mosses.

parenchyma   tissue consisting of uniformly thin-walled cells.

peat   deposits of incompletely decomposed plant material, often mainly of *Sphagnum.*

pendant   a growth form in which long gametophores droop downward from the substratum.

pendent   drooping downward.

perennial   persisting as a growing plant for many years.

perianth   a protective chlorophyllose tube that surrounds the archegonia, derived from fused leaf bases, characterizing jungermannialean hepatics.

perichaetium (-a)   the cluster of leaves with the enclosed female sex organs.

periclinal   parallel to the circumference of the surface.

perigonium (-a)   the cluster of leaves with the enclosed male sex organs.

perigynium (= involucre)   a sleeve of thallus tissue that surrounds and protects the archegonia.

peristome   the circumference of the opening of a moss sporangium, usually with teeth.

Permian   a period of geologic time beginning approximately 280 million years ago and lasting 55 million years.

phaneropore   pores or stomata exposed on the surface.

phenolic   a type of compound where at least one hydroxyl group is attached to an aromatic unit (e.g., flavonoids are phenolics).

photoperiodism   response to duration and timing of dark and light periods.

phototropic   sensitive to light.

phragmoplast   microfibrils parallel to the spindle axis of a dividing cell after the chromosomes have moved to the poles; concerned with formation of the cell plate between dividing cells.

phycoplast   an assemblage of microtubules perpendicular to the spindle and at the equator of the cell after the chromosomes have moved toward the poles.

phyllid (-ia)   a term sometimes used in prefernce to "leaf."

phyllodioicous   with dwarf male gametophores epiphytic on the female gametophore.

phytosterol   steroid molecule that is restricted to plants.

pinnate   branches originating in a regular pattern in a single plane on two sides of a stem.

pit   a natural recess or perforation in the cell wall.

plasma membrane   outer surface membrane that bounds the protoplast in a cell.

plasmodesmata   minute protoplasmic threads that extend through openings in cell walls and connect protoplasts of adjacent cells.

Pleistocene   an epoch of geologic time beginning 2.5 million years ago and persisting up to 5000 years ago, during which the most recent continental glaciations occurred.

pleurocarpous   a moss in which the female sex organs are borne on reduced lateral branches; thus the gametophore bears several lateral sporophytes.

plications   longitudinal pleats.

polar   with poles; used to describe a spore in which the apex of the three-angled face forms one pole, and the rounded face opposite forms the other pole.

polypeptide   a sequence of amino acids linked together by peptide bonds.

polyploidy   variations in chromosome number involving more than the diploid number of complete chromosome sets.

postical   on the undersurface.

precursor   a structure that initiates the development of a more elaborate structure.

productivity   accumulation of energy and nutrients by green plants.

progenitor   the ancestral type that initiates a lineage.

protonema (-ta)   the filamentous juvenile stage that precedes the formation of the gametophore (= first hair).

proximal   the surface closest to the point of attachment.

pseudodichotomous   equal forked branching achieved through the destruction of the apical cell.

pseudoparaphyllium (-a)   small irregularly shaped leaflike structure that arises around branch buds on the stems of some mosses.

pseudoperianth   an involucre that resembles a perianth, but is of thallus tissue, and forms mainly after sporophyte development.

pseudopodium   an elongated mass of leafless gametophore tissue that pushes the sporophyte beyond perichaetial leaves (in *Andreaea* and *Sphagnum*).

pteridophytes   a term used to indicate ferns and their allies.

pyrenoid   an organelle associated with the chloroplast of some hornworts.

pyriform   in the shape of a pear; i.e., with an expanded portion having an elongate neck.

Q   **quadrant**   a mass of four cells arranged around a central point.
quadrate   square.

R   **radial**   arranged symmetrically in more than two rows around a central axis.
recurved   curved under; used to describe leaf margins that curve downward, away from the stem apex.
reduction   a term used to indicate that a plant has lost features that were assumed to have been present in its ancestor.
reniform   the shape of a kidney.
reticulate   forming a network.
retort cell   a cell with a pore at the upper end, terminating a short projecting neck, as in cortical cells of branch stems of some species of *Sphagnum*.

revolute   strongly recurved.

rhizoid   filamentous structures that affix bryophytes to their substratum.

rhizomatous   possessing a creeping stem.

rostrate   with a long snout.

S   **saprophyte**   a plant that gains its nutrient directly from non-living organic matter, usually through the mediation of fungi.

secund   used to describe leaves in which the apices all point in one lateral direction.

sesquiterpenoid   a terpenoid built up of three isopreme units (see terpenoid).

sessile   affixed directly, without a stalk.

seta   the stalk of the sporophyte.

siliceous   a substratum in which silica (or sand) is the predominant substance.

silicicole   growing on siliceous substrata that tend to be acidic.

simple leaves   leaves that have no lobes.

simple pore   a pore that is directly encircled by epidermal cells a single cell in thickness.

slime pore   stomatelike pore in the thallus of some hornworts.

somatic   referring to the differentiated cells that form the body of an organism.

specialized   a term used to imply that an organ or organism has undergone changes that accommodate it more effectively to given conditions.

sphagnol   a chemical substance extracted from *Sphagnum*.

sporangium (-a)   the sac that contains the spores that originate through meiosis.

sporogenous   describes the layer in a sporangium that gives rise to the spores.

sporophyte   the part of the life cycle that is diploid (2n) and produces the spores through meiosis.

stem-calyptra   a sheathing structure that protects the developing embryo, constituted of both archegonial and extensive numbers of stem cells.

stereid   thick-walled cells of narrow diameter that offer support to organs of mosses.

steroid   a terpenoid based upon six or more isoprene units (see terpenoid).

substratum (-a)   the medium upon which an organism grows.

succubous   a term used to describe leaf arrangement in jungermannialean hepatics is which the lower border of a leaf covers the upper border of the leaf immediately below it as viewed from the dorsal surface.

symbiosis   an intimate and extended association between two or more organisms of different species.

symplast   held within the cell protoplast or lumen.

T   **tannin**   a complex phenolic polymer restricted to plants, used for tanning leather.

terpenoid   a general form for a compound formed of a sequence of five-carbon units termed isoprenes.

tetrad   a group of four; used to describe the four spores that are the final product of meiosis of the spore mother cell.

tetraploid   an organism that has four sets of chromosomes in its sporophytic cells.

thallus   a flattened gametophore in which no leaf-like organs dominate the structure.

thallus-calyptra   a sheathing layer formed of both archegonium and thallus tissue that protects a developing sporophyte.

thin-layer chromatography (TLC)   a technique of chromatography in which dissolved substances are separated on a thin layer that differentiates them on the basis of their differing solubilities.

thylakoid   the site of the light-trapping reactions of photosynthesis.

tissue   any of the cellular structures that make up the organs of a plant.

trabecula (-ae)   slender strands of supportive cells that cross air chambers, e.g., in moss sporangia.

Triassic   a period of geologic time beginning 225 million years ago and lasting 35 million years.

trigone   a three-angled structure; the thickened bulging cell wall that characterizes the corners of some cells in bryophytes.

triradiate   with three lines that radiate from a central point.

tuber   a specialized vegetative structure in a gametophore; in hepatics and hornworts, designating an apical cell surrounded by protective layers of dead cells, which allow the apical cell to survive through unfavorable conditions for a time; in mosses, used to designate rhizoidal gemmae.

turbinate   the shape of a toy top, with the expanded part basal.

turf   thickly matted; used to describe gametophores that are densely crowded together with upright shoots.

turgor   pressure exerted on the cell walls by the fluid within, which gives the cell rigidity.

U   **underleaf**   leaves on the ventral surface of the stem (= amphigastrium).

uniseriate   a filament made up of a single linear series of cells.

unistratose   a layer one cell in thickness.

V   **vacuole**   a space or cavity within the cytoplasm, filled with cell sap.

vector   any means by which diaspores are transported.

vegetative   growth through mitotic cell divisions to produce further tissue.

venter   the expanded portion of the archegonium that encloses the egg.

ventral   on the undersurface.

viability   ability to germinate.

vicariad   closely related species derived from a common ancestral population divided by geographic isolation.

vitta   a row of cells differing in shape from the remainder of the leaf cells in the central part of a leaf of some jungermannialean hepatics.

W   **weft**   a growth form in which gametophores are loosely interwoven.

Z   **zygote**   the first cell of the sporophyte, formed as a result of fusion of the sperm with the egg.

Appendix A

Collecting Bryophytes and Processing for Study

WHERE TO COLLECT

A most important aspect of collecting bryophytes is learning where these plants grow and when they are in the best condition to study. Although bryophytes occupy a wide diversity of habitats, some sites yield a greater variety than others. In a humid temperate climate, for example, there may be lush thick carpets of bryophytes on the forest floor; yet in these same sites the bryophytic diversity is often very limited.

Possibly the sites yielding the greatest floristic richness are humid forest-margined canyons of water courses. Here many habitats suitable for bryophytes are not rapidly colonized by the larger, more rapidly growing seed plants. In such canyons epiphytic bryophytes tend to be well represented. Open, high-elevation forests also are often rich in bryophytes, particularly in humid climates, whether at temperate or tropical latitudes. Such forests near coastal areas are also rich in bryophyte diversity and coverage.

The bryophyte coverage and diversity in arctic and alpine areas is often considerable. Unfortunately many of the bryophytes that occur in these regions grow in complex intermixed populations. Furthermore, sporophyte production is often erratic in such areas, and absence of sporophytes often presents difficulties in accurate determination.

WHEN TO COLLECT

Spring is the season when many bryophytes, especially hepatics, mature their sporophytes. Another season with a burst of sporophyte production is late summer to autumn. The sporophytes persist in mosses many months after they mature, but in hepatics the sporophytes generally disintegrate soon after the spores are shed. The sporophytes are present in most hornworts throughout the season of active growth, which varies in different climates. In temperate areas late spring to early summer is the best time to collect sporophyte-bearing material in a condition that makes the material determinable with a reasonable amount of effort.

COLLECTING TOOLS AND METHODS

Bryophyte collecting is extremely simple and usually requires readily available and modest equipment. Some bryophytes adhere very firmly to their substratum, and so a tool is needed to remove the specimen with or without its substratum. Often a knife is suitable for scraping bryophytes from rock or bark or for removing bark with the adhereing bryophyte. Probably a more efficient tool is a carpenter's wood chisel. A hammer and cold chisel are especially effective in removing the rock substratum with its specimen. Either a knife or chisel can be used to cut or peel patches of bryophyte colonies from their substratum of earth or decaying wood.

If a specimen has a good deal of soil adhering to it, the soil should be washed loose if the bryophyte is sufficiently large. This may pose particular difficulties with hepatics, where sporophytes can be seriously damaged upon washing.

Aquatic bryophytes often have considerable extraneous matter attached to them. This should be rinsed off as carefully as possible, without injuring the specimen, before the specimen is dried. In such aquatic bryophytes, especially mosses, the excess water should be carefully squeezed (or pressed) out between the hands, Particular caution must be taken not to break the brittle stems of some bryophytes, as for example, *Sphagnum*.

After collection, specimens are usually teased loose from each other, cleansed of extraneous material, including loose fragments of soil, flattened out, and exposed in a packet (made of newspaper) to air dry. Most bryophytes should not be placed under great pressure when they are being dried. Such pressure often ruptures sporophytes and destroys some of the critical morphologic features of the specimen. It may make some specimens (especially *Sphagnum*) difficult to dissect.

Specimens, dependent upon the size of the bryophyte, should be made so that they fit into a 10 × 15 cm packet, if this is possible without damage to the specimen. The size of the packet is enlarged to accommodate larger specimens.

OBSERVATIONS ON FRESH LIVING MATERIAL

Many hepatics possess a number of features that should be noted while the specimen is alive. Leafy hepatics, in particular, should be examined microscopically to record the average number, size, and morphology of the complex oil bodies in each cell of the leaf. It is easier to record the structure of thallose hepatics in fresh than in dried material. The nature of air pores and air chambers is especially important, and cross-sections can be made freehand with a razor blade, and the observations recorded.

It is advisable to make freehand sections on the stage of a binocular dissecting microscope. Lowest magnification is usually sufficient. The bryophyte should be placed upon a clean microscope slide and a few drops of water added (if necessary) to keep the sections moist. Sectioning is accomplished by holding the bryophyte with fine forceps with one hand, while a sharp single-edge razor blade (or scalpel) is used to cut thin sections. The sections are retained, while the larger fragments are removed, and a cover-slip is added. Sketches of the thallus structure can be made at this point, and any notes concerning structure can be recorded. Permanent slides can be made with Hoyer's solution (see p. 397), but this mounting medium bleaches out any color and destroys cell contents. Such a permanent slide, when dry, is sometimes very useful for reference in critical determination of future collections.

LABEL INFORMATION

The label that identifies a specimen should contain the following information:

1. Identity of the specimen (genus and species and the author for the species). Subspecific identity is included with any specimen that possesses such names.
2. Habitat, including substratum, elevation (when known), and general vegetation type in which it occurs.
3. Geographic locality, including geographic name of country, state, county or district, nearest politically named locality with its approximate latitude and longitude. Abbreviations should not be used.
4. Collector's name, date collection was made, collection number.
5. Determiner's name, and date determination was made.
6. It is often useful to indicate that the material has sporophytes (c. fr.)

A sample label is presented in Fig. A-1. The label is affixed to the outer flap of the packet.

FIGURE A-1

> ## BRYOPHYTES OF CANADA
> *Bryum argenteum* Hedw.   c.fr.
> on exposed gravel of rooftop
> University of British Columbia campus,
> Point Gray, Vancouver area
> BRITISH COLUMBIA. 49°10′N, 123°10′W.
> Coll. W. B. Schofield No. 2764   June 6, 1982
> Det. W. B. Schofield   June 6, 1982.

PACKETING

Bryophyte packets for collecting can be made out of newspaper. Colored newspapers sometimes pigment the specimen, so should be avoided. If large amounts of material are collected, the packet can be made to accommodate the amount, but attention must always be made to the ultimate size of the specimen to be accommodated in the permanent packet. Thus, when possible without damaging the specimen, material should be divided into appropriate-sized clumps that can be accommodated by a 10 × 15 cm packet.

Permanent packets should be made of a high rag content paper. Such packets are durable over a long period of time and do not discolor with age. Cheaper paper tends to discolor and become brittle in time. Newspaper packets, for example, often become brittle in less than 10 years time.

Packet folding is illustrated in Fig. A-2.

FILING AND STORAGE

It is often important to put a stiff card in the packet. This prevents the specimen from damage in handling. Specimens in their packets can be filed upright, like file cards. The packet size can be readily accomodated by shoe boxes, an inexpensive means of maintaining a collection. A filing system is designed that allows the user to retrieve the specimens most effectively, since the researcher often needs to refer to named material for comparison when determining an unknown species. A small collection is satisfactorily arranged alphabetically under genus, then alphabetically for species under each genus. As one becomes better acquainted with bryophytes, it is often more convenient to file material by family, then by genus and species alphabetically within each family.

Closed boxes tend to keep most insect pests from destroying material, but silverfish and cockroaches sometimes destroy sporophytes of specimens. For permanent preservation and storage, well-prepared specimens should be presented to national, state, or university herbaria. This makes the material easily available to international and national users and preserves valuable records of distribution.

FURTHER READING

Conard, H. S., S. Flowers, P. M. Patterson, and R. E. Wynne. 1945. The bryophyte herbarium. A moss collection: preparation and care. *Bryologist* **48**:198–202.

Flowers, S. 1973. *Mosses: Utah and the West,* pp. 68–73. Provo: Brigham Young University Press.

Schuster, R. M. 1953. Boreal Hepaticae: a manual of the liverworts of Minnesota and adjacent regions. *Am. Midl. Nat.* **49**:259–263.

FIGURE A-2

Steps in folding a packet. The dimensions given are for a permanent packet. For a field collection packet of newspaper, the same procedure is followed, but with a change in dimensions of the packet. (1) Size and folding lines indicated. (2) Fold up lower $\frac{1}{3}$ of sheet. (3) Fold inward the two sides. (4) Fold down front flap. (5) Final packet, with position of label noted. (From Ireland, 1982.)

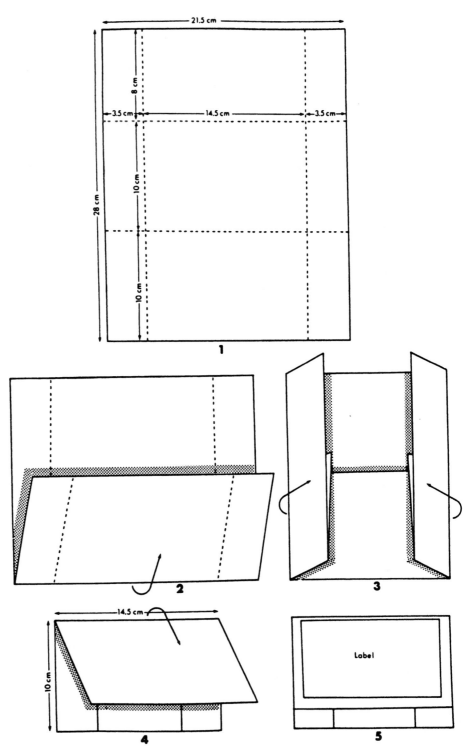

Appendix B

Comparison of Bryophyte Classes

Hepaticae	*Anthocerotae*	*Musci*
	Sporophyte	
Calyptra at base of mature sporophyte	"Calyptra" at base of mature sporophyte; occasionally also apical on young sporophyte	Calyptra at apex of mature sporophyte
First zygotic division transverse	First zygotic division longitudinal	First zygotic division transverse
Seta of thin-walled cells	Seta absent	Seta usually with stereids
Seta usually short-lived	Seta absent	Seta persistent
Sporangium usually opening by longitudinal lines	Sporangium usually opening by longitudinal lines	Sporangium usually opening by operculum
Peristome teeth absent	Peristome teeth absent	Peristome teeth usually present
Sporangium jacket without stomata	Sporangium usually with stomata	Sporangium usually with stomata
Sporangium jacket usually with transverse or radial thickenings across cells	Sporangium jacket usually without special thickenings across cells	Sporangium jacket without special thickenings across cells
Elaters usually present	Elaters usually present	Elaters absent
Columella absent	Columella usually present	Columella usually present
Spores dispersed over brief period	Spores usually dispersed over extended period	Spores usually dispersed over extended period
Growth determinate	Growth usually indeterminate	Growth determinate

Hepaticae	*Anthocerotae*	*Musci*
	Gametophyte	
Protonema usually very reduced	Essentially no protonema	Protonema usually extensive
Thallose or leafy	Thallose	Leafy
Gametophore often bilaterally symmetrical	Gametophore rarely bilaterally symmetrical	Gametophore rarely bilaterally symmetrical
Thallus sometimes with air chambers and dorsal pores	Thallus never with true air chambers, pores ventral	Not thallose
Chloroplasts numerous in each cell	Chloroplast often single in each cell	Chloroplasts numerous in each cell
Oil bodies usually complex	Oil bodies simple	Oil bodies simple
Hydroids and leptoids absent	Hydroids and leptoids absent	Hydroids often present, leptoids rare
Rhizoids usually smooth and unicellular; thallose taxa often with pegged rhizoids	Rhizoids always smooth usually unicellular	Rhizoids smooth or externally papillose, multicellular
Leaves usually in 3 ranks	Not leafy	Leaves usually in more than 3 ranks
Leaves often lobed	Not leafy	Leaves rarely lobed
Leaves never costate	Not leafy	Leaves often costate
Leaf cells usually iso-diametric	Not leafy	Leaf cells often elongate
Leaf cells often trigonous	Not leafy	Leaf cells rarely trigonous
Perianth or involucre often around archegonia	Archegonia embedded in thallus	Perichaetium surrounding archegonia
Archegonia discrete organs	Archegonia not discrete organs	Archegonia discrete organs
Paraphyses absent	Paraphyses absent	Paraphyses usually present

Appendix C

Stains for Revealing Pores in Sphagnum

The stain to be used is a 1–2% aqueous solution of methylene blue, gentian violet, or crystal violet. It can be added directly to the wet leaves of the dissected material on a microscope slide. The stain can be kept in a dropping bottle. A single drop is usually sufficient to stain material. If the amount is excessive, the excess moisture (including the stain) can be removed from the slide by placing the edge of a paper towel against the water's edge on the slide; then a drop of clear water is added to replace the moisture. These dyes can permanently dye clothing and fingers, so should be used with caution. They are also poisonous.

Appendix D

Freehand Sectioning Bryophyte Material for Anatomical Study

Freehand sectioning of material is most easily accomplished by the use of a single-edged sharp razor blade. Material should be cleared of soil particles and placed upon a microscope slide before sectioning. To acquire thin sections it is best to view the material, during sectioning, through the lowest power of a dissecting microscope.

If material is dry, it can be moistened by immersing it in boiling water or by adding water to the specimen on the slide. A wetting agent is highly efficient in rehydrating material. Aerosol OT (dioctyl sodium sulfosuccinate), available from the Fisher Scientific Company (U.S.A), is highly effective; a stock solution can be made readily with a dilution of approximately 1:100 in water. A similar and reasonably effective wetting agent can be made with a dilution from a detergent, but the proportions need to be established through trial and error. Such a wetting agent or boiling is useful only for anatomical studies of dead material, since these procedures destroy the living contents of cells.

Even fresh living material should be dissected in a drop of water sufficient to keep it moist. The sections are also easier to retain in water than on a dry slide.

Transverse sections of thalli, stems, and leaves can be cut freehand. A certain amount of experience quickly enables one to make very effective sections. For sections of stems, the leaves are removed with fine tipped forceps. The leaves of mosses can be stripped off by holding the tip of the leafy shoot with one pair of forceps, and with another pair of forceps stripping off the leaves by gently pinching the leafy shoot and and stripping it downward. With fragile leaves each leaf must be stripped off by pulling it downward carefully. Such a procedure is usual for leafy hepatics. The leaves can be removed to another slide for study, and sections can be made of the stem with a razor blade. Again the stem must be held on the surface of the slide by a pair of fine tipped forceps, and with the other hand sections are cut by holding the razor blade vertically over the stem and at right angles to its length; short chopping motions are used to cut several transverse sections. Cross-sections of thalli and leaves are made in the same way. In sectioning some leaves it is simpler to leave them attached to the stem.

FURTHER READING

Wagner, D. A. 1981. Pohlstoffe, a good wetting agent for bryophytes. *Bryologist* **84**:253.

Appendix E

Mounting Medium for Permanent Slides

Although Hoyer's solution results in distortion in the leaves of some bryophytes, it is useful for most. The chloral hydrate in the solution also causes bleaching of the cell contents. Since the solution dries very slowly, the slides prepared using it must be stored horizontally for a very extended period. The great advantage of Hoyer's solution is that specimens which are dissected and thoroughly moistened in water can be transferred directly to Hoyer's solution before a cover slip is added to make the final slide.

The formula for the solution, as given by Anderson (1954), is:

Distilled water	50 cc
Gum arabic (U.S.P. Flake)	30 g
Chloral hydrate	200 g
Glycerin	20 cc

The ingredients are mixed in that order at room temperature. Flake gum arabic is recommended since it goes into solution more rapidly than other forms. Furthermore, fewer bubbles usually result if this form is used. An electric rotary magnetic mixer tends to reduce the number of bubbles as well. The final mixture does not need to be filtered. After the final mixture is produced it should be allowed to stand for several hours before it is used. This permits most of the bubbles to disappear. It should be stored in air-tight bottles.

FURTHER READING

Anderson, L. E. 1954. Hoyer's solution as a rapid permanent mounting medium for bryophytes. *Bryologist* **57**:242–244.

Appendix F
Some Simple Methods for Culturing Bryophytes

METHOD I

For culture of small bryophytes the method of Berrie (1951) utilizes glass specimen tubes in which a cylinder of ordinary notepaper forms the substratum to hold the bryophyte. Small samples of the plant are removed from their substratum and placed upon a damp sheet of paper, which has been cut to a size that is accommodated by the length and circumference of the specimen tube. Two or three drops of water are placed in the bottom of the tube. The paper is made into a tube with the specimen facing inside, and this is placed into the specimen tube, which is then corked. The tubes are placed near a window that receives no direct sunlight. Every few weeks it is necessary to add a few drops of water to each tube to keep the specimen moist. The paper with the plant attached can be removed readily from the glass tube for examination. This method is useful for creeping mosses and hepatics, especially those that grow affixed to bark or rock. The notepaper should be washed to remove any obvious impurities before the specimen is placed upon it. It has been possible to grow such plants successfully for up to a year.

METHOD II

Another simple method was noted by Richards (1947). He used the natural media from which the bryophytes were collected. Such cultures can be started from whole tufts or shoots or from spores, gemmae, or fragments. For bryophytes that grow on soil, it is important to sterilize the soil medium in an autoclave before the bryophytes are "planted." This avoids some contamination (especially by fungi and algae) for a time. Environmental conditions for the cultures should attempt to match the natural environment where the bryophyte was collected. Porous pans or flower pots standing in an inch of water tend to be satisfactory to hold the soil for most soil-growing bryophytes. To keep the humidity high it is often necessary to cover pots with a sheet of glass, which has the additional advantage of reducing the number of air-carried contaminating organisms that frequently enter such cultures. Diffuse light is usually preferable, but it may be necessary to determine suitable light conditions by trial and error. To maintain pure cultures it is often necessary to weed out other species that enter. In temperate regions, best growth of bryophytes tends to be in spring and autumn, but in tropical areas growth can occur at most seasons when moisture conditions are favorable.

METHOD III

Another method for culturing bryophytes that grow on soil is to use inert material as the substratum, to which nutrient solutions can be added. Quartz sand can be thoroughly washed in a jet of water to remove extraneous matter, subjected to dilute HCl for 24 hr to remove carbonates, then washed thoroughly in tap water, and finally in distilled water. This sand is then autoclaved at 95°C. Calcareous sand can be washed in a similar way, but without the acid bath. The liquid nutrient medium is the same as in method IV.

The sand is then placed into glass petri dishes and molded into small domes so that it can be accommodated by the depth of the dish. The dome in the center of the dish allows for drainage of that portion of the substratum, upon which the bryophyte is placed. The liquid medium is added to saturate the sand. Such bryophytes can be transplanted readily without damage.

METHOD IV

Proskauer (1969) recommended a vermiculite substratum watered with a standard culture solution. The vermiculite is placed in glass dishes 9 cm in diameter and 5 cm high, then covered with petri dish lids. The vermiculite substratum is autoclaved in the half-filled dishes before culturing is begun. The substratum is then wetted with the culture solution, which contains, per liter of distilled water:

0.2 g	NH_4NO_3
0.1 g	$MgSO_4 \, 7H_2o$
0.1 g	$CaCl_2$
0.1 g	KH_2PO_4

This solution, as stock solution, is made up at 10 times the concentration stated, autoclaved, and diluted with distilled sterile water as needed. When cultures are planted, clean gametophore fragments should be used, in order to avoid any unnecessary algal or fungal contamination. An agar medium can be made utilizing the same culture solution.

METHOD V

In culturing spores of Jungermanniales to note sporeling development, Nehira (1966) used a liquid medium of one-half strength Knop's solution. This is made up of:

$NaNO_3$	0.375 g
$CaCl_2 \cdot 6H_2O$	0.125 g
KH_2PO_4	0.125 g
$MgSO_4 \cdot 7H_2O$	0.125 g
KCl	0.06 g
$FeCl_3$, 3%	1 drop
distilled H_2O	1000 ml

The pH of the medium is kept at 6.0–7.0. The medium is autoclaved preceding its use. Approximately 20 ml of the medium is placed in a petri dish. Spores are sewn on this liquid medium in the petri dish, and then the dish is closed. The dishes are then placed near a window under diffused light, avoiding direct exposure to sunlight.

FURTHER READING

Basile, D. V. 1972. A method for surface-sterilizing small plant parts. *Bull. Torrey Bot. Club* **99**:313–316.

———— 1975. A comparison of some macronutrient media used to culture bryophytes. *Bryologist* **78**:403–413.

Berrie, G. H. 1951. Culture of small bryophytes. *Trans. Br. Bryol. Soc.* **1**:485.

Bopp, M. 1983. Developmental physiology of bryophytes, in Schuster, R. M. (ed.), *New Manual of Bryology,* Vol. I, pp. 276–324. Nichinan: Hattori Botanical Laboratory.

Nehira, K. 1966. Sporelings in the Jungermanniales. *J. Sci. Hiroshima U. Series B., Div. 2* **11**:1–49.

Proskauer, J. 1969. Studies on Anthocerotales VIII. *Phytomorphology* **19**:52–66.

Richards, P. W. 1947. The cultivation of mosses and liverworts. *Trans. Br. Bryol. Soc.* **1**:1–3.

Schelpe, E. A. C. L. E. 1952. Techniques for the experimental culture of bryophytes. *Trans. Br. Bryol. Soc.* **2**:216–219.

Schneider, M. J., P. D. Voth, and R. F. Troxler. 1967. Methods for propagating bryophyte plants, tissues, and propagules. *Bot. Gaz.* **128**:169–174.

Zander, R. H. 1979. Regenerated herbarium material for biosystematic and cytological studies. *Bryologist* **82**:323.

———— 1979. Techniques for study of Pottiaceae. *Taxon* **28**:643–644.

Appendix G

Squash Techniques for Cytological Study of Chromosomes of Mosses

MEIOSIS

Plants bearing sporangia which are fully hydrated prior to or about the time of coloring of the annulus or peristome should be collected. The stage of sporangium development in which meiosis occurs varies in species so a range of sporangial development is needed. If necessary, plants can be kept in plastic bags or boxes at temperatures around 14–16°C for 1–2 weeks to allow sporangia to mature.

Place the sporangium in (1) a drop of fixative (1:3 glacial acetic acid:absolute ethanol) or (2) directly into stain (aceto-orcein or aceto-carmine). Remove the operculum and squash the sporangium gently to allow the columella to be forced out with the sporocyte mass attached. If it comes out easily and the mass is translucent (*not* yellow or opaque or green) it may be at a suitable stage. Remove as much as possible of the columella leaving sporocytes on slide. If prefixed, as in (1), add a second drop of fixative, allow to almost dry, then a third drop. Just as it evaporates add a drop of stain. If not prefixed, as in (2), add a second drop of stain. Cover with a cover slip, gently tap the surface to spread the cells (avoid sideways movement), and check for correct stage. Additional tapping and firm pressure through blotting paper may be needed to squash the cells and spread the chromosomes. Check under 40× objective and apply further pressure as needed. Only practice can help you judge how much.

If prefixed, cytoplasm will clear quickly, leaving chromosomes visible. If not prefixed, contents may intrude for some time, and the slide will not be as useful for study until it is made permanent.

To make permanent slides of the material. A piece of dry ice (small if only a few slides, larger if many) is needed. Fragments are often available from biochemistry departments. Cylinders of CO_2 can also be used if regular and continual supplies are needed. Freeze the slide and cover slip. Using a razor blade, release the cover slip. Place both slide and cover slip into a shallow dish of absolute ethanol for 1–5 min with upper surfaces uppermost.

Remove from the ethanol the slide with cells on upper surface. To the part of the slide from which the cover slip was removed, add Euparal while still moist, and cover with a new cover slip. Mount the original cover slip as well. Different mounting media are available, and the technique varies slightly for each. This technique is suitable for both mitotic and meiotic counts. Some small loss of cells may occur, but the technique gives a permanent record for future study. Hoyer's solution should not be used as a mounting medium, since it bleaches out the stain.

MITOSIS

Ramsay (1982) gives a summary of this procedure. [The procedure of Inoue and Iwatsuki (1976) gives the best results for centromere location.] Fix stem apex (1–2 mm) in 1:1:1 ethanol:glacial acetic acid:chloroform for 3 hr at 18°C. Soak in 45% acetic acid for a few minutes. Stain in 2% aceto-orcein for 10 hr at 15°C. Dissect in fresh aceto-orcein on a slide. Cover with a cover slip. Heat to a about 100°C on a hot plate. (Do not boil.) Tap on the cover slip to spread out the cells. Seal with Valap immediately, or examine, make drawings, and later make a permanent slide as set out above. (Hepatics may be teated in the same manner.)

ACETO-ORCEIN STAIN

Dissolve 2.2 g orcein in 100 ml of glacial acetic acid by gentle boiling and cool. This is a stock solution and should be diluted to 45% with distilled water when used.

FURTHER READING

Darlington, C. D., and L. F. LaCour. 1976. *Handling of Chromosomes* (6th ed.). London: George Allen & Unwin. (Recipes for stains and general techniques are available.)

Dyer, A. F. 1979. *Investigating Chromosomes*. London: Arnold. (Useful for interpretation.)

Inoue, S., and Z. Iwatsuki. 1976. A cytotaxonomic study of the genus *Rhizogonium* Brid. (Musci). *J. Hattori Bot. Lab.* **44**:137–145.

Ramsay, H. P. 1982. The value of karyotype analysis in the study of mosses. *J. Hattori Bot. Lab.* **53**:1–71.

—— 1983. Cytology of mosses, in Schuster, R. M. (ed.), *New Manual of Bryology*, Vol. I, pp. 149–221. Nichinan: Hattori Botanical Laboratory.

Vaarama, A. 1964. Notes on certain details of the karyological technique with mosses. *Portugaliae Acta Biol. (Ser. A)* **8**:81–94.

Appendix H
Collecting Material for Study of Major Evolutionary Lines of Bryophytes

ANDREAEIDAE

Andreaea is usually in sporophyte-bearing condition in early spring or late autumn in temperate regions. In tropical latitudes it is found only at high elevations or is entirely absent; thus material should be obtained from a temperate region. Fortunately, dried material is effective for study, and rehydration is easy. Antheridial plants usually bear perigonia at the same season that sporophytes mature. They are most easily found in dioicous species.

SPHAGNIDAE

Sphagnum is readily available in most parts of the world, and nearly any species is suitable for study of the gametophore. Since specimens of species related to *S. palustre* are large, they are very useful for study. Sporophytes are intermittently available, but in the Northern Hemisphere they can be found most readily later in the summer. Antheridia are frequently available in the spring and are betrayed by the deeper color of the somewhat swollen tips of divergent branches in and below the capitulum. For gametophore study at a later time, material can be collected and air-dried, then hydrated preceding study.

TETRAPHIDAE

The genus *Tetraphis* is widely distributed in the Northern Hemisphere, and often occurs in abundance on rotten wood in forests. Protonematal flaps can be found on rotten wood as well, but are absent in material where the leafy gametophores are obvious. For tropical regions and the Southern Hemisphere, material should be obtained from correspondents in the Northern Hemisphere. Fortunately, dried material can be hydrated readily for effective study. Sporophyte-bearing material can be collected during most of the year, but in summer and autumn the material usually has many sporophytes. Gemmiferous shoots are abundant during seasons when sporophytes are immature or absent.

POLYTRICHIDAE

Polytrichum juniperinum and *P. commune* are very widespread in the temperate and polar portions of both the Northern and Southern Hemispheres, and these species exhibit most of the characteristic features of both gametophore and sporophyte. In Australasia and New Guinea some species of *Dawsonia* are readily available and are also easy to examine. Sporophytes are best in late spring or early summer. The complex peristome of *Dawsonia* is atypical of Polytrichidae. Material of *Pogonatum* is available in most parts of the world and illustrates most features of the subclass. Perigonial plants with immature antheridia are usually available at the same season that sporangia mature.

BUXBAUMIIDAE

Material for *Buxbaumia* is rarely available in sufficient quantity for class examination. *Diphyscium,* on the other hand, is widely distributed and often abundant in the Northern Hemisphere, but scattered and rare in the Southern Hemisphere. Sporophyte-bearing material is usually in good condition in late summer and can be collected and air dried for class use later. It is readily hydrated and illustrates most major features of both sporophyte and gametophore. Careful examination of the material usually reveals both young perichaetial and young perigonial plants.

BRYIDAE

For study of this subclass, both pleurocarpous and acrocarpous mosses should be chosen to illustrate both haplolepidous and diplolepidous peristomes. Most specimens can be collected, air dried, and stored for later study when rehydrated. Mature sporophytes are readily available in most areas in spring to autumn, dependent on the species. It is best to choose species from the local flora, but a number of genera are sufficiently widespread that they can be recommended: *Bryum, Barbula, Bartramia, Brachythecium, Ceratodon, Dicranella, Dicranum* (or *Dicranoloma*), *Fissidens, Funaria, Grimmia, Leucobryum,*

Mnium, Plagiothecium, Pohlia, Thuidium. Perigonia are usually available at the time when sporangia mature.

ARCHIDIIDAE

The genus *Archidium,* although widespread in much of the Northern Hemisphere and scattered in the Southern Hemisphere, is often difficult to obtain in suitable condition. It grows on exposed damp soil and is barely visible to the naked eye. It is probably best treated as a demonstration from herbarium material.

CALOBRYALES

Although *Haplomitrium* is widely distributed, it is often difficult to obtain in quantities sufficient for class study. In New Zealand and Southeast Asia it is locally abundant and is best collected at the time for study. For *H. hookeri,* of the Northern Hemisphere, sporophyte-bearing material is usually available in summer, and antheridial plants are usually in good condition at the same time. In Southeast Asia, *H. mnioides* is in best condition at the same season, and the plants are large enough for easy study. In New Zealand and South America, *H. gibbsiae* is in excellent condition in the spring.

Takakia lepidozioides is in good condition much of the year, but it is rare except in Pacific North America, where it is locally abundant, but confined mainly to areas of difficult access. After collecting, material can be refrigerated in an ordinary household refrigerator for long periods of time. It is best to study living material, but herbarium material, when rehydrated, reveals most morphological details.

JUNGERMANNIALES

This order shows extraordinary diversity in gametophores, and material for study should reflect this. Some local genera should be included, but some widespread genera can be recommended which illustrate this diversity: *Bazzania, Calypogeia, Cephalozia, Frullania, Herbertus, Jungermannia, Lejeunea, Lepidozia, Marsupella, Plagiochila, Porella, Radula.* In the Northern Hemisphere *Scapania* and *Lophozia* should be included as well, and in the Southern Hemisphere *Geocalyx, Goebelobryum,* and *Schistochila* should be added. Material must be studied soon after it is collected in order to see the nature of the oil bodies. Sporophytes tend to be best represented in the spring in temperate regions. Gemmae are usually abundant in the autumn and spring in temperate regions.

METZGERIALES

This order shows considerable diversity. In temperate areas of the Northern Hemisphedre the genera *Pellia, Fossombronia, Metzgeria, Aneura, Pallavicinia, Blasia,* and *Riccardia* are usually available. Sporophytes are present in many of these in the spring. In tropical and Southern Hemisphere latitudes, some of these genera are available, including *Metzgeria, Pallavicinia, Aneura,* and *Riccardia,* but the genera *Treubia, Symphyogyna,* and *Podomitrium* should be included if available. Sporophytes tend to be available in spring or late winter. For Australasia *Hymenophytum* is usually available in good condition in spring and should be studied. These should be studied soon after they are collected.

SPHAEROCARPALES

This order contains plants of small size (*Sphaerocarpos*) or difficult to find in a submerged aquatic habitat and of extremely local occurrence (*Riella*). *Sphaerocarpos* is usually in good condition in spring and is found on fine-textured soils of garden margins, trail margins, and banks. Even when air-dried the sporophyte-bearing material retains most critical features, so herbarium material can be used to demonstrate its appearance. *Riella* is probably best treated using herbarium material for demonstration.

MONOCLEALES

Material of *Monoclea* is available in New Zealand and in central and South America. Vegetative thalli are present throughout the year, but sex organs appear in late summer. Sporophytes are available in spring. The genus grows in humid forest, especially along streams, but also along paths. Material should be studied when fresh.

MARCHANTIALES

Most genera of this order bear mature sporophytes in spring to summer. Some genera are widely distributed in the world, including *Riccia, Marchantia,* and *Lunularia.* The latter two genera are common greenhouse weeds. In some areas, especially the tropics, *Dumortiera* is common; *Conocephalum* is widespread in the Northern Hemisphere. For study of thallus structure, *Marchantia* is excellent, and *Riccia* shows another type of diversification. Material must be studied when freshly collected. In nature most genera are found in open sites on mineral soil. *Conocephalum* is often on humid, somewhat shaded rock faces, especially near streams. *Riccia* is found, especially in spring, on moist mineral soil and on shores of pond margins.

ANTHOCEROTALES

The genera *Anthoceros* and *Phaeoceros* are widely distributed and usually have sporophytes in spring until autumn. Most species grow on earth, especially on somewhat shaded banks, and tend to occur in sites that remain humid for extended periods. *Dendroceros,* an epiphytic species, is often in good condition for study during the winter. This is predominantly a tropical genus. Material must be studied soon after collection to observe the structure as well as plastid details.

Appendix I

Keys for Determining Subclasses and Orders of Bryophytes

The following keys can be used to determine the main subclasses and orders as they are treated in this text. Sporophytes are needed to determine several, and a number of genera cannot be placed readily by using such a simplified artificial key (see notes at end).

1. Gametophores with apparent leaves and stems _____ 2
1. Gametophores with an obviously flattened thallus _____ 16

2. Leaves lobed _____ Jungermanniales (Hepaticae)
2. Leaves unlobed _____ 3

3. Leaves with central costa or several costae _____ 23
3. Leaves without costae _____ 4

4. Leaf cells containing complex oil bodies _____ 5
4. Leaf cells lacking complex oil bodies _____ 10

5. Gametophores with a rhizomatous system, but which lack rhizoids _____ Calobryales (Hepaticae)
5. Gametophores usually with rhizoids _____ 6

6. Rhizoids multicellular, with oblique cross-walls to cells _____ 10
6. Rhizoids unicellular, or rarely with perpendicular cross-walls in the knobbed tip of the rhizoid _____ 7

7. Sporophytes with colorless seta when mature _____ 8
7. Sporophytes lacking seta when mature _____ 9

8. Lateral leaves transversely or obliquely inserted; leaves often in three distinct rows ———————————— Jungermanniales (Hepaticae)
8. Lateral leaves longitudinally inserted; leaves never in three distinct rows ———————————————— Metzgeriales (Hepaticae)

9. Sporangium spherical, enclosed within a calyptra and a unistratose sleeve of gametophore tissue when mature ——— Sphaerocarpales (Hepaticae)
9. Sporangium horn-shaped, not enclosed within a calyptra or unistratose sleeve when mature ———————— Anthocerotales (Anthocerotae)

10. Sporangium without columella, spores usually fewer than 32 ——————
———————————————————— Archidiidae (Musci)
10. Sporangium with columella, spores always more than 32 ——————— 11

11. Sporangium borne on pseudopodium when mature ——————— 12
11. Sporangium lacking pseudopodium ———————————— 13

12. Sporangium opening by longitudinal lines ——————— Andreaeidae (Musci)
12. Sporangium opening by operculum ——————— Sphagnidae (Musci)

13. Sporophytes with rigid seta composed in part of stereid cells ——— 24
13. Sporophytes with seta that collapses readily, composed mainly of thin-walled cells, never of stereids ———————————— 14

14. Leaves in three rows ——————— Jungermanniales (Hepaticae)
14. Leaves in two rows ———————————— 15

15. Lateral leaf base attachment longitudinal on stem ——————————
———————————————————— Metzgeriales (Hepaticae)
15. Lateral leaf base attachment transverse or oblique to length of axis ———————————————— Jungermanniales (Hepaticae)

16. Thallus with distinct dorsal pores ——— Marchantiales (Hepaticae)
16. Thallus lacking dorsal pores ———————————— 17

17. Thallus with internal air chambers or made up of longitudinal columns of cells with vertical air spaces between them ———————————
———————————————————— Marchantiales (Hepaticae)
17. Thallus lacking internal air chambers ——————— 18

18. Sporangium lacking seta ———————————— 19
18. Sporangium with seta ———————————— 20

19. Sporophyte cylindric, elaters usually multicellular ——————
———————————————— Anthocerotales (Anthocerotae)

19. Sporophyte spherical, elaters absent _____ Sphaerocarpales (Hepaticae)

20. Sporophytes on specialized carpocephalum _____ Marchantiales (Hepaticae)
20. Sporophytes not on carpocephalum _____ 21

21. Rhizoids arranged perpendicular to surface of thallus and also parallel to length of thallus _____ Monocleales (Hepaticae)
21. Rhizoids only perpendicular to thallus or not apparent _____ 22

22. Sporophyte borne on a short leafy branch with a perianth _____ Jungermanniales (Hepaticae)
22. Sporophyte never on short leafy branch with perianth _____ Metzgeriales (Hepaticae)

23. Leaves with dorsal lamellae; sporophytes usually with multicellular un-jointed peristome teeth _____ Polytrichidae (Musci)
23. Leaves without dorsal lamellae; sporophytes usually with jointed peri-stome teeth _____ 24

24. Sporangium with operculum and bearing four multicellular unjointed peri-stome teeth _____ Tetraphidae (Musci)
24. Sporangium lacking peristome teeth or with more than four peristome teeth _____ 25

25. Sporangium without operculum, opening by longitudinal lines _____ Andreaeidae (Musci)
25. Sporangium never opening by longitudinal lines, usually with operculum _____ 26

26. Peristome teeth in a single concentric circle or absent _____ 27
26. Peristome teeth in more than one concentric circle _____ 29

27. Epiphragm present _____ Polytrichidae (Musci)
27. Epiphragm absent _____ 28

28. Peristome consisting, at least in part, of a pleated cone of teeth, fused along most of their length _____ Buxbaumiidae (Musci)
28. Peristome absent or teeth jointed and separated along most of their length _____ Bryidae (Musci)

29. Endostome a pleated cone, exostomial teeth sometimes obscure _____ Buxbaumiidae (Musci)
29. Endostome never a pleated cone, exostomial teeth usually apparent _____ Bryidae (Musci)

Note: A number of genera are difficult to determine using such a simplified key. The following genera are highly distinctive and can be distinguished as follows:

1. *Buxbaumia* of the subclass Buxbaumiidae (Musci) has a highly distinctive sporophyte, but the gametophore is not visible with the naked eye. The endostome, which consists of a distinctive pleated cone, is unique to this subclass.

2. *Treubia* of the Metzgeriales (Hepaticae) has an unusual leafy thallus in which the lateral leaves are longitudinally oriented on the stem, but there are two rows of dorsal smaller leaves (or scales) that are transversely attached; each of these is associated with a lateral leaf and might be mistaken for a lobule of a jungermannialean hepatic. In the leaves of *Treubia*, however, there are scattered cells, each of which contains a single complex oil body; these give the gametophore a white-dotted appearance unique to this genus.

3. It is necessary to section fresh green thalli of Marchantiales in order to view the internal anatomy with the air chambers. In some genera, including *Riccia*, the air chambers may be obscure.

4. In older thalli of some Anthocerotae, some internal chambers may be found. These are mucilage chambers and not air chambers. If each of the cells of the thallus has a single lenticular chloroplast, this immediately marks it as belonging to the Anthocerotae. The horn-shaped sporophytes are unique to the class.

Appendix J

Manuals for Determination of Bryophytes

There is, for the mosses, the monumental work, in German, of V. F. Brotherus (1924-1925), *Musci,* in A. Engler's *Die Natürlichen Pflanzenfamilien,* which describes all families and genera of mosses of the world known to him at that time. There are keys to genera for each family, and keys also to species. Most genera are illustrated. Although greatly in need of modernization and expansion, this book is invaluable. A similar treatment for Hepaticae and Anthocerotae by V. Schiffner (1909) was produced before the period of careful, taxonomic hepaticological studies of the mid-20th century and is consequently less useful. It does contain keys to the genera known at that time and illustrates many. There are no other works that attempt to treat all bryophytes of the world. The full citations of these publications are as follows:

Brotherus, V. F. 1924-1925. *Musci,* in Engler, A. (ed.), *Die Natürlichen Pflanzenfamilien,* Band 10 and 11. Leipzig: W. Engelmann.

Schiffner, V. 1909. *Hepaticae & Anthocerotae,* in Engler, A. (ed.), *Die Natürlichen Pflanzenfamilien I.* Teil. 3. Abt. I Hälfte, pp. 3–141. Leipzig: W. Engelmann.

The following manuals or publications treat bryophytes of various parts of the world and can be used to determine genera and species in most cases.

TEMPERATE AND ANTARCTIC SOUTHERN HEMISPHERE

Allison, K. W. 1964. Revised key to the moss genera of New Zealand. *Tuatara* 12:157–184. (Keys to all genera known from New Zealand at that time.)
——— and J. Child. 1971. *The Mosses of New Zealand.* Dunedin: University of Otago Press. (Guide to common mosses; keys, illustrations.)
——— 1975. *The Liverworts of New Zealand.* Dunedin: University of Otago Press. (Guide to common hepatics and hornworts; keys, illustrations.)

Greene, S. W. (ed). 1974. *A Synoptic Flora of South Georgian Mosses.* London: British Antarctic Survey. (The first parts of the flora of the region; includes some genera of Bryidae and all Polytrichidae; illustrations, keys, descriptions.)

———, D. M. Greene, P. D. Brown, and J. M. Pacey. 1970. *Antarctic Moss Flora.* London: British Antarctic Survey. (Includes *Andreaea, Pohlia, Polytrichum, Psilopilum,* and *Sarconeurum;* keys, illustrations, descriptions.)

Hodgson, E. A. 1963–1964. Revised generic keys to the hepatic flora of New Zealand. *Tuatara* **11**:195–207. Ibid. **12**:1–13. (Keys to all genera known from New Zealand at the time.)

Menéndez, G. H. Hässel de. 1962. *Estudio de las Anthocerotales y Marchantiales de la Argentina.* Opera Lilloana **7**. (In Spanish; keys to genera and species, illustrations.)

———, and S. S. Solari. 1975. *Flora criptogámica de Tierra del Fuego. Vol. I. Bryophyta Hepatopsida.* Buenos Aires. (In Spanish; the first fascicle treating Calobryales and the families Vetaformaceae and Balantiopsaceae of the Jungermanniales; keys, illustrations, descriptions.)

Robinson, H. 1975. *The Mosses of Juan Fernandez Islands.* Smithsonian Contrib. Bot., Vol. 27. (Keys to genera and species.)

Sainsbury, G. O. K. 1955. *A Handbook of the New Zealand Mosses.* Roy. Soc. N. Z. Bull., Vol. 5. (Keys to species of most genera known from New Zealand at that time; illustrations.)

Vitt, D. H. 1974. A key and synopsis of the mosses of Campbell Island, New Zealand. *N. Z. J. Bot.* **12**:185–210. (Keys.)

TEMPERATE AND ARCTIC NORTHERN HEMISPHERE

Note that works describing local floras have been excluded when more comprehensive manuals are available that include a broader geographic area.

Abramova, A. L., L. I., Savicz-Ljubitakaja, and Z. N. Smirnova. 1961. *Handbook of the Mosses of Arctic U.S.S.R.* Moscow: Acad. Sci., U.S.S.R. (In Russian; keys and descriptions to families, genera, and species; illustrations.)

Agnew, S., and M. Vondrácels. 1975. A moss flora of Iraq. *Feddes Repert.* **86**:341–489. (Keys and illustrations to most species.)

Arnell, S. 1956. *Illustrated Moss Flora of Fennoscandia. I. Hepaticae.* Lund: C.W.K. Gleerup. (Keys and descriptions to genera and species; illustrations.)

Augier, J. 1966. *Flore des Bryophytes.* Paris: Paul Lechavalier. (In French; keys to families, genera, and species of bryophytes of France; illustrations.)

Bardunov, L. V. 1969. *The Handbook of the Mosses of Central Siberia.* (In

Russian; keys and descriptions to families, genera, and species; illustrations.)

Berghen, C. V. 1955–1957. *Flore Générale de Belgique. Bryophytes. Vol. 1. Hepatics.* Brussels: Jardin Botanique de l'État. (In French; keys and descriptions to families, genera, and species; illustrations.)

Bilewsky, F. 1965. Moss-Flora of Israel. *Nova Hedw.* 9:335–434. (Keys and descriptions to genera and species; illustrations.)

Casares-Gil, A. 1919, 1923. *Flora Ibérica Briofitas. Part 1, Hepáticas. Part 2, Musgos.* Madrid: Mus. Nac. Cienc. Nat. (In Spanish; keys and descriptions to genera and species; illustrations.)

Chen, P. C. 1963, 1978. *Genera Muscorum Sinicorum.* Peking. (In Chinese; keys and descriptions to families and genera; illustrations.)

Choe, Du-Mun. 1980. *Illustrated Flora and Fauna of Korea. Vol. 24, Musci. Hepaticae.* Ministry of Education. (In Korean; keys and descriptions of species; illustrations.)

Chopra, R. S. 1975. *Taxonomy of Indian Mosses.* New Delhi: CSIR, Bot. Mgr. 10. (Keys and descriptions to classes, subclasses, families, genera, and species; illustrations.)

Crum, H. A. 1983. *Mosses of the Great Lake Forest* (3rd ed.) Ann Arbor: Univ. of Mich. Herb. (Keys and descriptions to genera and species; illustrations.)

——— and L. E. Anderson. 1981. *Mosses of Eastern North America.* 2 vols. New York: Columbia University Press. (Keys and descriptions to genera and species; illustrations.)

Demaret, F., et al. 1959–1968. *Flore Générale de Belgique. Vol. 2, Bryophytes. Vol. 3, Mosses.* Brussels: Jardin Botanique de l'État. (In French; keys and descriptions to families, genera, and species; illustrations; incomplete.)

Dixon, H. N., and H. G. Jamieson. 1924. *The Student's Handbook of British Mosses* (3rd ed.). Eastbourne: V. Sumfield.

Flowers, S. 1961. *The Hepaticae of Utah.* Univ. of Utah Biol. Ser. 12:1–108. (General chapters, keys and descriptions to families, genera, and species; illustrations.)

——— 1973. *Mosses: Utah and the West.* Provo: Brigham Young University Press. (General chapters, keys and descriptions to families, genera, and species; illustrations.)

Frye, T. C., and L. Clark. 1937–1947. *Hepaticae of North America.* Univ. of Washington Publ. Biol. 6:1–1018. (Keys and descriptions to families, genera, and species; illustrations.)

Gams, H. 1950. *Kleine Kryptogamenflora von Mitteleuropa. I. Die Moos and Farnpflanzen (Archegoniaten)* (3rd ed.), 186 pp. Jena: G. Fischer. (In German; keys.)

Gangulee, H. C. 1969–1980. *Mosses of Eastern India and Adjacent Regions.* Calcutta: P. K. Ghosh. (Keys and descriptions to genera and species; illustrations.

Grout, A. J. 1928–1940. *Moss Flora of North America, North of Mexico,* 3 vols. Newfane, Vermont, publ. by author. (Keys to families, genera, and species; illustrations.)

Howe, M. A. 1899. *The Hepaticae and Anthocerotes of California. Mem. Torrey Bot. Club* 7:1–208. (Keys to species; illustrations.)

Instituto Botanico Boreali-Occidentali. 1978. *Flora Tsinlingensis.* Peking, China, Tom. III Bryophyta I. (In Chinese; keys and descriptions to genera and species of mosses; illustrations.)

Ireland, R. R. 1982. *Moss Flora of the Maritime Provinces.* Nat. Mus. Canada Publ. Bot. 13. (Keys and descriptions to genera and species; illustrations.)

Iwatsuki, Z., and M. Mizutani. 1972. *Coloured Illustrations of Bryophytes of Japan.* Osaka: Hoikusha Publ. Co. (In Japanese; keys and descriptions to genera and species; illustrations.)

Landwehr, J., and J. J. Barkman. 1966. *Atlas van de Nederlandse Bladmossen.* Amsterdam: Koninklijke Nederlandse Natuurhistorische Vereniging. (In Dutch; illustrations and brief comments to species of mosses.)

———, S. R. Gradstein, and H. van Melick. 1980. *Atlas Nederlandse Levermossen.* Amsterdam: Koninklijke Nederlandse Natuurhistorische Vereniging. (In Dutch; illustrations and brief comments to species of hepatics and hornworts.)

Lawton, E. 1971. *Moss Flora of the Pacific Northwest.* Nichinan: The Hattori Botanical Laboratory. (Keys to families, genera, and species; illustrations.)

Liao-Ning Provincial Agriculture and Soil Research Institute. 1969. *Flora Muscorum Chinae Boreali-orientalis.* Peking. (In Chinese; keys and descriptions to families, genera, and species of mosses; illustrations.)

——— 1981. *Flora Hepaticarum Chinae Boreali-orientalis.* Peking. (In Chinese; keys and descriptions to families, genera, and species of hepatics and hornworts; illustrations.)

Limpricht, K. G. 1890–1904. *Die Laubmoose Deutschlands, Österreichs und der Schweiz.* Leipzig: K. Kummer. (In German; keys and descriptions to genera and species; illustrations.)

Lye, K. A., and E. V. Watson. 1968. *Moseflora.* Oslo: Universitets-vorlaget. (In Norwegian; keys to genera and species of bryophytes of Norway; illustrations.)

Macvicar, S. M., and H. G. Jamieson. 1926. *The Student's Handbook of British Hepatics.* Eastbourne: V. V. Sumfield. (Keys and descriptions to genera and species; illustrations.)

Mönkemeyer, W. 1927. Die Laubmoose Europas, in Rabenhorst, L. (ed.), *Kryptogamen-Flora von Deutschland, Österreich und der Schweiz,* Band IV. Leipzig: Akad. Verlag. (In German; keys and descriptions to families, genera, and species; illustrations.)

Mueller, K. 1951–1958. Die Lebermoose Europas, in Rabenhorst, L. (ed.), *Kryptogamen-Flora 6* (3rd ed.) Leipzig: Akad. Verlag. (In German; general chapters, keys, and descriptions to families, genera, and species; illustrations.)

Noguchi, A. 1976. *Handbook of Japanese Mosses.* Tokyo: Hokuryukan Publ.

Co. (In Japanese; general chapters, keys to genera and species; illustrations.)

Nyholm, E. 1954–1969. *Moss Flora of Fennoscandia. II. Musci.* 6 fasc. Lund. (Keys and descriptions to families, genera, and species; illustrations.)

Pavletic, Z. 1968. *Flora Mahouina Jugslavije.* Zagreb: Inst. Bot. Sveucilista Zagrebu. (In Jugoslavian; keys and descriptions to families, genera, and species of mosses; illustrations.)

Pilous, Z. 1948. *Nase Mechy. Illustrovany Klie Kurcovani mechu ceskoslovenskych.* (In Czech; keys and descriptions to families, genera, and species of mosses; illustrations.)

Rejment-Growchowska, I. 1966. *Flora Polska Watrobowce.* Tome I. Warszawa: Polska Akad. Nauk, Inst. Bot. (In Polish; keys and descriptions to familes, genera, and species of hepatics; illustrations.)

Schljakov, R. N. 1976, 1979, 1980, 1981. *Bryophyta-Liverworts of the Northern U.S.S.R. Anthocerotae; Hepaticae: Jungermannidae-Metzgerianeae.* Leningrad: Publishing House "Science." (In Russian; keys, descriptions, and illustrations to genera and species.)

Savich-Lyubitskaya, L. I., and Z. N. Smirnova. 1970. *Handbook of the Musci of the U.S.S.R. Acrocarpous Mosses.* Moscow: Bot. Zurn. (In Russian; keys and descriptions to families, genera, and species; illustrations.)

Schuster, R. M. 1966–1980. *The Hepaticae and Anthocerotae of North America, East of the Hundredth Meridian,* 4 vols. New York: Columbia University Press. (General chapters, keys, and descriptions to families, genera, and species; illustrations).

Singh, V. B. 1966. *Bryophytes of India II. Marchantia I.* (Key to species; illustrations.)

Smith, A. J. E. 1978. *The Moss Flora of Britain and Ireland.* Cambridge: Cambridge University Press. (Keys and descriptions to genera and species; illustrations.)

Srivastava, K. P. 1964. *Bryophytes of India. I. Ricciaceae. Bull. Nat. Bot. Gard.* **104**:1–103. (Keys and descriptions to genera and species; illustrations.)

Watson, E. V. 1969. *British Mosses and Liverworts.* Cambridge: Cambridge University Press. (Keys and brief descriptions to species; illustrations.)

Weber, W. A. 1973. *Guide to the Mosses of Colorado.* Occas. Papers Inst. Arctic Alpine Res. 6. (Keys to families, genera, and species.)

AUSTRALIA AND OCEANIA

Bartram, E. B. 1933. *Manual of Hawaiian Mosses.* B. P. Bishop Museum, Honolulu, Bull. 101. (Keys and descriptions to genera and species; illustrations.)

Catcheside, D. G. 1980. *Mosses of South Australia.* South Australia: Government Printer. (General chapters, keys to genera and species; illustrations.)

Rodway, L. 1916. *Tasmanian Bryophyta Vol. II Hepatics*. Hobart: Roy. Soc. Tasmania. (Keys and descriptions to species and genera.)

Scott, G. A. M., and I. G. Stone. 1976. *The Mosses of Southern Australia*. London: Academic Press. (Keys to genera and species; illustrations.)

Whittier, H. O. 1976. *Mosses of the Society Islands*. Gainesville: University of Florida Press. (Keys and descriptions to families, genera, and species; illustrations.)

———, H. A. Miller, and C. E. B. Bonner. 1963. Musci, in *Bryoflora of the Atolls of Micronesia. Beih. Nova Hedw.* **11**:7–41. (Keys and descriptions to genera and species; illustrations.)

TROPICAL AND SUBTROPICAL LATITUDES

Abeywickrama, B. A. 1959. The genera for the liverworts of Ceylon. *Ceylon J. Sci. (Bio. Sci.)* **2**:33–81. (Keys to genera; illustrations.)

——— 1960. The genera of the mosses of Ceylon. *Ceylon J. Sci. (Biol. Sci.)* **3**:41–122. (Keys to genera; illustrations.)

Bartram, E. B. 1939. *Mosses of the Philippines. Philip. J. Sci.* **68**:1–437. (Keys and descriptions to families, genera, and species; illustrations.)

——— 1949. *Mosses of Guatemala. Fieldiana: Botany* **25**:1–442. (Keys and descriptions to genera and species; illustrations.)

Berghen, C. vanden. 1972. Hépatiques et Anthocérotées, in Symoens, J. J. (ed.), *Exploration hydrobiologique du Luapula*, Vol. VIII, Fasc. 1. (In French; keys and descriptions to families, genera, and species; illustrations.)

Bizot, M., and T. Pócs. 1974. East Africa bryophytes. *Act. Ac. Paed. Agriensis. N. S.* **12**:383–449. (Keys, descriptions, illustrations.)

Chuang, C-C. 1973. A moss flora of Taiwan exclusive of essentially pleurocarpous families. *J. Hattori Bot. Lab.* **37**:419–509. (Keys to families, genera, and species.)

Crum, H. A., and W. C. Steere. 1957. *The Mosses of Puerto Rico and the Virgin Islands*. N.Y. Acad. Sci. Scientific Survey of Puerto Rico and the Virgin Islands **7**:395–599. (Keys and descriptions to orders, families, genera, and species; illustrations.)

——— and E. B. Bartram. 1958. A survey of the moss flora of Jamaica. *Bull. Inst. Jamaica, Sci. Ser.* **8**:1–90. (Keys to genera and species.)

Fleischer, M. 1904–1924. *Die Musci der Flora von Buitenzorg (Zugleich Laubmooseflora von Java)*. 4 vols. Leiden: E. J. Brill. (In German; keys and descriptions to genera and species; illustrations.)

Florschütz, P. A. 1964. *The Mosses of Suriname. Part 1*. Leiden: E. J. Brill. (Keys and descriptions to genera and species; illustrations incomplete.)

Gangulee, H. C. 1969–1980. *Mosses of Eastern India and Adjacent Regions*. Calcutta: P. K. Ghosh. (Keys and descriptions to genera and species; illustrations.)

Kashyap, S. R. 1929. *Liverworts of the Western Himalayas and the Punjab Plain*. 2 vols. Lahore, India: University of Punjab. (Keys and descriptions to genera and species; illustrations.)

Magill, R. E. 1981. Bryophyta: Part 1, fasc. 1. Mosses, in Leisterner, D. A. (ed.), *Flora of Southern Africa*. (Keys to families, genera, and species of mosses; further fascicles to be published; illustrations.)

Potier de la Varde, R. 1936. *Mousses du Gabon. Mem. Soc. Sci. Natar. Math. Cherbourg* **42**:1–270. (In French; keys; descriptions; illustrations.)

Robinson, H. 1967. Preliminary studies on the bryophytes of Colombia. *Bryologist* **70**:1–61. (Keys to species of selected genera of bryophytes; illustrations.)

Sim, T. R. 1926. *The Bryophyta of South Africa. Trans. Roy. Soc. South Africa.* **15**:1–475. (Keys and descriptions to genera and species of bryophytes; illustrations.)

Sloover, J. L. de. 1973–1979. Note de bryologie africaine. *Bull. Jard. Bot. Nat. Belg.* **43**:333–348. Ibid. **45**:103–124, 131–135, 237–271, 313–321. Ibid. **46**:379–385, 427–447. Ibid. **47**:31–48, 155–181. Ibid. **49**:393–408. (In French; keys to species of some genera of mosses; illustrations.)

Spruce, R. 1885. *Hepaticae of the Amazon and of the Andes of Peru and Ecuador. Trans. Proc. Bot. Soc. Edinburgh* **15**:1–588. (In Latin; descriptions to families, genera, and species; illustrations.)

Wijk, R. van der. 1958. Precursory studies on Malaysian mosses II. A preliminary key to the moss genera. *Blumea* **9**:143–186. (Keys only.)

Index

Page numbers in *italic* indicate illustrations. Please refer to Contents for detailed list of material contained in each chapter.

ERRATA
Introduction to Bryology
July 2001

Page 13: paragraph 3, line 7 (Fig. 2-2, P) not (Fig. 2-20, P)

Page 34: Fig. 4 - 2 (E) add after rhizoids "chloroplasts not shown"

Page 40: Fig 4 - 7 Dehiscence "in", not "is"
          4 - 7 (A) "operculum", not "opeculum"

Page 47: Suzuki "combination", not "contribution"

Page 53: paragraph 2, line 4 "there", not "these"
          paragraph 3, line 8 "any", not "an"

Page 60: paragraph 2, line 4 "clothe", not "close"

Page 66: paragraph 4, line 5 "branches", not "banches"

Page 67: Fig. 6 - 9 col. (columella), is not labelled

Page 69: paragraph 1, line 4 "spores", not "spore"

Page 80: paragraph 3 add after *Diphyscium* "and *Theriotia*"
          Add at end of paragragh "*Muscoflorschuetzia* lacks peristome teeth"

Page 102: paragraph 1, line 7 "away", not "independently"
          line 13 delete "in the clone"
          paragraph 4, line 3 "antheridia", not "anteridia"

Page 107: Fig 8 - 18 C and D are reversed in the illustrations

Page 109: Fig 8 - 19 in legend, the final B should be D

Page 122: Fig 9 - 2 (C-E) "*A. clavatum*", not "*A. stellatum*"

Page 164: under 3: "Succubous", not "Succulous"
          Fig 13 - 11 add at the end of the legend:
          Note: Figs A,C,E,G,I,K are lateral views,
          Figs B,D,F,J,L are dorsal views

Page 200: Fig 15 - 4 under A: "archegonia", not "archengonia"

Page 209: Fig. 16 - 2 C and D are reversed in the figure

Page 210: paragraph 5, line 5 "inner", not "outer"

Page 222: Fig 17 - 12 "Schematic", not "Schmatic"

Page 225: Fig. 17 - 17 "archegoniophore", not "archgeoniophore"

Page 239: paragraph 6: "condition", not "conidtion"

Page 275: paragraph 1, line 4 "*Physcomitrium*", not "Physomitrium"

Page 277: Fig 21 - 1 Add to legend:
    I. *Bryum billiardieri* n = 10
    This is an Australian species
    J. *Racomitrium lanuginosum* n = 14
    specimen from British Columbia, Canada
    K. *Macromitrium* sp. n = 9
    Note tiny chromosome (arrow) specimen
    from New Guinea.
    L. *Papillaria amblyacis* n = 11
    Not one small bivalent separates early (arrows)
    Specimen from Australia
    M. *Cyathoporum bulbosum* n = 5. Specimen from Australia.
    N. *Hypopterygium rotulatum* n = 9
    Note the small chromosome pair separates early
    (arrow). Specimen from Australia
    O. *Rhytidiopsis robusta* n = 12
    Specimen from British Columbia, Canada

Page 320: Paragraph 4, line 5 "condition", not "conditon"

Page 376: glossary complicate bilobed
    delete "may be" and substitute "folds over the other longitudinally"

Page 377: glossary "diaspore", not "dispore"

W.B. Schofield
Department of Botany
University of British Columbia
Vancouver, B.C. Canada V6T1Z4

CPSIA information can be obtained at www.ICGtesting.com
Printed in the USA
BVOW082355110612

292330BV00007B/113/P